Functional Textiles and Clothing

Abhijit Majumdar · Deepti Gupta ·
Sanjay Gupta
Editors

Functional Textiles and Clothing

Springer

Editors
Abhijit Majumdar
Indian Institute of Technology Delhi
New Delhi, India

Deepti Gupta
Indian Institute of Technology Delhi
New Delhi, India

Sanjay Gupta
World University of Design
Sonepat, Haryana, India

ISBN 978-981-13-7720-4 ISBN 978-981-13-7721-1 (eBook)
https://doi.org/10.1007/978-981-13-7721-1

This Springer imprint is published by the registered company Springer Nature Singapore Pte Ltd.
The registered company address is: 152 Beach Road, #21-01/04 Gateway East, Singapore 189721, Singapore

Preface

Textile materials, be it fabric or home furnishing, are primarily chosen based on their aesthetic appeal and comfort. However, the perception about textiles is changing as they are expected to possess some functionality apart from fulfilling their basic purpose. A new domain of functional textiles has emerged in the last few decades which encompasses the areas like protective textiles, medical textiles, industrial textiles, sports textiles, automotive textiles, and packaging textiles. All these materials must have specific functional features to meet the end-use requirements. Therefore, material selection, product design, and manufacturing are rather complex and therefore challenging in case of functional textiles.

Functional textiles and clothing sector has emerged as the growth area in recent times. Recent reports peg the global market to reach USD 4.72 billion by 2020, at a CAGR of 33.58% between 2015 and 2020, on the strength of commercial market applications in healthcare, sports, fitness, fashion, military, and automotive sectors. India, as the second largest producer and fourth largest exporter of textile and apparel goods in the world, has already identified this area under its technical textiles push a decade back. With its domestic apparel market showing signs of maturing and a healthy growth rate, India is looking forward to make a place for itself in this high-value, high-tech textile and apparel sectors.

Looking at the emerging needs of industry and academia and to cope with the impending challenges of designing and manufacturing of functional textiles, an international event entitled 'Functional Textiles and Clothing Conference 2018' was organized by Indian Institute of Technology Delhi during February 9–11, 2018, in collaboration with World University of Design, Sonipat, and PSG College of Technology, Coimbatore. The conference provided an interdisciplinary platform to leading scientists, academicians, researchers, research scholars, entrepreneurs, and market stakeholders to have in-depth exchanges on the recent scientific developments, research results, cutting-edge technologies, innovations, trends, concerns, supply chain issues and challenges and opportunities in the field of functional textiles and clothing.

This book contains selected papers presented in the conference. The papers are organized under eight sections, namely protective textiles, medical textiles, smart textiles, textile chemical processing, garment and accessory design, testing and characterization, supply chain management and sustainability, and traditional textile art and crafts. Interested readers may contact the editors or the authors in case they have any query.

New Delhi, India — Abhijit Majumdar
New Delhi, India — Deepti Gupta
Sonepat, India — Sanjay Gupta

Contents

Part I Protective Textiles

Thermal Resistance of Leather and Membranes for Summer Desert Military Footwear Under Different Climate Conditions 3
Dragana Kopitar, Jadranka Akalovic and Zenun Skenderi

Woven Military Fabrics from the Aspect of the Microbial Barrier Permeability 11
Ivana Schwarz, Beti Rogina-Car and Ruzica Brunsek

Development of Thermal Insulative Nonwoven Fabric Through Advance Material Application 21
Rohit Naik and Arup Rakshit

Investigation on Sound Absorption Characteristics of Nonwoven Coir Mats 37
G. Thilagavathi, A. Muralikrishnan, N. Muthukumar and S. Neelakrishnanan

Part II Medical Textiles

Functionalized Silk for Surgical Suture Applications 49
S. Viju, L. Marian Shilpa and G. Thilagavathi

Effect of Domestic Laundering on Removal of Bacterial Contamination from Nurses' White Coats 67
Priyanka Gupta, Nilanjana Bairagi and Deepti Gupta

Part III Smart Textiles

Hybrid Cover Yarn's Element Orientation and Its Impacts on Mechanical/Tensile Behavior of Conductive Yarns and Fabrics 77
Ali Asghar, Mohd Rozi Ahmad, Mohamad Faizul Yahya, Syed Zameer Ul Hassan and Muhammad Kashif

Facile Metallization Technique of Textiles for Electronic Textile Applications 91
Md. Momtaz Islam, Musa Ali Reza, Dewan Murshed Ahmed, Md. Abdullah Al. Mamun and Hasan Shahariar

Development of Smart Textiles for Medical Care 101
Arindam Basu, Saurab Jain and V. S. Khoiwal

Photoluminescent Printed Fabrics: An Innovative Solution to Natural Nightlight 107
Richa Sharma and Nilanjana Bairagi

Development and Characterization of Metal Woven Electric Heating Fabrics 119
N. Muthukumar, G. Thilagavathi, T. Kannaian and S. Periyasamy

Part IV Textile Chemical Processing

Mechanical Properties of the Silk Degummed with Citric Acid and Ultrasound 131
Ruzica Brunsek, Ivana Schwarz and Mirta Than

Photocatalytic Decolorization of Rhodamine B Dye Solution Using TiO_2 Coated Cotton Fabric 139
Anu Mishra and Bhupendra Singh Butola

Chemical Modification of Indian Yak Fibre for Development of Jute/Yak Fibres Blended Warm Textile 151
Kartick K. Samanta and A. N. Roy

Statistical Optimisation of Nano-Zinc Oxide-Based Fire-Protective Finish on Jute Fabric 167
Ashis Kumar Samanta, Reetuparna Bhattacharyay (Roy), Arindam Bagchi and Ranjana Chowdhuri

A Study on the Efficiency of Lavender Microcapsules on Silk/Lyocell Blended Fabrics 193
Mariyam Adnan and J. Jeyakodi Moses

Part V Garment and Accessory Design

Method Development for Modeling, Designing, and Digital Representation of Outdoor and Protective Clothing 205
Muhammad Awais and Sybille Krzywinski

Safer Custody Clothing: Designing for Female Prisoners at Risk of Self-Harm 219
Jane Ledbury, Nicholas Hall, Barbara Shepherd and Laura Parker

Deeds Not Words, an Exploration into the Women's Suffrage Movement in Ulster Through the Fashion and Freedom Project 229
Alison Gault

Dynamic Anthropometry for Investigation of Body Movement Comfort in Protective Jacket 241
Inga Dabolina, Eva Lapkovska and Ausma Vilumsone

Development of Training Modules for Visually Impaired for Rehabilitation in Garment Manufacturing Units 261
Megha Gupta and Ritu Mathur

Part VI Testing and Characterization

Thermo-Physiological Comfort and Microbial Properties of Different Textile Raw Materials and Structures 285
Dragana Kopitar, Beti Rogina-Car and Zenun Skenderi

Handle Assessment of Knitted Mattress Fabrics Treated with Flame Retardant Finishes Using Fabric Touch Tester Device 295
Atiyyah Binti Haji Musa, Benny Malengier, Simona Vasile and Lieva Van Langenhove

Development of a Smoothness Tester for Fabrics 307
M. S. Parmar, Nidhi Sisodia and Maheshwar Singh

Thickness Loss of Handmade Carpets After Dynamic Loading 321
Shravan Kumar Gupta, Kamal Kanti Goswami and Abhijit Majumdar

Part VII Supply Chain Management and Sustainability

A New Collaborative Model for Demand-Driven Supply Chains: A Case Study on Textile Industry 339
Ke Ma, Sébastien Thomassey and Xianyi Zeng

Customer Analytics in Fashion Retail Industry 349
Chandadevi Giri, Sebastien Thomassey and Xianyi Zeng

Recent Developments in Recycling Silk Saris 363
S. Nivedita and Gargi

Awareness of Green Manufacturing in Apparel Industry 371
Ankur Saxena and Ajit Kumar Khare

Sustainable Production by Modifying Reduction Clearing in Polyester Dyeing 383
S. D. Kiruthika, R. Ugamoorthi, C. Venkatachalapathi and S. Ramarethinam

Part VIII Traditional Textile Arts and Crafts

Crafting Lives: Redefining Culture and Artisan Lives Through the Revival of Crafts in the State of Punjab, India 407
Simrita Singh and Anu H. Gupta

Moving Lights as Moving Spaces: Reinterpreting Traditional Bamboo *Chik* Making 427
Shubhra Singh

An Ergonomic Perspective of Uttarakhand Weavers 445
A. Goel and R. Garbyal

About the Editors

Abhijit Majumdar is a Chair Professor in the Department of Textile Technology, Indian Institute of Technology, Delhi. He has done his M. Tech in Textile Engineering and MBA in Technology Management from IIT Delhi, and his Ph.D in Production & Operations Management from Jadavpur University, Kolkata. His research interests include fabric manufacturing technologies, protective textiles, soft computing based modeling and sustainable operations and supply chain management. He has edited 2 books, and authored 2 books, 1 monograph, 6 book chapters and 95 research articles in refereed journals. He is a recipient of Outstanding Young Faculty Fellowship (2009–2014) of IIT Delhi, Teaching Excellence Award (2015) of IIT Delhi and Gandhian Young Technological Innovation Award (2017).

Deepti Gupta is a Professor in the Department of Textile Technology, IIT Delhi. She has more than 30 years research experience and her research interests include textile chemistry product design and development, antimicrobial finishing, anthropometrics and garment sizing, design and engineering of functional clothing. She is a member of various governmental, professional and industrial committees and has edited 4 books, and authored 1 book, 5 chapters and 87 research articles in refereed journals.

Sanjay Gupta is the Vice Chancellor of the World University of Design, Sonepat, India, and was previously the Dean of National Institute of Fashion Technology and the Founding Dean of the School of Design at GD Goenka University. He has done his Ph.D. in textile technology from IIT Delhi. He was a UNDP fellow to Fashion Institute of Technology (FIT), New York and a Visiting Professor at École Nationale Supérieure des Arts et Industries Textiles (ENSAIT), France. He was involved in the key initiative of bringing fashion schools worldwide under the umbrella of the International Foundation of Fashion Technology Institutes (IFFTI). He has over 100 publications and, over 30 presentations in national/international conferences and seminars. He has also been a member of committees & bodies at national/international level like ASSOCHAM, CII, IFFTI, BIS, NISTI besides being on the editorial boards of scientific journals.

Part I
Protective Textiles

Thermal Resistance of Leather and Membranes for Summer Desert Military Footwear Under Different Climate Conditions

Dragana Kopitar, Jadranka Akalovic and Zenun Skenderi

Abstract The desert climate is characterized by a big gap between day and night temperatures. In summer dry season, daytime temperatures can approach 45 °C and drop to 15 °C during the night. Aim of the paper is to investigate thermal resistance of footwear layers in military footwear for dry desert season under different climate conditions focusing on the temperature difference between day and night. Thermal resistance was determined using the Sweating Guarded Hotplate equipment according to the standard ISO 11092 and under climate condition known in deserts, respectively, 15 and 30 °C with relative humidity of 40%. Increasing the temperature, thermal resistance of leather for collar have tendency to decreases, while thermal resistance of leather for vamp and quarter have tendency to increase. The thermal resistance differences are not great but different tendencies of thermal resistance change is visible. It can be concluded that leather thickness influences on thermal resistance change under temperatures range of 15–30 °C. Behavior of the membranes under different temperatures is various. With the temperature increase, thermal resistance of two-layered membrane for tongue decreases for 5.9%. Thermal resistance differences, in temperature range of 15–30 °C, between minimum and maximum values for quarter and lining membranes are 17% for quarter and 13% for lining. Thermal resistance of fabrics for special uses, as protective footwear, should take in consideration climatic conditions under which fabrics will be used.

Keywords Thermal resistance · Military footwear · Climate conditions · Cattle velour suede leather

D. Kopitar (✉) · Z. Skenderi
Faculty of Textile Technology, University of Zagreb, Prilaz baruna Filipovica 28a, 10000 Zagreb, Croatia
e-mail: dragana.kopitar@ttf.hr

J. Akalovic
Faculty of Textile Technology, University of Zagreb, Study in Varazdin, Hallerova aleja 6, 42000 Varazdin, Croatia

A. Majumdar et al. (eds.), *Functional Textiles and Clothing*,
https://doi.org/10.1007/978-981-13-7721-1_1

1 Introduction

The desert climate, also known as an arid climate, is frequently characterized by three different climates; "cool" dry season where daytime temperatures peak between 35 and 45 °C and fall to 10–15 °C at night. Daytime temperatures can approach 45 °C during the "hot" dry season and drop to 15 °C during the night. During the rainy season, temperatures can range from 35 °C in the daytime to 20 °C at night [1]. Most of the time, desert is very dry, not humid, but after the desert has a good amount of rainfall, the desert becomes very humid. Summers in semi-arid deserts are long and have average temperatures between 21 and 27 °C where temperatures usually stay below 38 °C. During the evening, temperatures in semi-arid deserts drop to around 10 °C. If the outside temperature is in the range 18–25 °C, most people without clothing would begin to feel uncomfortable. If the outside temperature is in the range 30–40 °C, for any length of time, it can be life threatening. From a physiological point of view, the human body feels comfortable at about 26 °C with insulation of 0.093 m^2 °C W^{-1} [2]. Keeping the feet warm during day and night, where the temperature gap in the desert is significant, could be a problem and could jeopardize thermal comfort. Generally, a recognized solution for footwear is building up thin breathable layers of insulation as footwear liner [3]. Aim of the paper is to investigate thermal resistance of footwear layers in military footwear for dry desert season under different climate conditions focusing on the temperature difference between day and night.

2 Materials and Methods

2.1 Materials

The research of thermal resistance under different climate conditions of fabrics incorporated in summer desert military footwear was carried out (Fig. 1).

Military footwear, with inbuilt thermoplastic toe cap, is made of natural sand-colored chrome tanned leather of two thicknesses, as well vegetable and water repellent finished leather. The lining is three-layered beige-colored water repellent/waterproof and vapor permeable membrane of following raw material composition: the first layer is 85% PA/15% PES fiber, the second layer is PTFE membrane where the third layer is made of 100% PA fiber. The footwear tongue is made of two-layered laminate in which front side is made of PA fiber, backside from PES fiber bonded with adhesive. The quarter is made of two-layered laminate in which front side is made of PA/PUR, backside made of PES fiber mutually bonded with adhesive.

Fig. 1 Summer desert military footwear

2.2 *Method of Thermal Resistance Measurement Under Steady-State Conditions*

Thermal resistance was determinate using the Sweating Guarded Hotplate equipment according to the standard ISO 11092 and under climate condition known in deserts. The sample to be tested is placed on the heated plate at 35 °C temperature with the conditioned air ducted to flow along and parallel to its upper surface with a speed of 1 m s^{-1}. For the determination of thermal resistance (R_{ct}), the air temperature is set at 20 °C, relative humidity of 65% and airspeed at 1 m s^{-1}. After reaching the test conditions and steady state, the recording of values can be started [4, 5].

The thermal resistance of the fabric is calculated according to the following equation [4]:

$$R_{ct} = \frac{(T_m - T_a) \cdot A}{H - \Delta H_c} - R_{ct0} \tag{1}$$

where: R_{ct} is the thermal resistance [m^2 °C W^{-1}], T_m is temperature of measuring unit [°C], T_a is the air temperature during testing [°C], A is the area of the measuring unit [m^2], H is the heating power supplied to the measuring unit [W], ΔH_c is the correction term for heating power for the measurement of thermal resistance, and R_{ct0} is the apparatus constant for the measurement of thermal resistance [m^2 °C W^{-1}].

Thermal resistance R_{ct} of the tested material is determined as the arithmetic mean of the values of three individual specimens (Fig. 2).

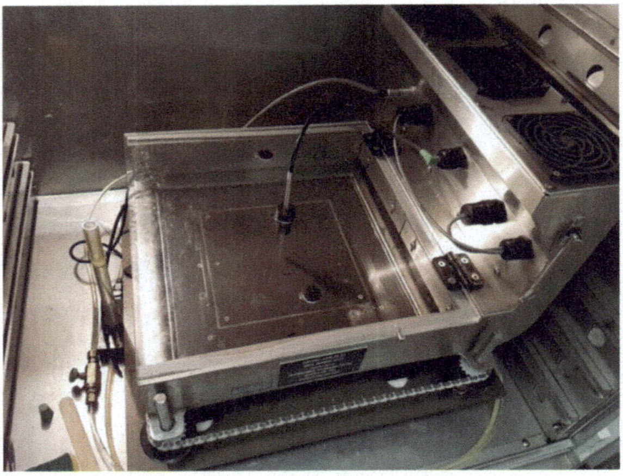

Fig. 2 Sweating guarded hot plate

Table 1 Parameters of material incorporated in summer desert military footwear

Samples			m (g m^{-2})	t (mm)
Cattle velour suede leather	Collar	M	719	0.95
		SD	0.59	0.02
Cattle velour leather	Vamp, quarter	M	1689	2.15
		SD	0.73	0.05
Two-layered membrane	Tongue	M	443	1.08
		SD	0.17	0.01
Two-layered membrane	Quarter	M	423	0.72
		SD	0.46	0.01
Three-layered membrane	Lining	M	230	0.53
		SD	0.09	0.01

Besides determination of thermal resistance under standard conditions, thermal resistance of samples was determined under summer desert temperatures (15 and 30 °C) with relative humidity setup at 40%.

3 Results and Discussion

Results of mass per unit area and thickness of fabrics incorporated in summer desert military footwear are presented in Table 1.

The thermal resistance tested according to the standard conditions and at temperatures of 15 and 30 °C with relative humidity at 40% is presented in Table 2 and Fig. 3.

Table 2 Thermal resistance of material incorporated in summer desert military footwear under standard and summer desert climate conditions

Samples			$R_{ct\_15°C}$ (m^2 °C W^{-1})	R_{ct} (m^2 °C W^{-1})	$R_{ct\_30°C}$ (m^2 °C W^{-1})
Cattle velour suede leather	Collar	M	0.0150	0.0149	0.0142
		SD	0.0004	0.0018	0.0008
		CV	2.68	12.21	5.49
Cattle velour leather	Vamp, quarter	M	0.0186	0.0187	0.0202
		SD	0.0013	0.0018	0.0003
		CV	7.04	9.60	1.64
Two-layered membrane	Tongue	M	0.0372	0.0360	0.0350
		SD	0.0017	0.0015	0.0002
		CV	4.60	4.13	0.47
Two-layered membrane	Quarter	M	0.0155	0.0169	0.0128
		SD	0.0009	0.0008	0.0002
		CV	5.79	4.87	1.48
Three-layered membrane	Lining	M	0.0146	0.0141	0.0165
		SD	0.0010	0.0013	0.0035
		CV	6.73	9.22	20.95

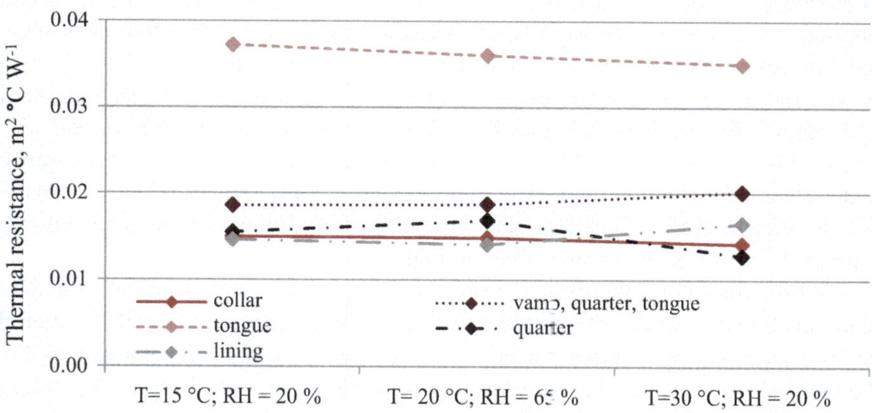

Fig. 3 Thermal resistance of military footwear layers under different climatic conditions

Comparing thermal resistance of cattle velour leather with different thicknesses and under different temperatures, different tendencies are visible (Table 2). With the increase of temperature (from 15 to 30 °C), thermal resistance of leather for collar (0.95 mm thickness) has tendency to decrease. Contrary to leather for collar, thermal resistance of cattle velour leather (2.15 mm thickness) intended for vamp and quarter has tendency to increase (from 0.0186 to 0.0202 m^2 °C W^{-1}) with temperature increase. Thermal resistance differences of cattle velour leather in temperature range from 15 to 30 °C are not great (5.3% decrease of 0.93 leather thickness and 8.60% increase of 2.15 mm leather thickness), but different tendencies of thermal resistance change are interesting. Since both cattle velour leathers have same finishing and quite difference in thickness (126%), it can be concluded that leather thickness influence on thermal resistance change under temperatures range of 15–30 °C.

The significance analysis of thermal resistance of cattle velour suede leather for collar and cattle velour leather for vamp, quarter, and tongue under different climatic conditions was conducted. The statistical analysis shows that the thermal resistance of two different types of leather under different climatic conditions is significant ($p = 0.001$).

Behavior of the membranes under different temperatures is various. Thermal resistance of two-layered membrane for tongue decreases for 5.9% (from 0.0372 to 0.0350 m^2 °C W^{-1}) with temperature increase.

Thermal resistance of two-layered membrane for quarter at standard condition is highest (0.0169 m^2 °C W^{-1}), while thermal resistance of three-layered membrane for lining is highest at 30 °C (0.0165 m^2 °C W^{-1}). Thermal resistance differences, in temperature range of 15–30 °C, between minimum and maximum values for quarter and lining membranes are 17% for quarter and 13% for lining.

The significance analysis of thermal resistance of three different membranes for tongue, quarter, and lining under different climatic conditions was conducted. The statistical analysis shows that thermal resistance under different climatic conditions of membrane for tongue and quarter ($p = 1.03 \times 10^{-4}$) and membrane for tongue and lining ($p = 2.73 \times 10^{-5}$) is significant. Thermal resistance under different climatic conditions of membrane for quarter and lining ($p = 0.99$) is not significant.

The thermal resistance values of materials incorporated in footwear measured under standard conditions show that thermal comfort of the footwear at 20 °C and 65% relative humidity is provided [6, 7]. The significance analysis of thermal resistance under standard condition as well as at 15 °C ($p = 0.99$) and 30 °C ($p = 0.94$) is not significant. We can conclude that summer desert military footwear in summer semi-arid desert climate is providing thermal comfort.

For providing good thermal comfort, the thermal resistance of all incorporated materials for summer desert military footwear at highest temperature (30 °C) should be lowest, respectively, and take away heat from feet. At lowest temperature (15 °C), the thermal resistance of the footwear layers should be highest, keeping feet warm and providing good thermal comfort.

4 Conclusion

The protective footwear, sometimes, is subject to very high or very low temperature, so it is necessary to evaluate capability of footwear to resist these two extreme situations.

The fabric incorporate in military footwear for summer dry desert climate conditions shows changes in thermal resistance under temperature range of 15–30 °C and 40% of relative humidity.

With the increase of temperature (from 15 to 30 °C) thermal resistance of leather for collar (0.95 mm thickness) has tendency to decrease, while thermal resistance of leather for vamp and quarter (2.15 mm thickness) has tendency to increase. The thermal resistance differences are not great (5.3% decrease of 0.93 leather thickness and 8.60% increase of 2.15 mm leather thickness), but different tendencies of thermal resistance change are visible. Since both cattle velour leathers have same finishing and quite difference in thickness, it can be concluded that leather thickness influence on thermal resistance change under temperature range of 15–30 °C.

Behavior of the membranes under different temperatures is various. With temperature increase, thermal resistance of two-layered membrane for tongue decreases for 5.9%. Thermal resistance of two-layered membrane for quarter at standard condition is highest, while thermal resistance of three-layered membrane for lining is highest at 30 °C. Thermal resistance differences, in temperature range of 15–30 °C, between minimum and maximum values for quarter and lining membranes are 17% for quarter and 13% for lining. We can conclude that studied summer desert military footwear in summer semi-arid desert climate is providing good thermal comfort.

For providing good thermal comfort, the thermal resistance for summer desert military footwear at highest temperature (30 °C) should be lowest and take away heat from feet, while at lowest temperature (15 °C) the thermal resistance should be highest keeping feet warm. Thermal resistance of fabrics for special uses, as protective footwear, should take in consideration climatic conditions under which fabrics will be used.

Acknowledgements The research is conducted from the project IP-2016-06-5278 with the support of the Croatian Science Foundation.

References

1. Food and agriculture organization of the United Nations, Forestry Department The arid environments, http://www.fao.org/docrep/t0122e/t0122e03.htm. Last accessed 2017/10/12
2. Williams, J.T.: Textiles for Cold Weather Apparel. Woodhead, Cambridge (2009)
3. Kuklane, K.: Footwear for Cold Environments Thermal Properties, Performance and Testing. Doctoral Thesis. National Institute for Working Life, Stockholm (1999)
4. ISO 11092 Textiles—Physiological effects—Measurements of thermal and water-vapour resistance under steady-state conditions (sweating guarded-hotplate test)
5. Operators Manual for Material Evaluation Hotplate Model SGHP-8.2 S/N 223-xx, Measurement Technology Northwest (2006)
6. Colak, S.M., Ozdil, N., Ekinci, M., Kaplan, O.: Thermophysiological comfort properties of the leathers processed with different tanning agents. Tekstil ve Konfeksiyon **4**(26), 436–443 (2016)
7. Frydrych, I., Sybilska, W., Wajszczyk, M.: Analysis of selected physical properties of membrane fabric influencing the utility comfort clothing. Fibres Text. East. Eur. **6**(17), 50–55 (2009)

Woven Military Fabrics from the Aspect of the Microbial Barrier Permeability

Ivana Schwarz, Beti Rogina-Car and Ruzica Brunsek

Abstract Functionality is of utmost importance for fabrics designed for military purposes, where woven fabric structural parameters and finishing processes greatly affect the definition and achievement of high set properties demands. Considering the application conditions of military fabrics and the high exposure to microorganisms from environment, as well as from the direct and indirect transfer between users, the possibility of contamination is unavoidable. Therefore, the aim of this paper is to determine whether and to what extent military fabrics have the properties of microbial barrier permeability. Determination of microbial barrier permeability was conducted according to the newly developed method. The most resistant forms of microorganisms were used, the bacterial endospores of apatogenic species: *Geobacillus stearothermophilus* and *Bacillus atrophaeus*. The analysis of the obtained results and the analysis of correlations between some relevant fabric properties confirm the extremely complex structural aspect of the woven fabrics. Tested woven fabrics designed for military purposes show a very good microbial barrier permeability with a different adhesion of microorganisms, conditioned by the complex woven fabric structure, based on which can be approached to further targeted functional designing of specific elements.

Keywords Microbial barrier permeability · Military purposes · Woven fabric structures

1 Introduction

Military fabrics must meet a number of high set demands, and it is crucial that clothing and related equipment are lightweight, compact and durable and have high performance. The purpose of military camouflage fabrics is not only to assimilate with the environment, but also to protect from various conditions and situations in

I. Schwarz (✉) · B. Rogina-Car · R. Brunsek
Faculty of Textile Technology, University of Zagreb, Prilaz Baruna Filipovica 28a, 10000 Zagreb, Croatia
e-mail: ivana.schwarz@ttf.hr

A. Majumdar et al. (eds.), *Functional Textiles and Clothing*,
https://doi.org/10.1007/978-981-13-7721-1_2

which body protection is of vital importance. Functional criteria of military fabrics should meet high camouflage and physical–mechanical and thermal requirements. Camouflage properties are achieved by specific patterns and colours application onto woven fabric, depending on the end-use conditions, with the aim of breaking the silhouette of the human body and achieving imperceptibility. Requirements of physical and mechanical properties set for these fabrics have a direct impact on psychological factors of individuals, which can cause psychological discomfort and instability that interfere with motivation and readiness to perform high-risk tasks. Military fabrics are produced in a combination of natural and synthetic fibres, which, with their specific properties and structural parameters, as well as additional chemical finishing (resistance to UV, water, heat, flame, wind), meets a wide range of requirements, all in order to achieve the needed properties [1–3]. Physiological (skin and temperature) and environmental variables (conditions of temperature, humidity and wind) are extremely important in achieving maximum comfort and safety while wearing a fabric. Its insufficient properties can cause a variety of negative phenomena (discomfort, climate imbalance, heat stress, the emergence of a large amount of nuisance, visual and cognitive disruption) affecting the person's mental and physical stability [4].

Another extremely important feature of the fabric for military clothing, regarding the exposure to microbial influences, is the ability of body protection and control of microorganisms. Therefore, the question is whether and to what extent military fabrics have microbial barrier properties.

The subject of this research is fabrics designed for certain specific purposes, with high requirements for the human body protection, in order to preserve mental and physical health. The use of fabrics for military purposes in real application conditions (in various areas of habitation and activity) represents indispensable exposure to a possible microbial contaminated environment and thus a danger of microorganism's penetration and body contamination. The reason for this is the fabric ability to retain moisture, which represents a possible cause of endangering the health of an individual. Textile materials are in constant contact with microorganisms, not only from skin but also from the environment. Textile material is a media that supports the microorganism's growth and development, because it provides them an environment rich in nutrients necessary for their survival (moist air or wet textile product). Microorganisms are fed with skin cells and various substances used in chemical finishing processes (and also dirt) [5–8]. The growth of microbes implies increasing their cells, but also increasing their number after multiplication, resulting in colony formation. Microorganism's growth and development depend on many factors, both physical and chemical, as well as about the environment in which they are located. Methods of microbial transmission are different, through direct and indirect contact with contaminated surfaces, contaminated environment (water) and even non-hygienic conditions [9]. In order to provide maximum body protection from microorganism penetration in real application conditions, the fabric should have sufficient microbial barrier properties.

2 Materials and Methods

The fabrics used in this research are designed and produced for military purposes in the Čateks, Croatia. The fabric samples contain equal ratios of natural and synthetic fibres (cotton/PA 6.6 and cotton/PES), with mass per unit areas from 220 to 280 g/m^2, and woven in different weaves (Fig. 1). After weaving process, the fabrics were subjected to dyeing process, using reductive dyes to print camouflage patterns ("forest" and "desert" patterns) depending on the fabric end-use conditions. To achieve the needed properties, the fabrics were treated with protective coatings of water and oil repellents, which, together with the properties achieved by the fabric structural parameters, provides exceptional characteristics of strength, durability, comfort and stability.

The relevant structural parameters of woven fabrics were determined in accordance with the standardized testing methods, based on which further analysis was carried out:

- mass per unit area—*m* (g/m^2)—determination of mass per unit length and mass per unit area, in accordance with the standard HRN ISO 3801:2003,
- fabric thickness—*t* (mm)—determination of thickness of textiles and textile products, in accordance with the standard HRN EN ISO 5084:2003,
- fabric density—*d* (threads/10 cm)—determination of the number of threads per unit length, in accordance with the standard HRN EN 1049-2-2003.

The following tests were also performed:

- determination of fabric air permeability, according to the standard HRN EN ISO 9237:2003

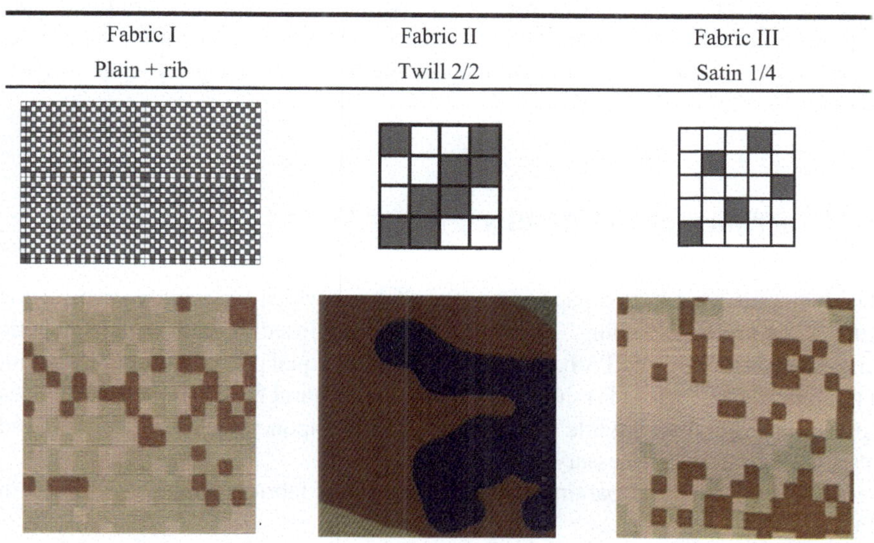

Fig. 1 Woven fabric samples used in the research

Fig. 2 Presentation of prepared samples for microbial barrier permeability testing

- determination of fabric breaking properties, according to the standard ISO 13934-1:1999
- determination of fabric thermal resistance ("hot plate"), according to the standard ISO 11092:1993
- determination of microbial barrier permeability according to the newly developed method. Tests were performed according to the following procedure: A test specimen of 22 cm × 22 cm is fixed to a ring-shaped device (Fig. 2) and placed into a transparent sterilization pouch. It is sterilized in the steam sterilizer Selectomat PL MMM (Münchener Medizin Mechanik) at 134 °C for 5 min. After the sterilization procedure under sterile conditions, the bacterial endospores of apatogenic species, in this case *Geobacillus stearothermophilus* 10^5 and *Bacillus atrophaeus* 10^6, are applied to the test area of the specimen. This is followed by incubation for 24 h. Afterwards, a CT3P agar print plate is used to take prints, first from the face side and afterwards from the back side. After taking prints, CT3P agar print plates are placed into a thermostat at 35 °C in order to perform incubation for 72 h, after which the number of bacterial colonies on the face and back sides of the fabric is read [10, 11].

3 Experimental and Results

The materials used in this paper are woven fabrics which are mainly used for the purpose of military clothing. The materials are composed of natural and synthetic fibres blends (cotton and PA 6.6, cotton and PES, in equal proportions of 50%) with a nominal yarn finesses (Tt) shown in Table 1. The natural fibre component provides comfort and flexibility, while the synthetic fibre component provides strength and resistance properties necessary for final application.

The basic structural parameters of tested woven fabric samples are shown in Table 1.

The difference of PA 6.6 and PES fibre properties, in a certain part, affects the yarn properties and thus the properties of finished woven fabrics. The values of standard

Table 1 Fabric structural parameters

Properties	Fabric I		Fabric II		Fabric III	
	Warp	Weft	Warp	Weft	Warp	Weft
Raw material	PA 6.6/cotton 50%/50%		PES/cotton 50%/50%		PA 6.6/cotton 50%/50%	
m (g/m^2)	220		250		280	
t (mm)	0.436		0.439		0.557	
Tt (tex)	17 × 2	40	20 × 2	20 × 2	30	50
d (thread/10 cm)	360	207	362	210	467	291

m—fabric mass per unit area (g/m^2), t—fabric thickness (mm), Tt—yarn finesses (tex), d—fabric density (thread/10 cm)

Table 2 Tested results of fabric breaking properties, thermal resistance and air permeability

Woven fabric properties		Fabric I		Fabric II		Fabric III	
		$\bar{x}$	CV (%)	$\bar{x}$	CV (%)	$\bar{x}$	CV (%)
F (N)	Warp	691.37	3.55	774.73	2.19	777.99	2.55
	Weft	441.30	2.22	500.14	1.96	845.01	5.00
ε (%)	Warp	43.17	5.47	18.67	5.58	32.33	6.25
	Weft	22.50	3.71	17.00	2.94	35.67	5.67
R_{et} (m^2 kW^{-1})		0.0185	3.9	0.0203	2.0	0.0127	1.3
R (mm/s)		22.244	4.2	35.972	6.3	10.220	4.6

F—breaking force (N), ε—elongation at break (%), R_{et}—thermal resistance (m^2 kW^{-1}), R—air permeability (mm/s), $\bar{x}$—mean value, CV—coefficient of variation (%)

strength range from 30 to 68 cN/tex for PA 6.6 and from 25 to 55 cN/tex for PES fibres. Apart from the raw material that carries the basic characteristics of the yarn, many other parameters, such as yarn structure and the fabric construction (density and weave), have an influence on the final woven fabric properties. The results of breaking properties, as well as thermal resistance and air permeability properties of tested fabric, are shown in Table 2.

Thermal resistance property of woven fabric used for military clothing is of utmost importance, regarding the extreme application conditions, where the influence of the environment (coldness in the winter and the heat in the summer days) affects the psychophysical condition of a person wearing a certain fabric. The results of tested woven fabrics show that the highest thermal resistance has Fabric II, with value of 0.0203 m^2 KW^{-1}, which is woven in a twill 2/2 weave with two-ply yarns (with more closed structure), preventing heat transfer through the yarn (Table 2). This means that such fabric has better properties of heat retention near the body, which is suitable for application during colder periods. Conversely, Fabric III shows the lowest thermal resistance with 0.0127 m^2 KW^{-1}, despite the fabric highest mass per unit area. The reason for this can be found in the structural characteristics of the fabric, which is woven with one-ply yarns, whose structure is more open and thus more susceptible to heat penetration. Furthermore, this Fabric III is woven in satin weave characterized by strong flotation, which additionally provides easier heat penetration between the threads.

Analysing the results of the air permeability tests (Table 2), it can be concluded that the structural parameters of the fabric, as well as the applied apertures, have influenced the low air permeability of the tested samples. The obtained results diametrically follow the thermal resistance values, where Fabric II provides the highest values of air permeability (35.972 mm/s) and Fabric III the lowest values (10.220 mm/s). This can be explained by structural parameters of Fabric III, i.e. in highest density of the warp and weft threads, which with its compactness affects the fabric cover and porosity, thus making the air transfer through such woven structure difficult.

The results of microorganism permeability through the fabric structure testing, carried out according to the newly developed method, are shown in Table 3. It is important to note that all tested samples were equally contaminated, with the same number of microorganisms (10^5 *Geobacillus stearothermophilus* and 10^6 *Bacillus atrophaeus*) applied on the face side of the fabrics. An interesting indication is that the number of microorganisms transmitted from the biological indicator to the face of the fabric differs significantly between the tested samples. This is caused by the fabric surface, conditioned by the fabric structural parameters and the applied apertures. Fabric III, woven in Satin weave (with a large flotations and a rare thread interlacing) retain on face side 992 CFU, Fabric II 540 CFU (45.6% less than Fabric III) and Fabric I (with the most frequent threads interlacing) even 82.5% less than Fabric III, only 174 CFUs. Despite the largest number of bacterial colonies on the face side of Fabric III, not one microorganism has passed through fabric to the back side, which shows its exceptional microbial barrier properties (Fig. 3). Furthermore, the remaining two samples, Fabric I and Fabric II, with a minimum number of only 1 CFU on the back side of fabrics, prove good fabric properties of microbial barrier permeability.

By analysing the ratio of applied and passed through CFUs from face to back sides of the fabrics, it can be concluded that there is a significant difference between Fabric I and Fabric II. The ratio of CFU on the face side and back side of Fabric II is 540:1, whereas this ratio in Fabric I is 174:1. Above mentioned shows that for the transition of 1 CFU on the back side of the Fabric I is required 174 CFU applied

Table 3 Results of microbial barrier permeability of tested fabrics after extreme contamination

Sample	The average number of bacterial colonies on the fabric face side (CFU)	The average number of bacterial colonies on the fabric back side (CFU)	Ratio (CFU)
Fabric I	174	1	174:1
Fabric II	540	1	540:1
Fabric III	992	0	–

CFU Colony-forming unit

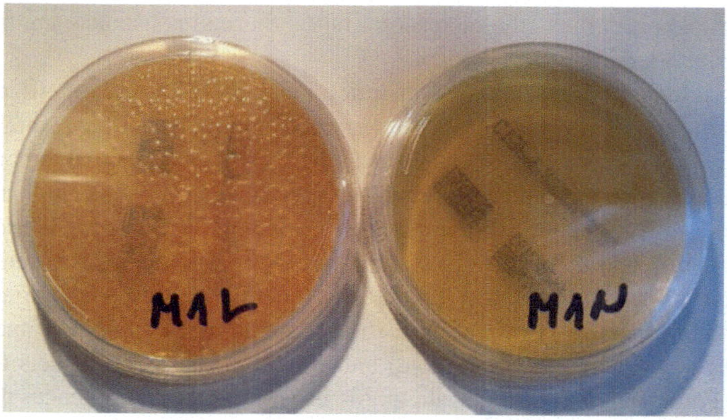

Fig. 3 Display of CT3P agar plate on the face and back sides of tested sample Fabric III

Table 4 Values of correlation coefficient of some significant fabric parameters

	d_{warp} (thread/10 cm)	d_{weft} (thread/10 cm)	t (mm)	R (mm/s)	R_{et} (m^2 KW^{-1})	CPU
CPU	−0.75819	−0.76798	−0.76170	0.27867	0.57745	1
R_{et} (m^2 KW^{-1})	−0.97015	−0.96637	−0.96883	0.94500	1	
R (mm/s)	−0.83749	−0.82912	−0.83453	1		
t (mm)	0.99999	0.99995	1			

on face side; while in Fabric II, the required amount of applied bacterial colonies is three times higher. This points to the conclusion that Fabric II has a better microbial barrier property, which can, once again, be explained by the fabric weave, regarding the approximately uniform structural fabric parameters. The twill 2/2 weave does not have abrupt or irregular thread interlacing, but the warp and weft threads are lined in identical proportions, with slight gradual shift, without any major bending. In contrast, the weaves with the most frequent interlacing and transitions from the face to the back sides of the fabric in a specific pattern (plain, rib) define more pronounced pores, allowing easier penetration of microorganisms through the fabric.

Correlation analysis was used to discover the degree of correlation between tested parameters, and the most interesting correlations, based on which significant conclusions were drawn, are presented in Table 4.

The air permeability property shows a very good correlation with the fabric thickness parameter (r = −0.83453), which is directly influenced by the fabric density, i.e. the density of warp and weft threads.

Statistical analysis of correlation indicators shows a strong dependent relation of thermal resistance property with the fabric thickness parameter (r = −0.96883). The results also show a strong correlation of thermal resistance with air permeability property (r = 0.94500), where the increase of air permeability (caused by decrease of fabric thickness) affects the increase of fabric thermal resistance, i.e. the ability of fabric heat retaining near the body.

Placing into correlation relations, the microbial barrier fabric properties with the fabric structural parameters indicate weaker correlation connections, which are still classified as good correlation ($r = -0.76170$). This is understandable considering that the microbe penetration is performed in a dry state, i.e. by mechanical penetration, which is affected by the fabric thickness parameter and thus the fabric density. Air permeability and thermal resistance properties are in poor correlation with the fabric microbial barrier property. The reason for this is, already mentioned, complex structural aspect of woven fabric, i.e. the constructional parameters and specific weave. The way of interlacing of warp and weft threads, which affects the fabric cover and compactness, as well as the pores structure, greatly influences the ability of adhesion and transition of microorganisms through the fabric.

All of the above findings are of utmost importance for the design process of woven fabric for specific purposes, which enable defining and achievement of the desired fabric properties depending on the end use, i.e. application conditions.

4 Conclusion

Based on the conducted research, it can be concluded that all of the relevant military fabric properties, including the microbial barrier property, are greatly affected, besides finishing processes, by the fabric structural parameters. The analysis of the obtained results and the analysis of correlations between some relevant properties confirm the extremely complex structural aspect of the woven fabrics. Tested fabrics for military purposes show very good microbial barrier properties, with the indicated differences, based on which can be approached to further targeted functional designing of specific elements.

Acknowledgements The financial support was provided within the framework, Support scientific research 2017, "Natural Fibres Agrotextile" (TP 12/17), financed by the University of Zagreb.

References

1. Wilusz, E.: Military textiles. Woodhead Publishing Limited, Cambridge, England (2008)
2. Sparks, E.: Advances in military textiles and personal equipment. Woodhead Publishing Limited, Cambridge, England (2012)
3. Winterhalter, C.A., Lomba, R.A., Tucker, D.W., Martin, O.D.: Novel approach to soldier flame protection. J. ASTM Int. **2**(2), 227–232 (2005)
4. Mijović, B., Skenderi, Z., Salopek, I.: Comparison of subjective and objective measurement of sweat transfer rate. Coll. Antropol. **33**(2), 315–320 (2009)
5. Thiry, M.C.: Prescription textile protection. AATCC Rev. **10**, 30–36 (2010)
6. Teufel, L., Redl, B.: Improved methods for the investigation of the interaction between textiles and microorganisms. Lenzing. Ber. **85**, 54–60 (2006)

7. Beck, W.C., Carlson, W.W.: Aseptic barriers. Arch. Surg. **112**(2), 2240–2244 (1981)
8. Borkow, G., Gabbay, J.: Biocidal textiles can help fight nosocomial infections. Med. Hypotheses **70**, 990–994 (2008)
9. Duraković, S.: Opća mikrobiologija. Prehrambeno-tehnološki inženjering, Zagreb (1996)
10. Rogina-Car, B., Budimir, A., Katović, D.: Microbial barrier properties of healthcare professional uniforms. Text. Res. J. **87**(15), 1860–1868 (2017)
11. Rogina-Car, B., Budimir, A., Turčić, V., Katović, D.: Do multi-use cellulosic textiles provide safe protection against the contamination of sterilized items? Cellulose **21**(3), 2101–2109 (2014)

Development of Thermal Insulative Nonwoven Fabric Through Advance Material Application

Rohit Naik and Arup Rakshit

Abstract In recent years, as a sway of high pace development in science and technology, people tend to have more aptitude towards using clothing for new functions, which ultimately contributes to opening of opportunities for further development and incorporation of new technologies along with novel materials. In this context, textiles are of fast decalescence or fast heat radiation media as far as comfort accountability of textile articles is concerned. The structure and texture of textiles play a very vital role in determining the thermal comfort level of the human body; hence, people need to obtain various advantages of the functional material design in order to improve the heat–moisture balance of textiles. Thermal comfort depends on the extent to which the clothing influences heat and moisture transport between the human body and environment. In order to come up with solution for this, the use of silica aerogels gains extreme attention due to their surprising properties and their existing potential applications in a wide of variety technological areas. Aerogel basically exhibits nanostructure which offers high porosity, high specific surface area, low density and outstanding heat insulation properties. This paper emphasis on development of thermal insulative PET nonwoven fabric with the application of silica aerogel which gives excellent thermal insulation with reduced bulk and weight which are generally enforced to insulative textile materials to induce functionality. Findings are supported by various analyses and testings followed by STATISTICA 6 software for ensuring statistical significance of all parameters. The developed product shows the potential to be used at various technical textile product ranges.

Keywords Silica gel · Thermal insulation · Nonwoven fabric

R. Naik (✉) · A. Rakshit
Veermata Jijabai Technological Institute, Matunga 400019, Mumbai, India
e-mail: naikrohitp@gmail.com

A. Rakshit
e-mail: akrakshit@vjti.org.in

A. Majumdar et al. (eds.), *Functional Textiles and Clothing*,
https://doi.org/10.1007/978-981-13-7721-1_3

1 Introduction

On sway of development in science and technology, people have increasing requirements on uses of clothing for new functions, which ultimately going to contribute towards release of new opportunities for further development and incorporation of new materials and technologies in the recent years. In the aspect of the comfort textiles, we hope that the textiles are of fast decalescence or fast heat radiation. The microstructure and texture of textiles play a very important role in determining the heat–moisture comfort level of the human body. Fabric, especially in cold environment, is designed to keep the body in its critical internal temperature by offering a specific rate of heat loss and works as a thermal barrier. Heat transfer through the textile occurred due to the mechanism of radiation, convection and conduction. Most of the methods for the determination of heat transfer through textile take into account all the three mechanisms as a whole. Considerable studies on heat and moisture transfer through woven and knitted fabrics structure have been conducted over the year [1–3]. In cold atmosphere, human body loses substantial amount of the heat through fabric, which is mostly attributed to conduction mode of heat transfer than other two modes viz. convective and radiative [4]. Designing lower thermal conductive fabrics involves the use of different fibre types, constructional parameters, special finishing and other techniques like membrane and composite [5]. In case of high-loft textile structures, method of incorporating hollow fibres has well established itself; which is practised for high-altitude clothing too. The study made by Yoshihiro et al. (2008) related to use of composite fabric/resin material to further reduce the thermal conductivity. Another way to achieve this is to make composite of fabric with the use of low thermal conductive material such as inorganic nanoporous silica sols.

Textiles material is discontinuous materials; in that, they are produced from macroscopic sub-elements (finite length fibres or continuous filaments). The discrete nature of textile fabric is generally a porous media that comprises two phases: solid (consisting of solid fibres) and gas (consisting of water vapour and dry air) that contribute directly to some of the transportation properties of the textiles, for example, thermal insulating characteristics, liquid absorption properties, softness and other tactile characteristics. Clothing functionality and performance have been a major concern over the years. Clothing combines both the functions: it acts as a barrier to the outside environment, and at the same time, it is the transporter of heat and moisture from the body to the surrounding environment; hence, it is essential to keep thermal balance between heat production and heat loss by maintaining transport of heat and moisture through the human–clothing–environment system [6]. The impact of clothing on thermal comfort depends on the extent to which the clothing influences heat and moisture transport between the human body and surrounding environment [7]. Fabric can be designed to render a specific rate of loss of insensible perspiration for assisting the skin in conserving essential levels of body fluids. Specific rates of heat loss are essential to keep the body at its critical internal temperature in a cold environment, so understanding of heat and mass flow through textile material becomes necessary. Now a days, there are many techniques used to design fabrics which works as a thermal barrier. Other forms which are used for this purpose can be enlisted as membranes, composite, fibre fills etc. To predict the degree of thermal insulation of textile materials, some theoretical models have been proposed through significant study on heat and moisture transfer in fabrics systems [8, 9]. Theoretical research on thermal properties is categorised as modes of heat transfer through textile materials, heat transfer through porous materials, and heat and mass transfer though textile materials. Based on woven geometry proposed by Peirce, researcher found heat loss by conduction and radiation at intersecting points, and by individual yarn whereas air in yarn pores, air trapped in fabric voids are treated as resistance to heat loss [10].

The main bottleneck which is associated with textile materials to be successful as thermal insulation materials can be enumerated as follows: Firstly, high loft or bulkiness of material so as to provide a predetermined amount of insulation by ensuring sufficient trapping of air. Secondly, the insulation depends on forced convection; such convective heat loss cannot be prevented by textile material. Third is that the textile alone cannot reach the level of thermal conductivity lower than 0.025 W/m K of air. Perhaps nano-fibres can do so, but still mass production and cost-effectiveness is a problem. Finally, use of such high loft material for thermal insulation becomes heavier and uneasy to manage especially when required to carry over a body. The proposed work aims at developing lightweight effective thermal insulation textiles in combination with nanoporous aerogel which provides the fundamental basis for the optimisation of material properties to achieve good performance of the clothing system along with that it creates numerous opportunities for research to unearth advanced fabrics/clothing suitable for various technical and allied applications.

2 Materials and Method

2.1 Material

Raw Material PET fibre is used for aerogel-reinforced fabric formation because polyester fibre is cost-effective, has low thermal conductivity, easily available in current market, excessively used in various technical textile applications and exhibits scope for bond formation with other chemicals or materials provided it is treated with appropriate chemicals along with proper experimental conditions like time, temperature, pressure, catalysts and heat.

Fabric Structure Thermal bonded nonwoven fabrics of 100% PET were chosen to see the combined effect of aerogel, different fabric GSM (gm/m^2) and structure of fabric on the thermal conductivity as it provides better structural suitability and geometrical adaptability for chemical treatments. In this concern, 100% PET nonwoven fabrics with different GSM (g/m^2) were prepared, including web formation by carding machine followed by thermal bonding process to produce thermal bonded nonwoven machine at DKTE COE for nonwoven, Ichalkaranji.

2.2 Methodology

Fabric Sample Preparation Thermal bonded nonwoven fabrics were subjected to chemical pretreatment in order to make the nonwoven fabric (especially PET material) suitable for application of silica sol. In this pretreatment process, first thermal bonded PET nonwoven fabric was subjected to 15% NaOH (pad) for 2 h which further followed by deionised water wash and drying. Furthermore, the fabric was made suitable for sol intake by treating with 1% APTES/toluene solution at room temperature followed by drying.

Aerogel Composite Synthesis [11]

Chemicals Tetraethylorthosilicate (>98%, TEOS), trimethylchlorosilane (>95%, TMCS), ethyl alcohol (>99%, EtOH), *N*-hexane (>99%) and deionised water were purchased from Sigma-Aldrich, India and used without any purification and time delay. *N*, *N*-Dimenthyl-formamide (>98%, DMF), hydrochloric acid (37%, HCl) and ammonia (25%, NH_4OH) were used by diluting 10-fold with deionised water.

Procedure One hundred and fifty millilitres of silica sol was prepared by a two-step acid/base catalysed sol–gel process. In the first step, solution "A" was prepared by mixing TEOS, EtOH, deionised H_2O and HCl in the molar ratio 1:7: 1: 1 × 10^{-5} and magnetically stirred for 30 min. The solution then refluxed for 24 h at room temperature. In the second step, solution "B" was prepared by mixing DMF, H_2O and NH_4OH and then added into the stock solution and stirred for 30 min. The final molar ratio of TEOS: EtOH: H_2O: DMF: HCl: NH_4OH is 1:7:2:0.25:10^{-5}:3.57 × 10^{-3}.

Synthesis of SiO_2 aerogel composites reinforced with nonwoven fabric: Aerogel was applied in the pre-synthesis stage into the nonwoven fabric by padding technique, and composites were obtained. The fabric/SiO_2 composite was prepared by padding silicic acid with 83% expression at 2.5 kg/cm^2 squeezing pressure. Monoliths gel starts forming after 3–4 min. The composite gels were kept at room temperature for 2 days for further solidification to form silica monoliths. These monoliths were aged in H_2O/EtOH (1:4, vol.) solution for 24 h and then TEOS/EtOH (1:4, vol.) solution to strengthen the gel network. The water and ethanol solvents in the pores of the wet gel were exchanged with isopropanol and *n*-hexane, respectively, after being immersed in a solution of 10% vol. The aged monolith is then transfer to TMCS/*n*-hexane mixture where it was kept at 35 °C for about 8 h for surface modification. The monoliths were washed in *n*-hexane for 32 h. The SiO_2 aerogel composites reinforced with nonwoven structure were synthesised by drying the monoliths at 70 °C for 12 h followed by further drying at 100 °C for 12 h. Table 1 gives an overall idea about the procedure followed for development of aerogel composite.

Table 1 Stepwise procedure to aerogel composite

S. No.	Step	Time	Temperature
1	Sol-Préparation {Acid-catalysation (solution A)}	30 min	Room
2	Solution a reflux	24 h	Room
3	Gel-Preparation Base-catalysation (Solution B)	30 min	Room
4	Fabric Pad with silicic acid (Gel formed inside fabric structure)	4–5 min	Room
5	Solidification of monoliths	48 h	Room
6	Ageing in EtOH	24 h	Room
7	Strengthening silica frame network in TEOS/EtOH	24 h	Room
8	Solvent exchange with Isopropanol, *N*-hexane	48 h	Room
9	Surface modification (hydrophobicity) with TMCS/*N*-hexane	8 h	35 °C
10	Monolith washing with N-hexane	48 h	Room
11	Drying	12 h	70 °C
12	Drying	8 h	100 °C

2.3 *Methods of Characterisation*

Finally, sample of 100% PET thermal bonded nonwoven of size 30 cm *30 cm was prepared with two different GSMs, i.e. 80 g/m^2 and 150 g/m^2. Samples are subjected to same type of pretreatments stated above, and sol is synthesised and applied on the sample with the help of padding technique to develop in situ aerogel. These samples are further subjected to various chemical treatments for solvent exchange listed in Table 1. Samples are then tested against untreated samples of same GSM in order to study the effect of aerogel application on various properties of nonwoven fabric. Main properties taken under consideration are thickness, thermal conductivity, specific thermal resistivity, air permeability and pore size and are shown in Table 2.

Apart from that sample was also analysed for their surface structure, functional group present, microscopic images by using scanning electron microscope (SEM) and Fourier transform infrared spectroscopy (FTIR).

Furthermore, the results obtained were statistically analysed by using STATISTICA-6 software for their level of significance. For various parameters factorial analysis of variance (ANOVA) test is carried out which gives the *P* value for analysing significance level (if *p* value less than 0.05 then difference is significant) along with that Regression Summary for Dependent Variable are also studied to obtain correlation coefficient (*R*). Graphs were also drawn by using STSTISTICA-6 software which shows clear trend of change in every parameter before and after application of aerogel material on nonwoven fabric.

Table 2 Testing standards and instrument utilised

Properties		ASTM Standard	Instrument
Physical properties	GSM (g/m^2)	–	GSM cutter
	Thickness (mm)	D1777-96	Thickness tester
Transfer properties	Air-permeability ($cm^3/cm^2/s$)	D737-04	TexTest (Fx3300)
	Porosity (%)	F316	PMI Capillary flow Porometer
	Thermal conductivity (W/m K)	C518	LASERCOMP Thermal conductivity tester
Resistive behaviour	Thermal resistance (m K/W)	C518	LASERCOMP Thermal conductivity tester
Miscellaneous	FTIR		FTIR 8400S
	SEM		

3 Result and Discussion

In Table 3, average results for various properties tested for silica/aerogel nonwoven fabric are summarised which reflects the significance of improvement in functionality of fabric after incorporation of aerogel into thermal bonded nonwoven PET fabric.

Table 3 Summarised average result of properties for Sio_2 aerogel/nonwoven composite fabric

	Before	After	% Change	*P* value	*R* value
80 GSM					
Thickness (mm)	6.881	3.022	56.10	0.000	0.995
Thermal conductivity (W/m K)	0.035	0.029	16.09	0.000	0.985
Specific thermal resistivity (m K/W)	28.137	33.534	19.18	0.000	0.993
Air permeability ($cm^3/cm^2/s$)	297.20	95.20	67.96	0.000	0.998
Mean pore size (Micron)	190.6	137.8	27.70	0.002	0.897
150 GSM					
Thickness (mm)	7.631	4.170	45.35	0.000	0.995
Thermal conductivity (W/m K)	0.033	0.028	14.35	0.000	0.985
Specific thermal resistivity (m K/W)	29.758	34.746	16.76	0.000	0.993
Air permeability ($cm^3/cm^2/s$)	263.20	72.80	72.34	0.000	0.998
Mean pore size (Micron)	179.6	116.4	35.18	0.002	0.897

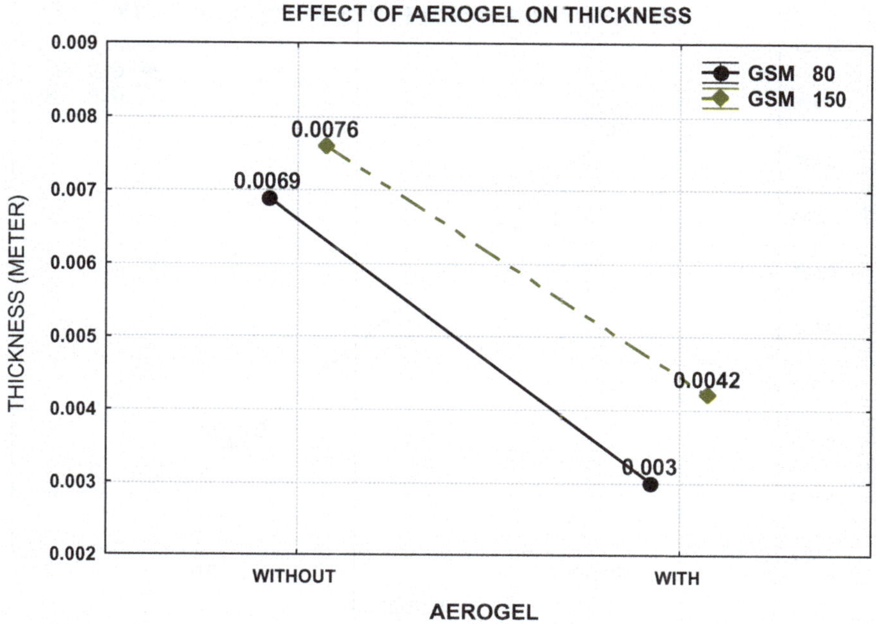

Fig. 1 Effect of aerogel on nonwoven fabric thickness

3.1 Effect of Aerogel on Thickness of Nonwoven Fabric

It is observed from Fig. 1 that for both types of fabrics, there is a significant reduction in thickness value after the application of aerogel. However, the PET nonwoven with GSM 80 shows 56.10% reduction in fabric thickness while fabric with 150 GSM gives 45.35% less thickness value. The reason behind this reduction is may be due to gelation of aerogel inside nonwoven structure which leads to stick fibre components in more closely packed condition. Second reason behind reduction in thickness value is heating provided during the drying stage of aerogel formation. That heat is going to evaporate all the excessive solvents present inside the nonwoven material as well as from the microstructure of aerogel material itself which results in recess of thickness.

3.2 Effect of Aerogel on Thermal Conductivity of Nonwoven Fabrics

It is clearly seen from the Fig. 2 that, after application of aerogel; thermal conductivity value shows significant decrease i.e. for 80 GSM; thermal conductivity falls from avg. value 0.0355 W/m K to avg. value 0.0298 W/m K. which is almost 16.09% decrease. While in case of nonwoven fabric of 150 GSM, it shows drop in thermal

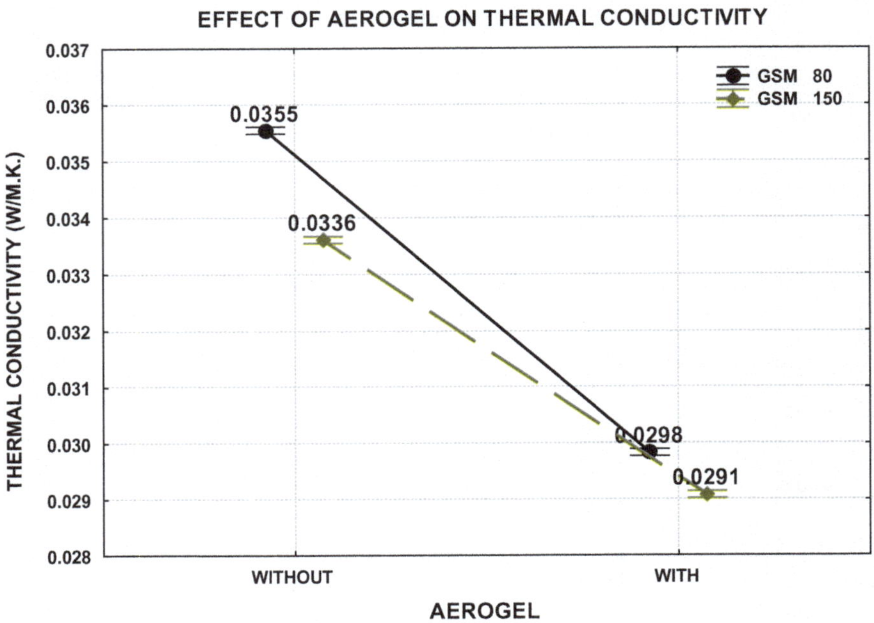

Fig. 2 Effect of aerogel on nonwoven fabric Thermal conductivity

conductivity by about 14.35%, i.e. decrease from 0.0336 to 0.0291 W/m K. Since both nonwoven fabrics irrespective of their GSM have same add-on of silica aerogel, their value of reduced thermal conductivity hardly differs.

Main reason behind this significant decrease is formation of in-situ silica aerogel which creates nanoporous structure inside the nonwoven fabric that encapsulates air and reduces thermal conductivity, which is called as "Knudsen effect" [12, 13].

3.3 *Effect of Aerogel on Pore Size and Air Permeability*

From the Figs. 3 and 4 we can easily predict the effect of aerogel on pore size and that to on air permeability of thermal bonded PET nonwoven fabrics of both GSM. After application of silica aerogel, combined effect of pore size and air permeability shows that for 80 GSM nonwoven fabric, as of mean pore size decreases from 190.6 μ to 137.8, i.e. about 27.7% decrease in mean pore size value; air permeability sharply reduces to 95.2 $cm^3/cm^2/s$ from 297.2 $cm^3/cm^2/s$, i.e. recess of 67.9%. similarly, in case of thermal bonded nonwoven fabric of 150 GSM, after application of silica aerogel; value of mean pore size decreases by 35.18%, i.e. from 179.6 to 116.4 μ which leads to have a significant reduction in air permeability value of nonwoven fabric by 72.34%.

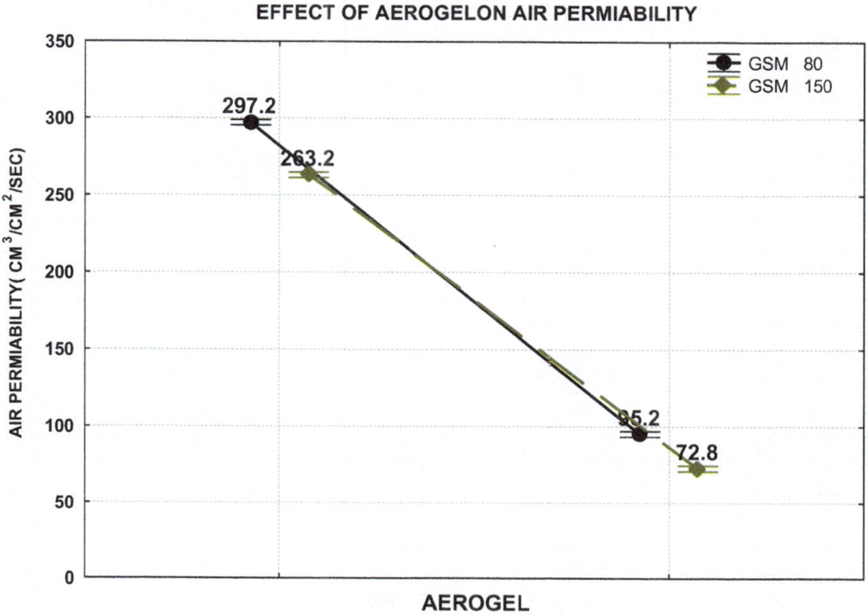

Fig. 3 Effect of aerogel on nonwoven fabric air permeability

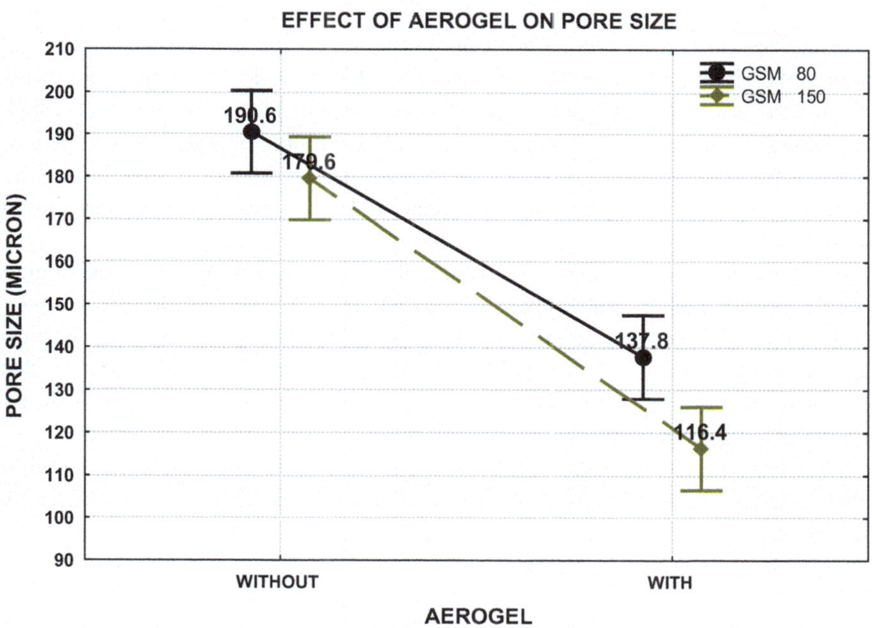

Fig. 4 Effect of aerogel on nonwoven fabric pore size

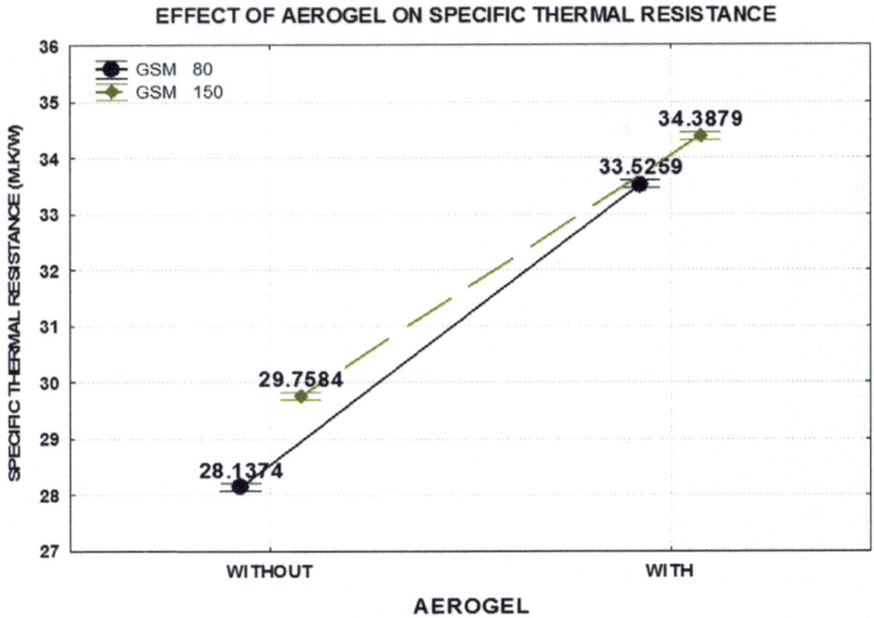

Fig. 5 Effect of aerogel on specific thermal resistivity of nonwoven fabric

Prime reason behind reduction in pore size is that the silica aerogel is in gel form, and as it gets dry in ambient pressure, its density reduces, shrinks and forms microparticles which stick and form bond on the surface of fibre and eventually block the open pore causing reduction in pore size relative to untreated samples. Pore size is one of the basic and microcharacteristics causes a significant change in transfer properties of fabric, so air permeability and thermal conductivity are not an exception for that.

Basically, air permeability is the function of specific surface area, and higher specific surface causes higher resistance to air flow. Aerogel fabric offers higher resistance to air flow because pure aerogel has higher specific surface area as compared to textile fabric as observed [13]. When aerogel combined with fabric, it adds its own surface area by reducing mean pore size resulting in lower air permeability of aerogel composite fabric.

Since these results show the significant reduction in pore size and air permeability value after application of aerogel, thermal conductivity also directly gets affected by these parameters.

3.4 *Effect of Aerogel on Specific Thermal Resistivity of Nonwoven Fabric*

From Fig. 5, it is observed that after application of silica aerogel, the nonwoven fabric with 80 GSM shows hike in specific thermal resistivity by 19.18%, i.e. specific thermal resistivity increases from 28.13 to 33.52 m K/W; same trend is shown in case of nonwoven sample of 150 GSM since its specific thermal resistivity increases from 29.75 to 34.38 m K/W, i.e. 16.76% increase in specific thermal resistivity value.

As specific thermal resistivity is a function opposite to thermal conductivity, it shows opposite trend in Graph 5. Main reason behind the increase in specific thermal resistivity value is reduction in mean pore size of nonwoven fabric which increases resistance to air flow and that to specific thermal resistivity value.

3.5 *SEM Analysis*

From Figs. 6 and 7, it is clearly observed that aerogel is successfully get synthesised inside the thermal bonded PET nonwoven fabric. Figure 7 clearly reflects that the free space between the fibres got filled with aerogel particles; at the same time, there is sufficient coating observed over fibre surface which is indication of good incorporation of aerogel particles inside fabric structure at the finest level.

(a) (b)

Fig. 6 **a** SEM image for untreated nonwoven fabric sample **b** SEM image for untreated nonwoven fabric sample

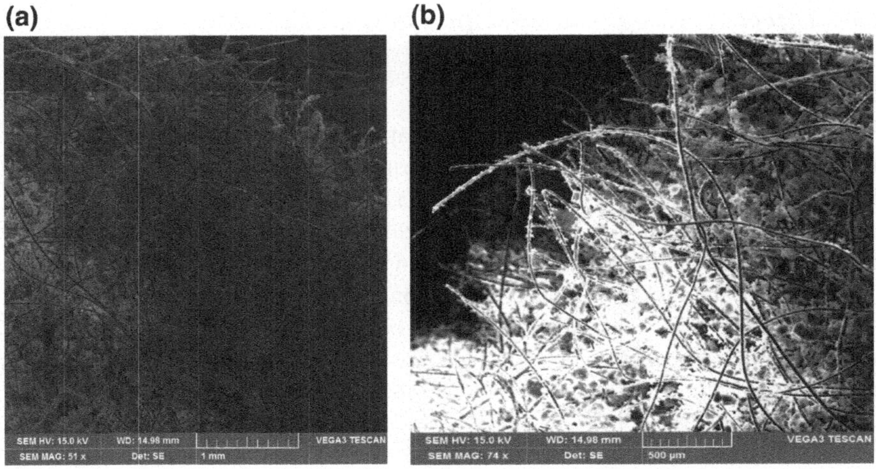

Fig. 7 **a** SEM image for silica aerogel treated nonwoven fabric sample **b** SEM image for silica aerogel treated nonwoven fabric sample

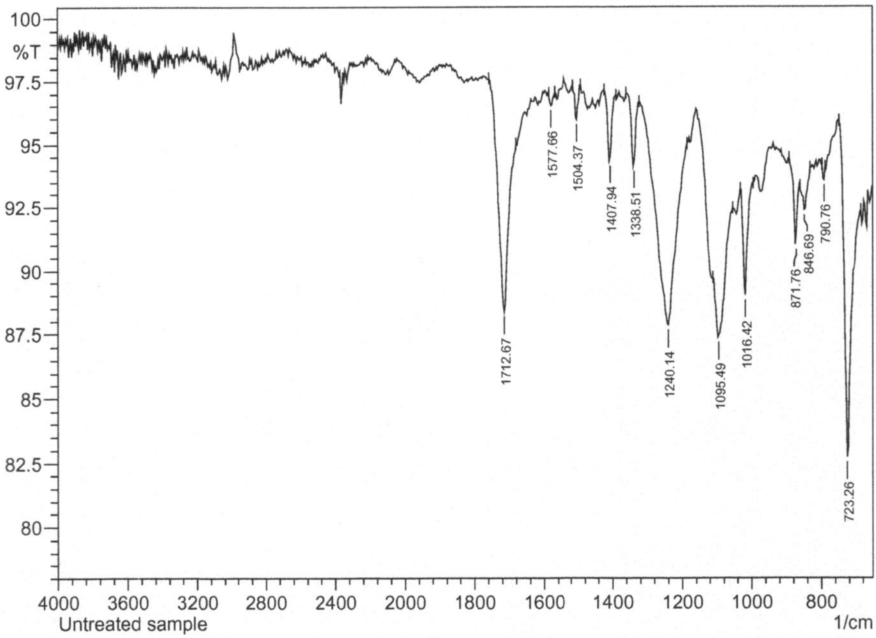

Fig. 8 FTIR image for untreated nonwoven fabric sample

3.6 FTIR Analysis

From Fig. 8, it is observed that at various transmitivity (T%) levels, graph shows different wave numbers range mainly in the following categories which clearly represent functional group of ester bonding. Hence, it is technically supported that untreated sample of nonwoven is of PET. Ester C=O stretch 1735–1750, strong C–C stretch 1000–1300, C=O very strong 1680–1820 (amides, ketones, aldehydes lower frequency for amides and carboxylic acid, esters) when C=O is conjugated [14] (Fig 9).

From Fig. 9 it is observed that the at various transmitivity (T%) levels, graph shows different wave numbers range mainly in the following categories which clearly represent functional group of Si–O–Si or Si–OR which is ultimately functional group of silica aerogel. Reference frequencies are as follows [8]

- Si–H silane 2100–2360 cm $^{-1}$ strong
- Si–OR 1000–100 cm^{-1} strong
- Si–CH_3 1250 ± 10 cm^{-1}

Apart from this as the graph contains both ester and silane functional groups, it is proven that there is successful in situ development of aerogel bonds with PET fabric

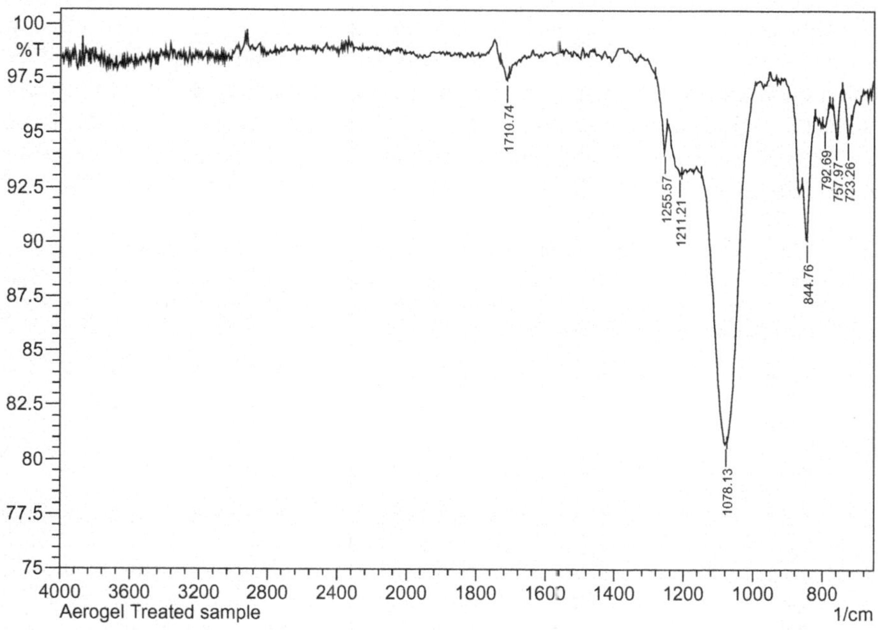

Fig. 9 FTIR image for silica aerogel treated nonwoven fabric sample

4 Conclusion

The present experimental work was aimed to reduce thermal conductivity of nonwoven fabrics by applying aerogel. Nonwoven fabric structures are produced on thermal bonded nonwoven machine by using different GSM (g/m^2) to understand the thermal behaviour and their interrelation. From this study, it is concluded that aerogel material is synthesised successfully in laboratory by sol–gel chemistry followed by ambient pressure drying method. In situ transition of silica aerogel with nonwoven fabric is carried out successfully in laboratory by sol–gel chemistry with ambient pressure drying method which gives reduced thermal conductivity along with significantly reduced loft and thickness which is bottleneck for all textile materials intended for thermal insulation articles. Polyester nonwoven fabric with 80 g/m^2 has thermal conductivity of 0.0298 W/m K. which is lowest amongst all samples with 16.09% reduction in thermal conductivity. Coupling of aerogel material along with textile substrate is successfully achieved which shows the presence of aerogel inside nonwoven structure along with coating on fibre surface which further shows the presence of respective functional groups (Si-OR or Si-O-Si) on nonwoven fabric technically supported by SEM and FTIR images. Aerogel–nonwoven composite fabric successfully achieves increase in specific thermal resistivity by about 16.76%. There is a significant reduction in air permeability by 72.35% and pore size by 34.18% in the fabric after the aerogel application for 150 g/m^2. While 67.96% reduction in air permeability with 27.70% reduction in mean pore size of 80 g/m^2 fabric. Reduction in thermal conductivity of nonwoven fabric is observed despite being a sharp reduction in thickness value which generally governs the air trapping and that to thermal conductivity. The developed product shows potential to be get used at various places like thermal protective middle layer fabric for insulative articles, jackets, car roofing and packing.

Acknowledgements The infrastructural and testing support received from DKTE COEs, Ichalkaranji, India for this research work, analytical testing support is received from SASMIRA, Mumbai, India and ICT, Matunga, Mumbai. Financial assistance rendered by textile manufactures department, VJTI, Matunga is gratefully acknowledged.

References

1. Woo, S.S., Shalev, I., Barker, R.L.: Heat and moisture transfer through nonwoven fabrics Part II: moisture diffusivity. Text. Res. J. **64**(4), 190–197 (1994)
2. Morris, G.J.: Thermal properties of textile material. J. Text. Inst. Trans. **44**(10), T449–T476 (1953)
3. Schuster, J., Heider, D., Sharp, K., Glowania, M.: Thermal conductivities of three-dimensionally woven fabric composites. Comp. Sci. Technol. **68**(9), 2085–2091 (2008)
4. Bhattacharjee, D., Kothari, V.K.: Prediction of thermal resistance of woven fabrics. Part II: mathematical model. J. text. Inst. **99**(5), 433–449 (2008)
5. Williams, J.T.: Textile for Cold weather apparel, 1st edn. Woodhead Publishing Series in Textile: 3–18 (2009)

6. Li, Y., Holcombe, B.V.: Mathematical simulation of heat and moisture transfer in a human-clothing-environment system Text. Res. J. **68**(6): 389–397 (1998)
7. Model, D.D., Tang, Tian, Song, Guowen, McDonald, Ancre: Characterizing the performance of a single-layer fabric system through a heat and mass transfer—Part I: heat and mass transfer model. Text. Res. J. **81**(4), 398–411 (2011)
8. Kothari, V.K., Bhattacharjee, D.: Prediction of thermal resistance of woven fabrics. Part I: mathematical model. J. text. Inst. **99**(5), 421–432 (2007)
9. Ismail, M.I., Ammar, A.S.A., El, Okeily M.: Heat transfer through textile fabrics: mathematical model. Appl. Math. Model. **12**(4), 434–440 (1988)
10. Liao, Yundan, Huijun, Wu, Din, Yunfei, Yin, Shuai, Wang, Moran, Cao, Anmin: Engineering thermal and mechanical properties of flexible fiber-reinforced aerogel composites. J. Sol-Gel. Sci. Technol. **63**, 445–456 (2012)
11. Bheekhun, N., Talib, R, Hassan, M.: Aerogels in aerospace: an overview. Adv Mater Sci Eng. 1–18 (2013)
12. Zhao, J.-J., Duan, Y.-Y., Wang, X.-D., Wang, B.-X.: Experimental and analytical analyses of the thermal conductivities and high-temperature characteristics of silica aerogels based on microstructures. J Phy D Appl Phy. **46** (2012)
13. Hilonga, A., Kim, J.-K., Sarawade, P.B.: Low-density TEOS-based silica aerogels prepared at ambient pressure using isopropanol as the preparative solvent. J. Alloy. Comp. **487**, 744–750 (2009)
14. William. R.: Infrared Spectroscopy. (2017). http://www2.chemistry.msu.edu/

Investigation on Sound Absorption Characteristics of Nonwoven Coir Mats

G. Thilagavathi, A. Muralikrishnan, N. Muthukumar and S. Neelakrishnanan

Abstract Due to higher CO_2 and other environmental issues, many researchers are working in the production of insulating materials made of natural and bio-based materials. This paper investigates the sound absorption characteristics of porous nonwoven mat made from coir fibre. The produced nonwoven samples were combined with a perforated panel to have hybrid configurations and characterized for acoustic properties. It was found that the developed porous coir mats had good sound absorption values at high-frequency range compared to low-frequency range. When Micro perforated panel (MPP) was placed in front of porous coir mat, the sound absorption characteristic was improved at low frequencies and some samples exhibited good sound absorption characteristics in both low and high frequencies.

Keywords Coir fibre · Impedance tube · Needle punching · Sound absorption

1 Introduction

Various musical programmes are conducted in the multipurpose concert halls. But the quality of the musical programme mainly depends on the acoustic atmosphere which includes volume of the hall, the surface area of various surfaces (walls, ceiling, etc.) and the absorption coefficient of the surfaces. For example, the required reverberation time for the classical music is 1.5–2 s and rock music is 0.8–1.5 s for empty halls. Some special musical programmes have low-frequency sound energy which requires a specific acoustic atmosphere in order to provide quality programme. A bass guitar typically has a low E string tuned to E1 or about 41 Hz. So the acoustic design and measurement range could in fact extend down to the 40 Hz frequency band.

Porous absorbing materials like glass wool, polyurethane foam and mineral fibre composites are commonly used for sound absorption applications. These materials

G. Thilagavathi (✉) · A. Muralikrishnan · N. Muthukumar
Department of Textile Technology, PSG College of Technology, Coimbatore, India
e-mail: thilagapsg@gmail.com

S. Neelakrishnanan
Department of Automobile Engineering, PSG College of Technology, Coimbatore, India

A. Majumdar et al. (eds.), *Functional Textiles and Clothing*,
https://doi.org/10.1007/978-981-13-7721-1_4

provide better performance for the sound having a medium and high frequency. Resonance absorbers like perforated plasterboard, perforated metal corrugated sheets and metal boxes provide better sound absorption for low-frequency sound. A single structure does not provide better sound absorption for a wide frequency range. Hence, sound-absorbing materials in the form of sandwich structure need to be constructed to provide sound absorption for a wide frequency range [1, 2].

Coir fibre is the natural organic fibre which is obtained from the outer shell of the coconut. The insulation characteristic of coir makes it an organic sound absorbent. Coir fibres are mixed with a binder to improve the functional and surface characteristics and used in building panels. Nonwoven fabrics have been widely used for acoustic applications and provide better performance for the sound having a medium and high frequency. The lower performance of nonwovens at low-frequency sound limits its applications in the automobile industry. Mohd et al. [3] experimented with coir-based samples both with and without binders and concluded that nonwovens made from raw coir fibres had better performance for both low- and medium-frequency sound waves compared to nonwovens made from coir fibres with binder. This behaviour is more remarkable with layer thickness. It was found that increasing coir fibre layer thickness increased the performance of sound absorption at low- and medium-frequency range. Also in another study, Mohd et al. [4] analysed the influence of compression on sound absorption performance of single layer coir panel and observed that the coir nonwoven panel made with higher compression rate had better sound absorption performance compared to uncompressed coir nonwoven panel.

Modern rhythmic music has a wide range of sound energy and requires good acoustic atmosphere to provide quality sound. Therefore, in recent years, multi-layer acoustic absorbers have been developed to absorb broadband noise. These materials are made by combining perforated plates and porous materials with airspace. The sound absorption performance of multi-layer acoustic absorbers depends on their manufacturing process [5–7]. Zulkifi et al. [8] studied the performance of coir fibre with perforated panel and results showed that coir fibre with perforated panel has higher sound absorption at low-frequency range. Increasing the sound absorption at low frequency can be a great contribution to noise control engineering [9]. Ayub et al. [10] developed a panel composed of coir fibre layer in different thicknesses 20, 35 and 50 mm with air gap backed by a rigid wall. Zulkifli et al. [11] studied the performance of sound absorber consisting of a perforated plate with porous coir backing.

The aim of present work is to develop hybrid fibrous panel from porous nonwoven coir mat and perforated panels for controlling low-frequency sound. The samples are planned to have various configurations like porous nonwoven coir mat, perforated panel with air gap, perforated panel backed by porous fibrous matt and perforated panel backed by porous fibrous mat and air cavity. The developed samples are characterized for its acoustic performances.

2 Materials and Methods

2.1 Fabrication of Nonwoven Coir Mats

Coir fibre is sourced from coir industries in Tamil Nadu, India. The opened fibres are passed through a needle punching machine where the fibres are made to entangle to form as a sheet. In order to get the structural integration, natural rubber polymer latex of 22% was sprayed over the coir mat and cured. Totally, six nonwovens were developed and the specifications of the developed nonwovens are given in Table 1.

Plywood of 3.4 mm thickness was purchased from local market with density of 67.27 kg/m^3 and perforations were made using a hand drill. The specifications of the developed micro-perforated panel (MPP) are given in Table 2. Porous coir mat samples, MPP and air gap are considered for making effective experimental set-up. Figure 1 shows the representation of various configurations of samples followed during measurements.

Table 1 Physical properties of porous nonwoven coir samples

Sample ID	Thickness (mm)	Weight (g/m^2)	Density (kg/m^3)	Porosity (%)
P1	6	900	150	86.9
P2	15	1650	110	90.4
P3	15	2625	175	84.8
P4	15	3150	210	81.8
P5	20	2200	110	90.4
P6	25	2750	110	90.4

Table 2 Physical properties of perforated panel

Sample ID	Material type	Thickness (mm)	Hole diameter (mm)	Hole spacing (mm)	Open area (%)
MPP1	Plywood	3.4	0.6	4	1.76
MPP2	Plywood	3.4	2	12	2.18
MPP3	Plywood	6.0	2	12	2.18
MPP4	Plywood	6.0	4	12	8.72
MPP5	Plywood	3.4	8	12	34.89
MPP6	Plywood	3.4	2	18	0.97
MPP7	Plywood	3.4	8	18	15.51
MPP8	Plywood	3.4	2	8	4.90
MPP9	Acrylic sheet	1.0	2	4	19.63

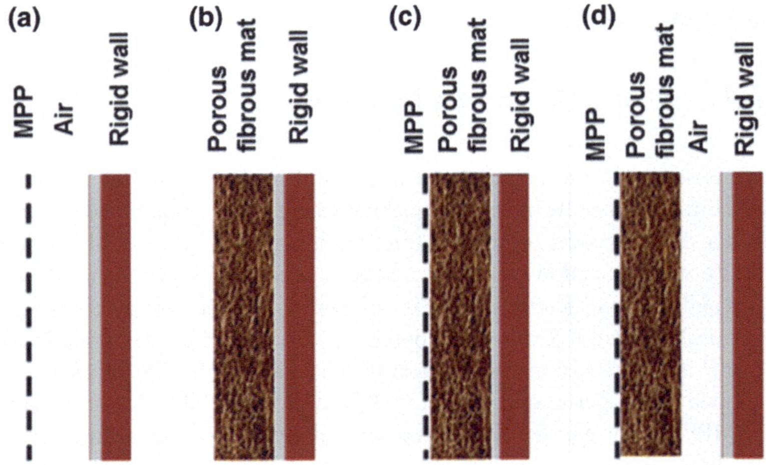

Fig. 1 Representation of various configurations, **a** MPP backed by air cavity (MPP+A), **b** porous absorber (P), **c** MPP backed by porous absorber (MPP+P) and **d** MPP backed by porous and air cavity (MPP+P+A)

2.2 Sound Absorption Characterization

The acoustic performance of the developed samples was evaluated using an impedance tube as per ASTM E 1050 standards. The impedance tube is made of a hollow cylinder consisting of loudspeaker and sample holder. Figure 2 indicates the positions of the microphone ports in the tube. The system of data acquisition and processing takes into account of these microphones as matched ones along with the other devices such as two analog signal conditioners and a two-channel Fast Fourier Transform (FFT) analyser. The microphones are connected to the individual channel

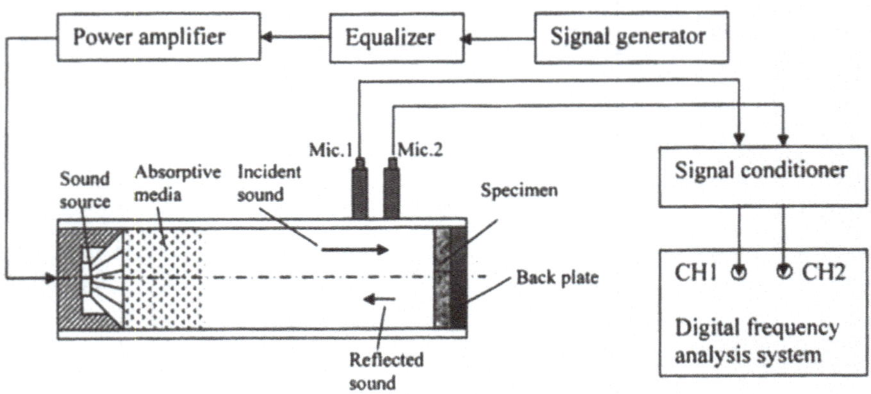

Fig. 2 Schematic diagram of an impedance tube set-up

of the analyser. The sound absorption coefficient is determined through the measured transfer function data using a microprocessor computer.

The sound absorption performance of a material is represented by Noise Reduction Coefficient (NRC) value which is an index obtained through as an average of sound absorption values at 250, 500, 1000 and 2000 Hz frequencies. The NRC values range from 0 (perfect reflection) to 1 (perfect absorption) to the nearest accuracy of 0.05.

3 Results and Discussion

3.1 Sound Absorption Characteristics of Porous Absorber

Figure 3 shows sound absorption characteristics of the developed porous absorber samples. The NRC values of porous samples (P1 to P6) are 0.08, 0.08, 0.11, 0.13, 0.12 and 0.18, respectively. In general, porous materials have minimum sound absorption characteristics at low frequencies (<1000 Hz) and have good sound absorption characteristics at higher frequencies. The maximum sound absorption coefficient among the samples is 0.13 for lower frequency (f < 1000 Hz) and 0.69 for higher frequency (f > 1000 Hz). This is the reason for the reduction in NRC values of porous materials.

The density of material plays a major role in sound absorption characteristics of any material. It was observed that the NRC values of the sample P2, P3 and P4 are 0.08, 0.11 and 0.13, respectively, and maximum absorption coefficient is 0.38 at

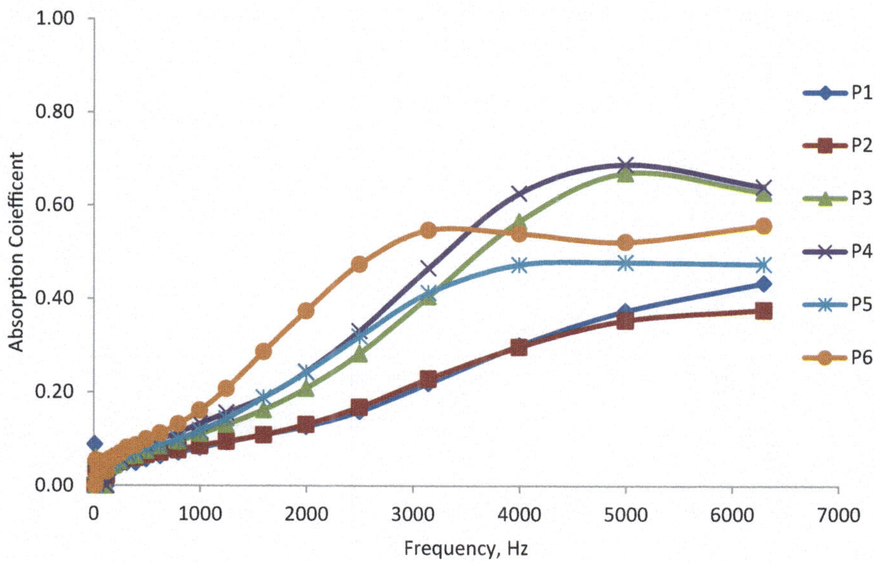

Fig. 3 Sound absorption characteristics of porous absorbers

6300 Hz, 0.67 at 5000 Hz and 0.69 at 5000 Hz, respectively. It was also observed that porous coir mat with density of 110 kg/m^3 (P2) has lowest sound absorption and porous coir mat with the density of 210 kg/m^3 (P4) has higher sound absorption for the entire frequency range. This is due to the presence of more number of coir fibres in the porous coir mat with the density of 210 kg/m^3 (P4). The loss in energy of sound wave is more when more number of fibres present in the porous layer. But there is no significant difference noticed in the absorption coefficient between the porous coir mat with the densities of 175 and 210 kg/m^3; this may be due to the lesser difference in porosity which results in minor difference in the air flow resistivity.

Thickness of the material is another factor which influences the sound absorption characteristics of any material. It was observed that the porous coir mat of thickness 25 mm (P6) has higher sound absorption coefficient values at low- and medium-frequency range compared to others. Low-frequency sound absorption has a direct relationship with thickness of material. When thickness of the material increases, the sound absorption characteristics increase and absorption curve also shifts towards lower frequencies. It is a well-known fact that if the porous absorber has thickness of one-tenth of the wavelength of the incident sound then it will have maximum sound absorption.

3.2 Sound Absorption Characteristics of Porous Absorber with MPP+Air Configuration

MPP backed by 50 mm air gap (MPP1+50 mm Air) has sound absorption only at low-frequency range of 630–1500 Hz and has the maximum sound absorption

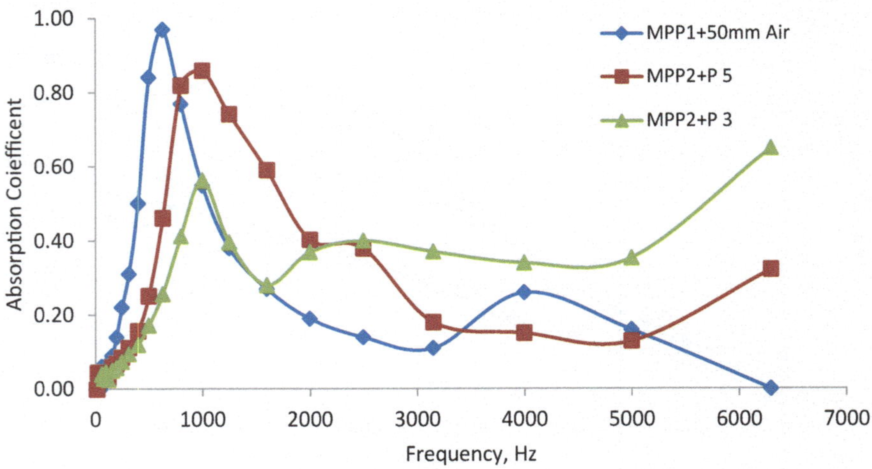

Fig. 4 Effect of porous absorber in MPP+Air configuration

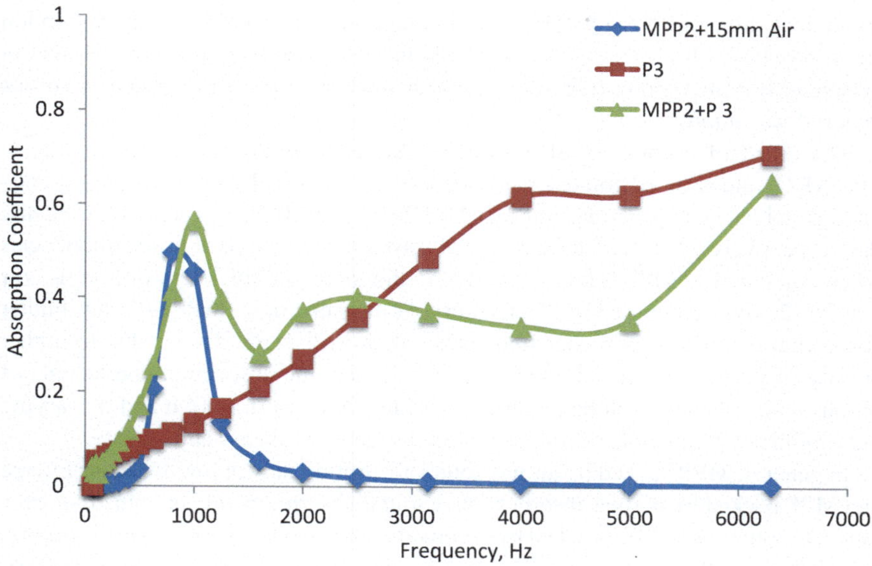

Fig. 5 Effect of MPP and porous absorber in MPP+Air configuration

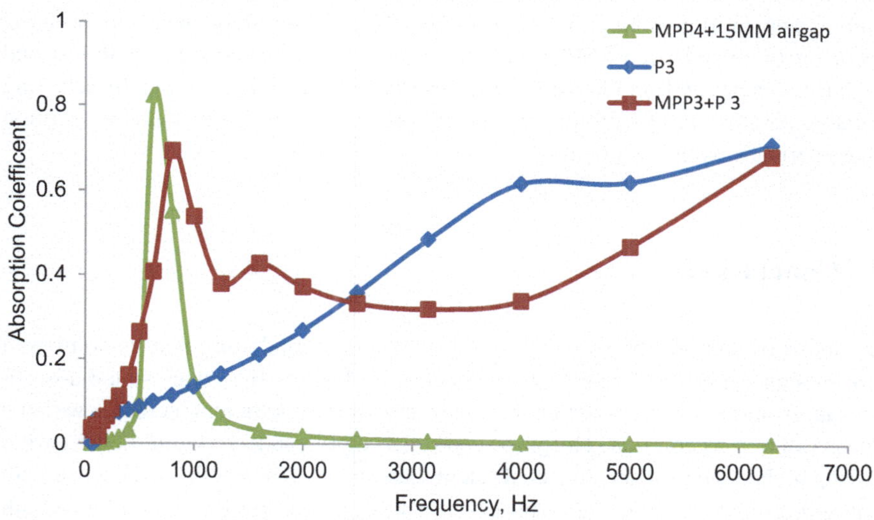

Fig. 6 Effect of MPP and porous absorber in MPP+Air configuration

coefficient value of 0.97 at 630 Hz. This configuration does not have any absorption characteristics at higher frequencies. While inserting porous nonwoven coir mat in between MPP and rigid wall, the absorption characteristics of MPP backed by porous material was improved.

The effect of porous absorber in MPP+Air configuration is shown in Fig. 4. The NRC values of perforated panel backed by air (MPP1+50 mm Air), perforated panel backed by porous coir mat MPP2+P3 and MPP2+P5 are 0.45, 0.30 and 0.4, respectively. MPP2+P5 have higher absorption coefficients at lower frequencies when compared to MPP2+P3; this is due to the increased thickness of porous coir mat P5 (20 mm) compared to P3 (15 mm). At the same time, MPP2+P3 has higher absorption coefficient at higher frequencies than MPP2+P5. This is due to higher density of porous coir mat P3 (175 kg/m^3). Hence, the absorption coefficients of porous materials may well be modified by adding MPP in front of it and by varying the significant parameters of porous material such as thickness and density.

In general, MPP backed by air has sound absorption only at low-frequency range and MPP backed by porous absorbers have sound absorption only at high-frequency range. In some cases, MPP backed by porous absorber showed good sound absorption at both low and high frequencies. This trend is shown in Figs. 5 and 6. It was observed that MPP2+P 3 and MPP3+P 3 have absorption trend of MPP2 backed by 15 mm air and MPP3 backed by 15 mm air gap at low frequencies, respectively, and have common absorption trend of porous absorber (P3) at higher frequencies. In this case, the acoustic impedance of MPP and porous absorber is just added in both low and high frequencies and act like wide-band absorber. Hence, it is clear that by selecting suitable parameters of MPP and porous absorber, the wide-band sound absorption characteristics would be obtained.

4 Conclusion

In this work, hybrid fibrous panel was developed by combining porous nonwoven coir mat and perforated panels and characterized for low-frequency sound absorption application. It was found that the sound absorption coefficient increases with the density of porous coir mat. The porous coir mats of 175 and 210 kg/m^3 density were having higher sound absorption coefficient than others. When MPP was backed with porous coir mat, the sound absorption characteristics improved as it combined both low-frequency and high-frequency absorption in some cases. The sound absorption characteristic of MPP backed by porous coir mat had two different behaviours. It had absorption characteristics of MPP backed air cavity for the low frequencies and absorption characteristics of porous coir mat for high frequencies. But the above trend was noticed only with high-density porous materials. The same configuration such as MPP backed with low-density porous coir mat did not have two-stage sound absorption characteristics. It had the resonance-type absorption curve, but the bandwidth of absorption was found to be wider. The results indicated that the developed coir fibrous mats can be used as sound-absorbing panels in auditorium and conference halls.

Acknowledgements The authors thank M/s PSG TECHS COE INDUTECH, Coimbatore for providing facilities to carry out this research.

References

1. Seddeq, H.S., Aly, N.M., Ali, Marwa A., Elshakankery, M.H.: Investigation on sound absorption properties for recycled fibrous materials. J. Ind. Text. **43**, 56–73 (2012)
2. Thilagavathi, G., Praveen, E., Kannaian, T., Sasikala, L.: Development of Natural Fibre Nonwoven for Application as Car Interiors for Noise Control. J. Ind. Text. **39**(3), 267–278 (2010)
3. Mohd Nor, M.J., Ayub, M.D., Zulkifli, R., Aminn, N., Fouladi, M.H.: Effect of compression on the acoustic absorption of coir fibre. Am. J. Appl. Sci. **7**(9), 1285–1290 (2010)
4. Mohd Nor, M.J., Ayub, M., Zulkifli, R., Amin, N., Fouladi, M.H.: Effect of different factors on the acoustic absorption of coir fiber. J. Appl. Sci. **10**, 2887–2892 (2010)
5. Delany, M.E., Bazley, E.N.: Acoustical properties of fibrous absorbent material. Appl. Acoust. **3**, 105–116 (1970)
6. Qian, Y.J., Kong, D.Y., Liu, S.M., Sun, S.M., Zhao, Z.: Investigation on micro-perforated panel absorber with ultra micro perforations. Appl. Acoust. **74**, 931–935 (2013)
7. Shoshani, Y., Yakubov, Y.: A model for calculating the noise absorption capacity of non woven fiber webs. Text. Res. J. **69**(7), 519–526 (1999)
8. Zulkifi, R., Mohd Nor, M.J., Mat Tahir, M.F., Ismail, A.R., Nuawi, M.Z.: Acoutstic properties of multi-layer coir fibres sound absorption panel. J. Appl. Sci. **8**(20), 3709–3714 (2008)
9. Zulkifli, R., Mohd Noor, M.J.: Noise control using coconut coir fiber sound absorber with porous layer backing and perforated panel. Am. J. Appl. Sci. **7**(2), 260–264 (2010)
10. Ayub MD, Mohd Nor MJ, Amin N, Zulkifi R (2009) A preliminary study of effect of air gap on sound absorption of natural coir fibre. In: Proceedings of the regional engineering postgraduate conference, Malaysia (2009)
11. Zulkifli, R.: Noise control using coconut coir fiber sound absorber with porous layer backing and perforated panel. Am. J. Appl. Sci. **7**(2), 260–264 (2010)

Part II
Medical Textiles

Functionalized Silk for Surgical Suture Applications

S. Viju, L. Marian Shilpa and G. Thilagavathi

Abstract In the present state of environmental consciousness, the modern quality necessities of a product not only stress on the inherent functionality but also a preparation method that is ecological. Hence, investigations on coating agents based on natural extracts for medical and healthcare textiles application are gaining global attention. In this work, a novel attempt has been made to coat the silk sutures with natural bioactive agents such as a*loe vera*, curcumin, and chitosan. The exhaustion method was adopted for the application of extracts in two different combinations, *aloe vera*, curcumin, and *aloe vera*, curcumin in combination with chitosan. Suture properties such as tenacity, knot strength, coefficient of friction, and antimicrobial property were tested. The process variables such as time, temperature, and concentration were optimized for better performance of the sutures by adopting the Box–Behnken response surface methodology. Sutures treated with *aloe vera* and curcumin in combination with chitosan had low values of coefficient of friction, i.e., 0.199 and maximum values of tenacity and knot strength of 39.86cN/tex and 30.56 cN/tex, respectively. The silk sutures treated with *aloe vera* and curcumin in combination with chitosan, under the optimized conditions exhibited a zone of inhibition of 6 mm for *Staphylococcus aureus* and 4 mm for *Escherichia coli* bacteria.

Keywords Silk · Suture · Tenacity

1 Introduction

Suture materials are normally used to adjoin tissues together after surgery. Some significant characteristic features of sutures include tenacity, knot strength, coefficient of friction, biocompatibility, and antimicrobial property. Another attribute which is desirable in the sutures is the antiscar property. Till date, there is no one suture substrates which can accomplish all the important characteristic features of sutures. The general practitioner should select the correct suture for the nature of

S. Viju (✉) · L. Marian Shilpa · G. Thilagavathi
Department of Textile Technology, PSG College of Technology, Coimbatore, India
e-mail: vijutext@yahoo.co.in

A. Majumdar et al. (eds.), *Functional Textiles and Clothing*,
https://doi.org/10.1007/978-981-13-7721-1_5

surgery that he is intended to perform as the requirements for suture support are diverse for different tissues. The general practitioner has many options for selection of sutures, and he may decide them based on his convenience and knowledge [1–3].

Silk is one of the widely used non-absorbable suture material in the medical field due to its unique features such as superior handling behavior, user-friendliness, knot security, and availability. Furthermore, it has been reported that bacteria prefer to adhere onto uncoated silk sutures and increase the risk of surgical site infections. Hence, coating with silicone or wax or antimicrobial agent is normally performed to improve the functional characteristics of silk sutures [3].

Most of the commercial antimicrobial sutures such as polyglycolic acid suture, monocryl plus and polydiaxanone plus are based on triclosan coating only. Recently, literatures recognize that the triclosan-coated sutures encompass considerable health-related issues [4]. Therefore, suitable coating substances are urgently needed.

For the purpose of improving the functional characteristics of silk sutures, different coating materials such as cynodon dactylon [5], thermomyces [6], poly (ε-caprolactone) and sulfamethoxazole trimethroprim [7], levofloxacin hydrochloride and poly (ε-caprolactone) [8], tetracycline hydrochloride [9], and chitosan [3] have been imparted onto silk filaments.

In a recent study, silk suture has been coated with chitosan and incorporated with a drug cynodon dactylon. When subjected to antimicrobial study, it is found that 1.7% chitosan and 7% drug at 60 °C have optimum antimicrobial efficiency with efficient bacterial reduction percentage against both *Escherichia coli* (*E. coli*) and *Staphylococcus aureus* (*S. aureus*) [5].

Parthiban et al. [6] applied natural fungal pigment, namely *Thermomyces,* onto silk sutures. They observed that the pigment concentration has a significant impact on antimicrobial property. The antimicrobial activity of the pigment treated suture increases against both *E. coli* and *S. aureus* bacteria with pigment concentration. Highest zone of inhibition 10 and 14 mm is observed against *E. coli* and *S. aureus,* respectively, at 2.5% pigment concentration.

In another study, poly (ε-caprolactone) (PCL) and sulfamethoxazole trimethroprim (SMZ) were coated onto braided silk sutures and their antimicrobial characteristics were studied. At high PCL concentrations, the zone of inhibition value of treated suture against *S. aureus* and *E. coli* bacteria was found to be higher [7].

Chen et al. [8] used levofloxacin hydrochloride and poly (ε-caprolactone), to develop antimicrobial silk suture structures with excellent physical characteristics and slow-release antibacterial efficiency. Circular braiding machine was used to braid the silk filaments; subsequently, antimicrobial treatment was given by two-dipping–two-rolling method alternatively before and after braiding. Zone of inhibition method was used to recognize the antibacterial activity of the treated suture material against *S. aureus* and *E. coli.* Durability of the coated suture material was also studied. The treated sutures exhibited slow drug release property and continuous antibacterial efficiency for more than five days.

Viju and Thilagavathi [9] studied the antimicrobial and physical characteristics of tetracycline hydrochloride (TCH)-treated silk sutures. They observed that TCH coating has no significant effect on some of the properties of sutures such as coefficient of friction, knot strength, and tenacity. However, antimicrobial property of the sutures changed considerably with TCH treatment. Furthermore, it was observed that with the increase in drug concentration, the antibacterial activity increases. At 1% TCH concentration, 32 and 25 mm zones of inhibition were observed against *E. coli* and *S. aureus*, respectively.

Silk-braided sutures were treated with chitosan and subsequently evaluated for its antimicrobial property by agar diffusion method (SN 195920). Furthermore, it was observed that with the increase in chitosan concentration, the antibacterial activity increases. At 3% chitosan concentration, 13 and 10 mm zones of inhibition were observed against *E. coli* and *S. aureus*, respectively [3].

Aloevera is a plant gives aloe gel, which has antimicrobial activity against different microbes and subdued the growth of *Mycobacterium*, *Trichophyton*, and *Bacillus subtilis* [10, 11]. Chitosan, a natural biopolymer, has several distinctive features like non-toxicity, wound healing ability, biodegradability, antimicrobial activity, and cationic nature. Chitosan is chemically named as beta-(1,4)-2-(amino)-2-deoxy-D-glycopyranose. In aqueous solution below pH 6.5, its protonated form (NH_3^+) acts similar to a cationic polyelectrolyte and reacts with negatively charged molecules, i.e., cell wall of the microbes and slow down their growth [12]. Curcumin is the dried tuber of the plant Curcuma domestica and chemically named as 1-7-bis (4-hydroxy-3-methoxy phenyl)- 1,6-heptadiene-3,5-dione. The presence of hydroxyl and methoxyl groups is supposed to be liable for the antimicrobial activity [13].

The antimicrobial efficiency of the natural extracts such as *aloe vera*, curcumin, and chitosan has paved way to coat these constituents individually and as in combination on silk sutures. Although numerous efforts have been performed to apply these materials as antimicrobial agents, its application as a coating on sutures and its influence on suture properties are not well documented in the literature. Hence, attempt has been made in this work to coat the silk suture with the natural extracts without compromising the other suture properties. The effects of the coatings on tenacity, knot strength, coefficient of friction, and antimicrobial characteristics were evaluated. The process parameters like the time, temperature, and concentration were optimized for enhanced performance of the silk sutures by employing a factorial design (Box–Behnken) for three-independent variables.

2 Materials and Methods

Bombyx mori silk filaments were received from Silk sericulture board, Bangalore, India. Chitosan polymer was received from Otto chemicals, India. *Aloe vera* and curcumin were procured from Herbs and Crops Overseas, India. Deionized distilled water was used throughout the experiment.

2.1 Degumming

Braided silk structures were dipped in 0.1% (w/v) Na_2CO_3 solution at 98–100 °C for thirty minutes. Subsequently, silk filaments were washed, dried, and conditioned.

2.2 Manufacture of Braided Silk Filaments

Braided silk filaments having a count of 130 tex were prepared using 16 degummed silk filaments using a braiding machine. The detailed method of manufacture has been described in our previous study [2].

2.3 Preparation of Chitosan Solution

Chitosan solution of concentration 1.5% (w/v) was made in 1.0% (v/v) acetic acid solution. The solution was subsequently subjected to stirring using the magnetic stirrer for 1 h at 60 °C.

2.4 Extraction of Aloe Vera, Curcumin, and Finishing Treatment

50, 75, and 100 g of *aloe vera* and curcumin powders each were soaked in 500-ml ethanol (non-polar solvent) for one day. Then, it was filtered and extracted using the rotary vacuum evaporator to give 10, 15, and 20%.

Ethanolic extracts of *aloe vera* and curcumin were coated individually and in combination by exhaustion method with liquor ratio of 1:20. The *aloe vera* and curcumin finished braided silk sutures were dried in room temperature. Further, they were treated with chitosan solution for 24 h at room temperature and then washed with 0.1M NaOH solution. Then, the samples are washed with excess water to get rid of the acetic acid. The silk filaments were subsequently dried at 100 °C for ten minutes.

2.5 Friction Measurement

A friction tester indigenously designed was used to measure the friction. Working principle of the friction instrument and test method has been elaborately discussed in a previously published paper [14]. Coefficients of friction of untreated and treated silk filaments were tested at a load of 100 g and at a sliding speed of 100 mm/min. For each sample, at least 10 readings were taken.

2.6 *Tenacity and Knot Strength Measurement*

Instron tensile tester was used to measure the tenacity and knot properties using ASTM standard D2101-82. Tenacity and knot strength were measured at an extension rate 9 mm/min and gauge length 150 mm. Surgeons knot was made for knot strength measurement [15].

2.7 *Scanning Electron Microscope*

Scanning electron microscope (SEM) was used to measure the surface characteristics of silk filaments.

2.8 *Evaluation of Antibacterial Activity*

Qualitative method (shake flask method) and quantitative method (agar diffusion method) were used to measure the antibacterial activity of the treated and untreated silk filaments.

2.9 *Box–Behnken Design*

In order to optimize the suture properties such as coefficient of friction, tenacity, and knot strength, Box–Behnken experimental design was used. In this study, a total of 15 experiments were conducted with the variables and levels according to the Box–Behnken response surface design (three-level design experiment) using the design expert software (version no. *9.0.4.1*). The particulars of process parameters used are shown in Table 1. In order to study the effect of process variables, regression equation and surface response curves were made.

Table 1 Particulars of process parameters

S. No.	Parameter	Units	(−1)	(0)	(1)
1	Time	Hours	0.5	1.0	1.5
2	Temperature	°C	30	55	80
3	Concentration	%	10	15	20

3 Results and Discussion

Finishing trials on the silk sutures with different combinations of the natural extracts were conducted under the 15 conditions given by Box–Behnken design which is shown in Table 2.

3.1 *Regression and Statistical Analysis*

ANOVA (analysis of variance) method is used to confirm the acceptability of the model. The regression equations acquired from the analysis of variance for each response are tabulated in Tables 3 and 4 They depict the relationship between the process variables such as time (X_1), temperature (X_2), and concentration (X_3) and their contributing effect on the responses (Y_1, Y_2, and i_3), the key properties of suture such as the coefficient of friction, tenacity and knot strength.

In both the above experimental trails, the values of p-value ("Prob > F") mentioned in Tables 3 and 4 associated with the dependent variables, coefficient of friction (CF), tenacity (TN), and knot strength (KS) are below 0.0500 which show that model conditions are found to be significant. Further, the value of R^2 given Tables 3 and 4 for the conducted experiments is greater than 0.6 which confirms that the designed model is valid. Also, Table 5 shows the Adjusted (Adj.) and predicted (Pred.) R^2 values from ANOVA table.

3.2 *Surface Plots: 2D Contour Plots*

3.2.1 Effect of the Process Parameters on Friction

Coefficient of friction is one of the important properties of the suture material. When compared to the monofilament sutures, the braided sutures have high frictional values. A higher coefficient of friction benefits knot security, resulting in a more secure knot. But at the same time, this high friction results in tissue drag and damage [16]. Hence, it is necessary to coat the sutures in order to reduce the drag.

While considering the impact of process variables as indicated in the below contour plots Figs. 1 and 2, it is noted that higher concentration of *aloe vera* and curcumin

Table 2 Experimental runs of Box–Behnken design and finish application on silk sutures with different extract combinations

Physical properties of sutures									
Sample	TE (hrs.)	TM (°C)	CN (%)	Aloe vera, curcumin in combination			Aloe vera, curcumin in combination with chitosan		
				CF	TN (cN/Tex)	KS (cN/Tex)	CF	TN (cN/Tex)	KS (cN/Tex)
1	0.5	30	15	0.426	32.44	23.03	0.226	37.22	28.05
2	1.5	30	15	0.418	32.77	23.45	0.224	37.58	28.17
3	0.5	80	15	0.467	28.34	19.48	0.305	29.41	21.54
4	1.5	80	15	0.465	28.47	19.01	0.235	29.66	20.32
5	0.5	55	10	0.436	31.81	21.89	0.235	36.56	26.99
6	1.5	55	10	0.432	31.85	22.03	0.231	36.96	27.12
7	0.5	55	20	0.408	34.84	24.02	0.199	36.62	30.12
8	1.5	55	20	0.396	35.39	25.32	0.203	39.86	30.56
9	1	30	10	0.438	31.42	21.32	0.237	36.23	27.09
10	1	80	10	0.464	28.58	19.97	0.277	30.03	21.01
11	1	30	20	0.341	36.03	26.74	0.217	39.48	29.67
12	1	80	20	0.449	29.84	20.67	0.259	32.01	23.01
13	1	55	15	0.44	32.92	22.34	0.227	37.27	26.36
14	1	55	15	0.416	33.04	23.98	0.237	38.91	27.99
15	1	55	15	0.439	31.01	23.06	0.211	36.29	28.73

TE—Time, TM—Temperature, CN—Concentration, CF—Coefficient of friction, TN—Tenacity, KS—Knot strength

Table 3 Empirical model of the silk sutures treated with *aloe vera* and curcumin in combination

Test	Regression equation	R^2-value	*p*-value
CF	$Y_1 = +0.43 - 3.250\text{E-}003\, X_1 + 0.028\, X_2 - 0.022\, X_3$	0.7076	0.0028
TN	$Y_2 = +32.32 + 0.13\, X_{1\text{-}2}.18\, X_2 + 1.56\, X_3 - 0.050\, X_1X_2 + 0.13\, X_1X_3 - 0.84\, X_2X_3 + 0.093\, X_1^2 - 1.91\, X_2^2 + 1.06\, X_3^2$	0.9670	0.0034
KS	$Y_3 = + 23.13 + 0.17\, X_{1\text{-}1}.93\, X_2 + 1.44\, X_3 - 0.22\, X_1X_2 + 0.29\, X_1X_3 - 1.18\, X_2X_3 - 0.37\, X_1^2 - 1.51\, X_2^2 + 0.56\, X_3^2$	0.9733	0.002

Table 4 Empirical model of the silk sutures treated with *aloe vera* and curcumin in combination with chitosan

Test	Regression equation	R^2-value	*p*-value
CF	$Y_4 = +0.23\text{-}9.000\text{E-}003\, X_1 + 0.021\, X_2 - 0.013\, X_3 - 0.017\, X_1X_2 + 2.000\text{E-}003\, X_1X_3 + 5.000\text{E-}004\, X_2X_{3\text{-}4}.000\text{E-}003\, X_1^2 + 0.027\, X_2^{2-4}.000\text{E-}003\, X_3^2$	0.8991	0.0465
TN	$Y_5 = + 37.49 + 0.16\, X_{1\text{-}3}.68\, X_2 + 1.40\, X_3 - 0.027\, X_1X_2 - 0.040\, X_1X_3 - 0.32\, X_2X_3 - 0.10\, X_1^{2-3}.92\, X_2^2 + 0.86\, i_3^2$	0.9784	0.0012
KS	$Y_6 = +27.69 - 0.066\, X_{1\text{-}3}.39\, X_2 + 1.39\, X_3 - 0.33\, X_1X_2 + 0.077\, X_1X_3 - 0.15\, X_2X_3 + 0.16\, X_1^{2-3}.34\, X_2^2 + 0.84\, X_3^2$	0.974	0.0019

Table 5 Adjusted (Adj.) and predicted (Pred.) R^2 values from ANOVA table

(a) *Aloe vera* and curcumin				(b) *Aloe vera*, curcumin and chitosan			
	Adj. R^2	Pred. R^2	Adeq. precision		Adj. R^2	Pred. R^2	Adeq. precision
CF	0.6278	0.4009	9.883	CF	0.7175	0.6417	8.129
TN	0.9076	0.9068	12.972	TN	0.9396	0.9083	13.745
KS	0.9253	0.8596	15.521	KS	0.9272	0.8396	12.914

decreases the frictional values. This may be due to the better coating of extracts onto the suture at higher concentration. Effect of temperature is also domineering, and increase in frictional values is obtained at higher temperatures. This may be due to the surplus activation of extracts at higher temperature and which results in uneven deposition of extracts onto the surface of the sutures. Thus, the friction value increases at higher temperature.

In the case of *aloe vera* and curcumin coating on the silk sutures, as the concentration of the coating substance increases, the friction value decreases. Yet a lower value of friction is necessary for better performance of the suture thread. Particle deposition of *aloe vera* and curcumin coating gives a high frictional value of 0.467. In case of commercial wax-coated silk suture, the coefficient of friction value is 0.225 [16]. Hence, the *aloe vera*-treated and curcumin-treated silk sutures are further coated with chitosan. When the sutures are coated with chitosan, the frictional values are found on a lower sale, i.e., 0.199. Lower values of friction are due to the

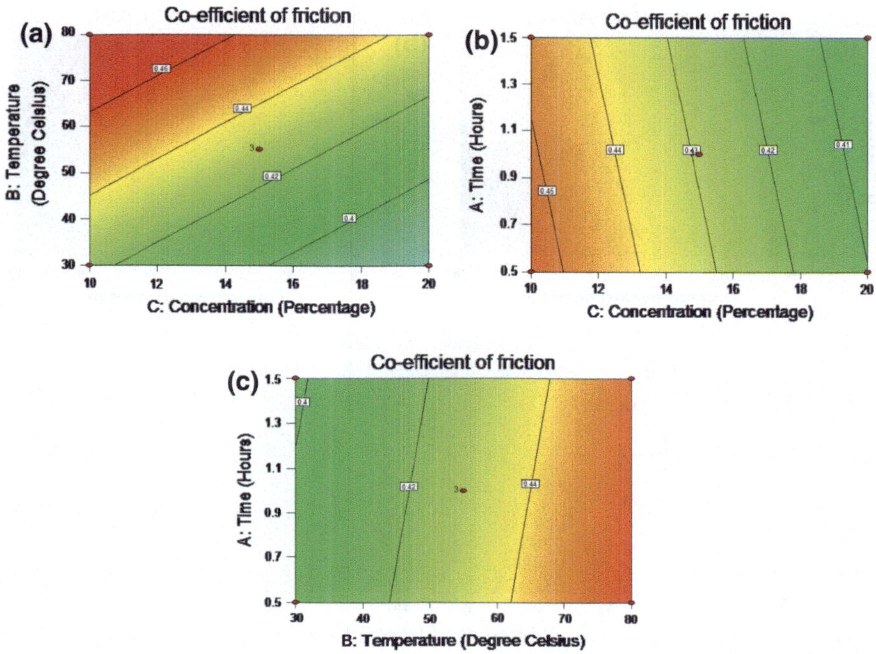

Fig. 1 Influence of **a** concentration and temperature, **b** concentration and time, **c** temperature and time on friction of the silk sutures treated with aloe vera and curcumin

formation of even film on suture surface. SEM photographs of untreated and treated silk sutures are shown in Fig. 3.

3.2.2 Effect of the Process Parameters on Tenacity

Surgeons usually select the suture with superior tensile properties. If the suture threads are very weak and the knot tying force is high as compared to the tensile strength, the suture can break easily during knot tightening [16].

Figures 4a and 5a illustrate the influence of concentration and temperature on tenacity when the time of the experiment remains constant. Maximum value of tenacity 36.03 cN/tex and 39.86 cN/tex is obtained at an extract concentration of 20% and low temperature level.

From Figs. 4a and 5a, it is noted that as the concentration increases, the tenacity increases in case of both silk sutures treated with *aloe vera* and curcumin and silk sutures treated with *aloe vera*, curcumin, and chitosan. This may be because, at higher concentrations, there is better binding of coating extracts onto silk-braided sutures, thus contributing to enhanced resistance to load. These findings are in track with the earlier results [17]. It is further noted that as the temperature increases, the tenacity of silk suture also increases. This may be due to the surplus activation of extracts

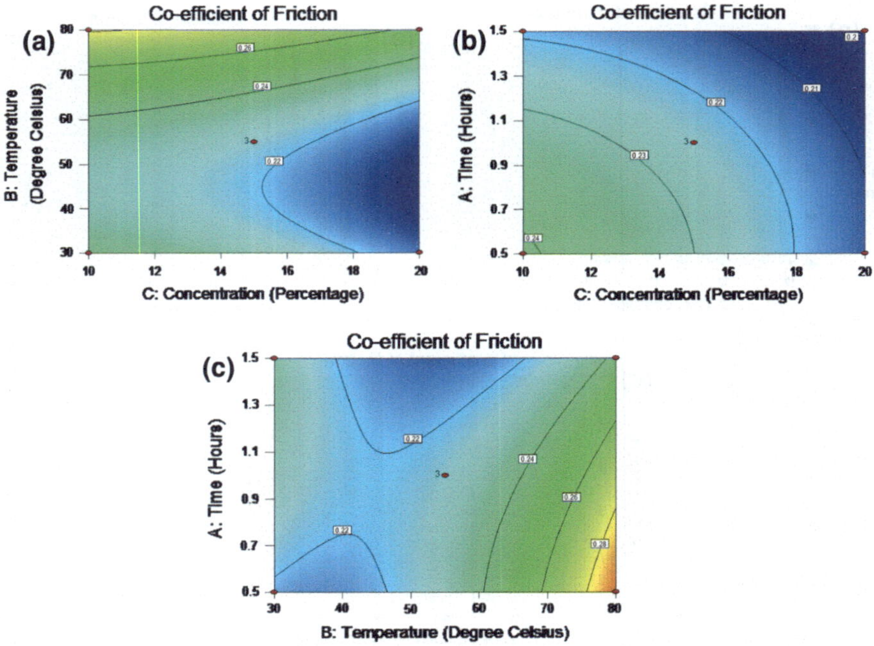

Fig. 2 Influence of **a** concentration and temperature, **b** concentration and time, **c** temperature and time on friction of the silk sutures treated with aloe vera, curcumin, and chitosan

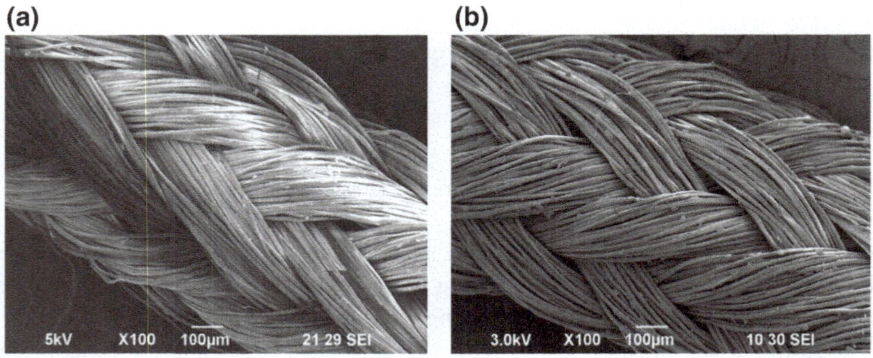

Fig. 3 SEM images of **a** untreated silk suture and **b** *aloe vera*, curcumin, and chitosan-treated silk suture

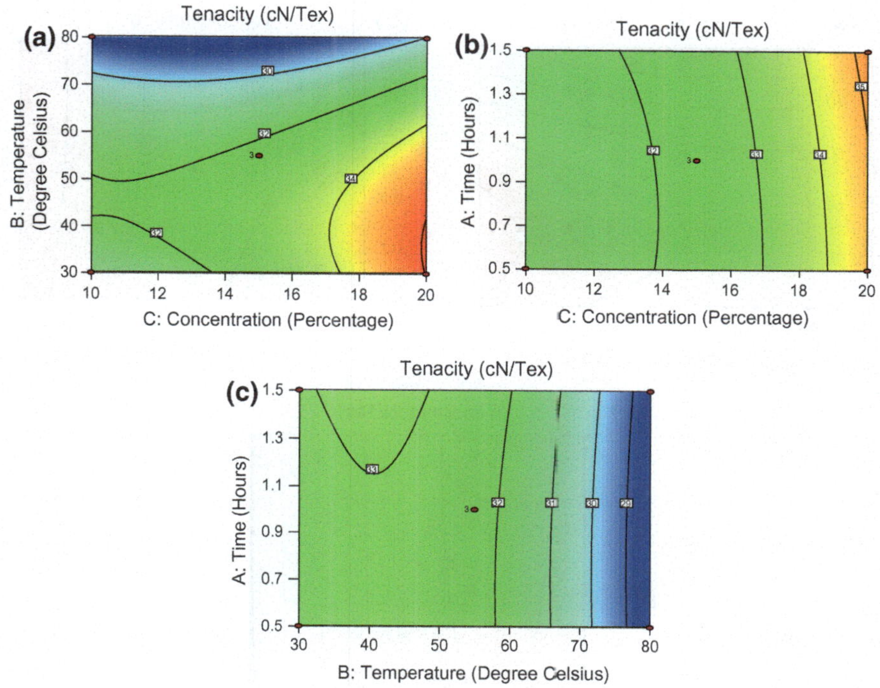

Fig. 4 Influence of **a** concentration and temperature, **b** concentration and time, **c** temperature and time on tenacity of the silk sutures treated with aloe vera and curcumin

at higher temperature and which results in uneven deposition of extracts onto the surface of the sutures. Thus, the tenacity values decreases at higher temperature.

Figures 4b and 5b indicate the influence of concentration and time on tenacity at a constant temperature. From Figs. 4b and 5b, it is noted the tenacity also increases with the concentration. Explanation for the similar trend has been discussed previously. When compared to the other variables, time may not be a significant contributing factor which affects the tenacity of the finished silk sutures. The possible reason might be due to the reactivity and fixation levels of the extracts onto the suture specimen reached the saturation point at 30-min time.

Figures 4c and 5c denote the change of tenacity with respect to the temperature and time at a constant concentration level. The maximum strength of 36.03 cN/tex and 39.86 cN/tex (from Table 2) is obtained at a temperature range of 30–55 °C and a treatment time of 1–1.5 h.

3.2.3 Effect of the Process Parameters on Knot Strength

Knot strength is yet another essential property required for selection of the suture material for wound closure.

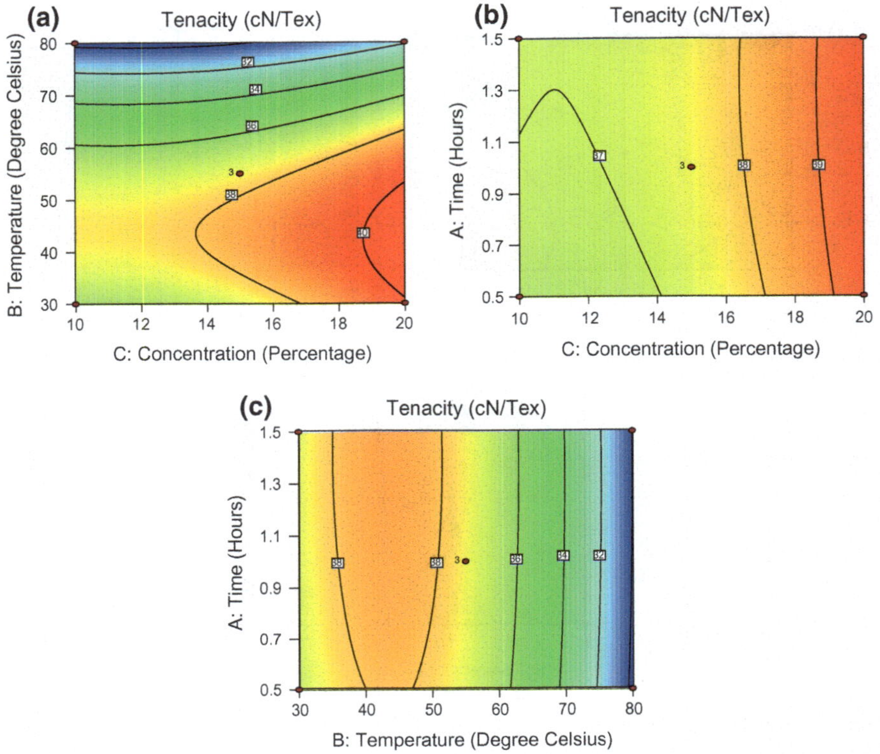

Fig. 5 Influence of **a** concentration and temperature, **b** concentration and time, **c** temperature and time on tenacity of the silk sutures treated with aloe vera, curcumin, and chitosan

The trend in the increase and the decrease of the knot strength as portrayed in the above contour plots (Figs. 6 and 7) is very much similar to tenacity of the sutures. From Figs. 6 and 7, it is observed that the tenacity is higher than the knot strength of the sutures, which confirms that the existence of knot reduces the tenacity. In general, the knot is considered to be the weakest part of the suture. Hence during knot strength test, the breakage happens at the knot area relatively than the suture strands, representing that the high stress concentration in the knot region [18]. The effect of concentration, temperature, and time on knot strength is very similar to that of tenacity.

When compared to the untreated silk sutures, the treated silk sutures showed better knot strength values. The maximum value of knot strength was 26.74 cN/tex and 30.56 cN/tex (from Table 2) which are obtained at a temperature range of 30–55 °C and extract concentration of 20%. This could be due to the reason that the coated silk sutures possess low friction values, and hence, the knot region is not fail easily during knot formation.

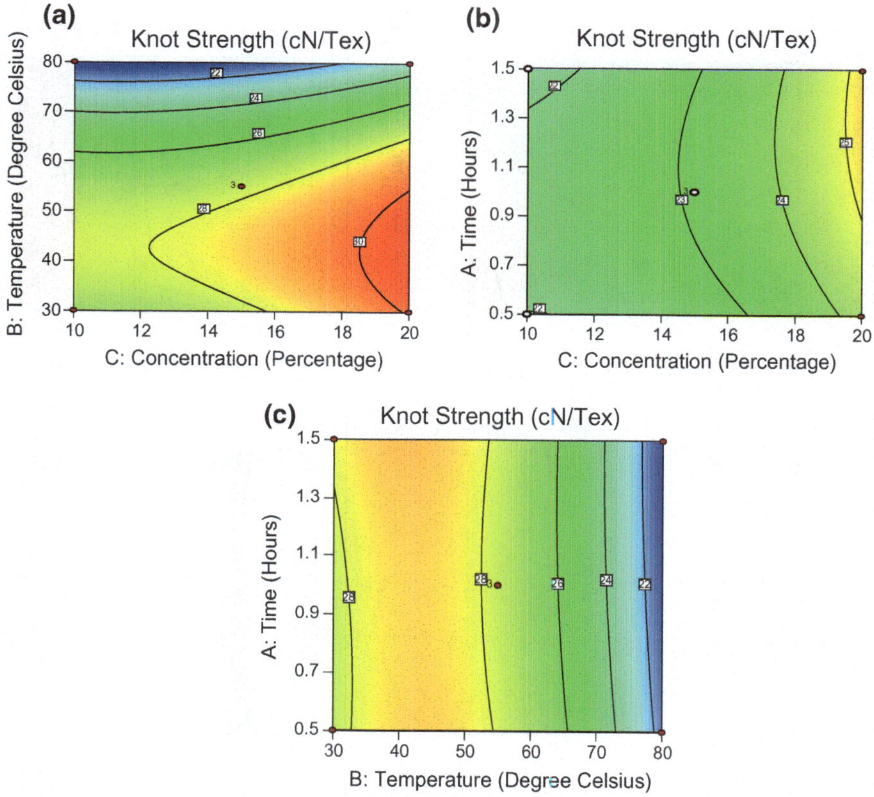

Fig. 6 Influence of **a** concentration and temperature, **b** concentration and time, **c** temperature and time on knot strength of the silk sutures treated with aloe vera and curcumin

Table 6 Optimization goal of the experiment

Response	Goal
Coefficient of friction	Minimize
Tenacity	Maximize
Knot strength	Maximize

3.3 *Optimization of the Model for the Responses*

The multi-objective optimization aims to achieve the best coating conditions for the coating the braided silk sutures without compromising the other suture properties like coefficient of friction, tenacity, and knot strength. Optimization goals of the project are indicated in Table 6.

Table 7 denotes the predicted values of the independent and dependent variables given by the model using the design expert software. Table 7a indicates the values for the suture samples treated with *aloe vera* and curcumin. Table 7b indicates the

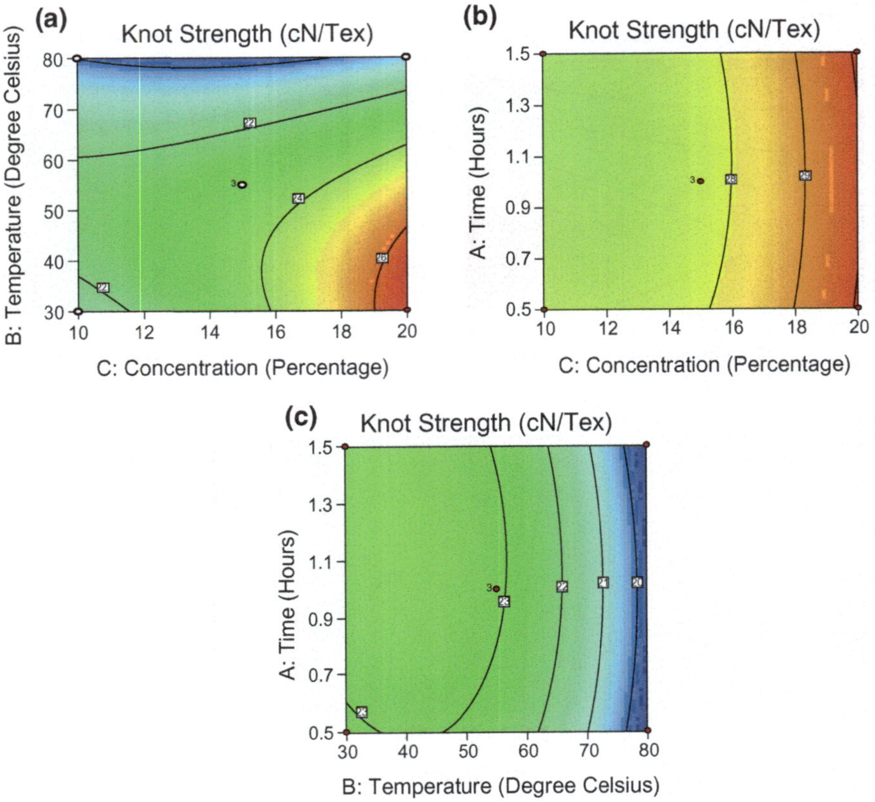

Fig. 7 Influence of **a** concentration and temperature, **b** concentration and time, **c** temperature and time on knot strength of the silk sutures treated with aloe vera, curcumin, and chitosan

values for the suture specimens treated with *aloe vera* and curcumin in combination with chitosan.

3.4 Antibacterial Activity

Based on the numerical optimization process carried out, it is observed that the suture properties like the coefficient of friction, tenacity, and knot strength are on an optimum level for the sutures treated with *aloe vera*, curcumin in combination with chitosan. As the aim of the work is to provide the antibacterial property to the suture material without compromising the other suture properties, the efficacy is assessed only for the sample treated under the optimized conditions given by the model.

The antibacterial activity of the optimized sample is assessed qualitatively by the agar diffusion method (SN195920:1992) against the standard test cultures, viz.

Table 7 Forecasted values given by the model

(a) Finish application of the sutures with *aloe vera* and curcumin in combination		(b) Finish application of the sutures with *aloe vera* and curcumin in combination with chitosan	
Independent variables	*Values*	*Independent variables*	*Values*
Time (hrs.)	1.5	Time (hrs.)	0.56
Temperature (°C)	30	Temperature (°C)	38
Concentration (%)	20	Concentration (%)	20
Dependent variables	*Values*	*Dependent variables*	*Values*
Coefficient of friction	0.376	Coefficient of friction	0.20
Tenacity (cN/tex)	36.44	Tenacity (cN/tex)	40.37
Knot strength (cN/tex)	27.04	Knot strength (cN/tex)	30.62

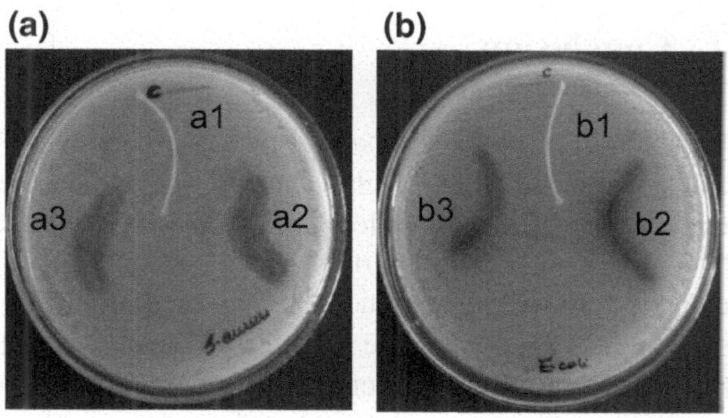

Fig. 8 **a** Zone of inhibition against *S. aureus* and **b** zone of inhibition against *E. coli*

S.aureus ATCC 6538 (gram-positive bacteria) and *E. coli* ATCC 11230 (gram-negative bacteria).

From Fig. 8a1 and b1, it is apparent that there is no zone formed for the untreated silk sutures, whereas in case of the treated samples, the zone of bacterial inhibition is clear, which emphasizes the antibacterial efficiency of the natural agents. This zone is formed due to the migration (leaching) of the coated substances through the agar medium. It is apparent that the activity of *aloe vera*, curcumin in combination with chitosan-treated sample is excellent at a higher concentration level of the bioactive agents. The zone is approximately 6 mm for *S. aureus* shown in Fig. 8a2, a3 and 4 mm for *E. coli* shown in Fig. 8b2, b3.

Aloe vera and curcumin in combination with chitosan are a potential suppressor for the bacterial growth [19]. It is noted that the presence of salicylic acid, anthroquinone, anthraacenone, and acemannon component in *aloe vera* helps to enhance antibacterial activity and also improved due to the presence of saponins, zinc, and amino acids in it [19]. Curcumin has tannins, which forms hydrogen bonding with hydroxyl groups of amino/hydroxyl groups of protein polymer. Tannins are generally accumulated in parts of the trees such as roots, leaves, wood, bark, or fruits.

The antibacterial activity of chitosan with silk could be due to ionic interaction between protonated amino groups of chitosan and carboxylate anions in silk [19].

The difference in activity among the gram-positive and gram-negative bacterium is owing to the difference in membrane features of bacteria. Generally, outer wall of gram-positive bacteria consists of peptidoglycan and acidic polysaccharides with numerous pores. This produces a higher proteolysis cleavage, which inhibits or destroys cellular functions on a higher level [20]. Conversely, in case of gram-negative bacterium, the cell wall consists of lipids, proteins, lipopolysaccharide (LPS), and an internal peptidoglycan cytoplasmic membrane. Therefore, here the outer membrane is usually of high molecular weight and does not allow the strange molecules to penetrate. This double-layer structure of the cell wall is the cause for the lower activity of the gram-negative bacterium [20].

4 Conclusion

Ethanolic extracts of the natural bioactive agents such as *aloe vera* and curcumin and aqueous solution of chitosan are applied on the silk sutures. Box–Behnken response surface methodology is adopted for optimization of the variables in relation to the physical properties of the suture. Statistical analysis is carried out by using the ANOVA technique where all models can be used for navigating the deign space with 95% confidence level. From the analysis, it is concluded that the sutures treated with *aloe vera* and curcumin in combination with chitosan had low values of coefficient of friction, i.e., 0.199, and maximum values of tenacity and knot strength 39.*86 cN*/tex and 30.56 cN/tex, respectively. As the main goal of this research is to provide the antibacterial activity without affecting the other suture properties, the silk suture treated under the optimized conditions is only verified for its antibacterial efficiency by the agar diffusion. Defined zone of inhibition 6 mm for *S. aureus* and 4 mm for *E. coli.* is obtained. With reduced number of experimental runs, substantial, logical and satisfactory results have been obtained, giving an endorsement for the use of the natural bioactive, eco-friendly, non-toxic agents, *aloe vera* and curcumin in combination with chitosan as coating substances for the silk sutures.

Acknowledgements One of the author Dr. S. Viju has been granted Career Award for Young Teachers (CAYT) by All India Council for Technical Education (AICTE), Government of India under Project Number [F.No.11-36/RIFD/CAYT/POL-1/2014-15]. The financial support received from AICTE for carrying out this work is gratefully acknowledged.

References

1. Pillai, C.K.S., Sharma, C.P.: Review paper: absorbable polymeric surgical sutures: chemistry, production, properties, biodegradability, and performance. J. Biomater. Appl. **25**(4), 291–366 (2010)

2. Viju, S., Thilagavathi, G.: Fabrication and characterization of silk braided sutures. Fiber Polym. **13**(6), 782–789 (2012)
3. Viju, S., Thilagavathi, G.: Effect of chitosan coating on the characteristics of silk-braided sutures. J. Ind. Text. **42**(3), 256–268 (2012)
4. Aiello, A.E., Clayton, E.M.R., Todd, M., Dowd, J.B.: The Impact of Bisphenol A and triclosan on immune parameters in the U.S. population, NHANES 2003-2006. Environ. Health Perspect. **119**(3), 390–396 (2011)
5. Sudha, D., Dhurai, B., Ponthangam, T.: Development of herbal drug loaded antimicrobial silk suture. Ind. J. Fibre Text Res. **42**, 286–290 (2017)
6. Parthiban, M., Thilagavathi, G., Viju, S.: Development of antibacterial silk sutures using natural fungal extract for healthcare applications. J. Text. Sci. Eng. **6**, 249 (2016). https://doi.org/10.4172/2165-8064.1000249
7. Pethile, S., Chen, X.J., Hou, D., Wang, L.: Effect of changing coating process parameters in the preparation of antimicrobial-coated silk sutures: an in vitro study. Fibers Polym. **15**(8), 1589–1595 (2014). https://doi.org/10.1007/s12221-014-1589-6
8. Chen, X., Hou, D., Tang, X., Wang, L.: Quantitative physical and handling characteristics of novel antibacterial braided silk suture materials. J. Mech. Behav. Biomed. Mater. **50**, 160–170 (2015)
9. Tummalapalli, M., Anjum, S., Kumari, S., Gupta, B.: Antimicrobial surgical sutures: recent developments and strategies. Polym. Rev. 1–24(2016). https://doi.org/10.1080/15583724.2015.1119163
10. Ammayappan, L., Jeyakodi Moses, J.: Study of antimicrobial activity of *aloevera*, chitosan, and curcumin on cotton, wool, and rabbit hair. Fiber Polym. **10**(2), 161–166 (2009)
11. Internet Health library. http://www.internethealthlibrary.com/Plant-Remedies/Aloevera.htm (2006)
12. Agarry, O.O., Olaleye, M.T., Michael, C.O.B.: Comparative antimicrobial activities of aloe vera gel and leaf. Afr. J. Biotech. **12**, 1413 (2005)
13. Liu, X.F., Guan, Y.L., Yang, D.Z., Yao, K.D.: Antibacterial action of chitosan and carboxymethylated chitosan. J. Appl. Polym. Sci. **79**, 1324 (2001). https://doi.org/10.1002/1097-4628
14. MahendraGowda, R.V., Mohanraj, S.: A novel approach to measure friction in textile fibre assemblies. Res. J. Text. Apparel. **12**(2), 30–38 (2008)
15. Ben Abdessalem, S., Debbai, F., Jedda, H., Elmarzougui, S., Mokhtar, S.: Tensile and knot performance of polyester braided sutures. Text. Res. J. **79**(3), 247–252 (2009)
16. Viju, S.: Development and characterisation of silk and Poly L-lactic acid based sutures. Ph.D thesis, PSG College of Technology, Anna University Chennai, Tamilnadu, India (2013)
17. Kavitha, T., Padmashini, R., Swarna, A., Gridev, V.R., Neelakandan, R., Senthilkumar, M.: Effect of chitosan treatment on the properties of turmeric dyed cotton yarn. Ind. J. Fibre Text. Res. **32**(1), 53–56 (2007)
18. Heward, A.G., Laing, R.M., Carr, D.J., Niven, B.E.: Tensile performance on nonsterile suture monofilaments affected by test conditions. Text. Res. J. **74**(1), 83–90 (2004)
19. Ammayappan, L., Jeyakodi Moses, J.: Study of antimicrobial activity of *aloevera*, chitosan, and curcumin on cotton, wool, and rabbit hair. Fibers Polym. **10**(2), 161–166 (2009)
20. Goy, R.C., Assis, O.B.G.: Antimicrobial Analysis of Films Processed from Chitosan and N,N,N-Trimethylchitosan. Brazilian J. Chem. Eng. **31**(3), 51–56 (2014)

Effect of Domestic Laundering on Removal of Bacterial Contamination from Nurses' White Coats

Priyanka Gupta, Nilanjana Bairagi and Deepti Gupta

Abstract *Objective* Effect of laundering in bacterial decontamination of nurses' white coats. *Methodology* Patches of sterilized polyester and polyester cotton blend fabric were stitched on a washed white coat, worn by nurses for two shifts. At the end of the second shift, one patch was removed and plated. The other patch was removed and plated after domestic laundering of the coat. Total microbial contamination on the patches was assessed in brain heart infusion (BHI) broth before and after laundering. *Findings* All white coats sampled in the study across different wards of the hospital were found to be contaminated even after laundering. After two shifts, the contamination on blend was 54% higher than polyester fabric. After the domestic laundering, the bacterial reduction was 76% on polyester and 81% on the blend. *Conclusion* Nearly 20% of the contaminants were retained on the coat after laundering, indicating that the nurses, in fact, are carrying potentially harmful infections both into the hospital and their homes. The current home-laundering practices followed in India are not sufficient for sterilization of the nurses' uniform; thus, laundering guidelines are required to completely decontaminate the coats.

Keywords Laundering · Decontamination · Uniforms · Hospital textiles · Infection control

---

P. Gupta (✉)
Department of Knitwear Design, National Institute of Fashion Technology Delhi, Hauz Khas, New Delhi 110016, India
e-mail: priyanka.gupta@nift.ac.in

N. Bairagi
Department of Knitwear Design, National Institute of Fashion Technology Bengaluru, Bengaluru 560102, Karnataka, India

D. Gupta
Department of Textile Technology, Indian Institute of Technology Delhi, Hauz Khas, New Delhi 110016, India

A. Majumdar et al. (eds.), *Functional Textiles and Clothing*,
https://doi.org/10.1007/978-981-13-7721-1_6

1 Introduction

There are a number of bacterial reservoirs in a hospital setting, and hospital textiles are one of them. The growth of micro-organisms on textiles inflicts a range of unwanted effects not only on the textiles itself but also on the wearer. These effects include the generation of unpleasant odour, stains and discoloration in the fabric, a reduction in fabric mechanical strength and an increased likelihood of contamination.

Surfaces in hospital environment get contaminated with pathogenic micro-organisms which usually come from patients, healthcare workers, visitors, or from external sources. Therefore, the hospital environment is considered as a reservoir of pathogenic micro-organisms and some of which can survive for a longer period of time. These pathogens can spread to patients through direct or indirect contact leading to the development of healthcare-associated infections (HAIs) [6].

The rising incidence of hospital infections is a matter of great concern in India. This growth is often attributed to the absence of effective infection control strategies in healthcare facilities. The studies are mostly focused on ineffective laundering practices followed by hospitals and nurses. Research finding of healthcare worker's uniform laundering is very limited, and the methods and study designs vary significantly, making positive correlations and definitive justifications difficult to confirm [7]. In one of the study, the occurrence of bacterial contamination of the linens in the hospital due to improper laundering was observed [5]. The linens in the hospital environment have more patient contact than healthcare workers' uniforms; hence, improper decontamination of the linens can have severe consequences for patients. The studies show the contradictions in the effectiveness of home laundering and industrial laundering of the scrubs.

The nature and frequency of laundering practices may have implications in the spread of infections. The laundering methods used for nurses' uniform are diverse in India. The two laundering methods which are used for washing the contaminated uniforms are industrial laundering and domestic laundering. The temperature, detergent, time of washing and drying conditions are different in both the laundering mechanisms. The lack of specific guidelines or policies for laundering in the literature suggests that the contamination arrives at the hospital due to improper handling of the uniforms during laundering. It has been reported that hospitals and other health organizations are unable to monitor the home-laundering practices of the healthcare workers' uniform and hence cannot prevent the problem [4]. Therefore, it is important to investigate the home-laundered scrubs as a possible source of contamination.

There are limited studies on the efficacy of domestic laundering methods practised in India in decontaminating soiled nurses' white coats. This paper investigates the effectiveness of domestic laundering method in eliminating bacterial contamination from nurses' white coats. Microbial contamination on polyester and blend fabric patches, worn for two shifts (12 h of exposure), was studied for home-laundered nurses' white coats, which is a normal practice in government hospitals in India.

2 Methodology

The study was undertaken in a 100-bedded government hospital in Delhi, India, during the month of April–May 2016. Two nurses from five different wards, namely paediatric, medicine, gynaecology, casualty and intensive care unit, participated in the study.

2.1 Sampling Method

Two kinds of fabric, namely polyester and polyester cotton blend fabric, were selected for the study in India based on the current uniform fabrics used for nurses' uniform. To investigate bacterial contamination in a hospital environment, bacterial growth was studied in brain heart infusion (BHI) broth for the total microbial load. Brain heart infusion broth is a highly nutritious general-purpose growth medium where all bacteria can grow. All chemicals and media used for bacterial sampling were procured from HiMedia Laboratories Pvt. Ltd., Mumbai.

The samples were taken from the abdominal region over the pockets of the nurses' white coat. The nurses were asked to get the laundered coat before the beginning of the shift. Fabric patch method was developed in the study to assess the bacterial load on nurses' white coats. A 20 cm × 10 cm fabric sample of 100% polyester was stitched to a 20 cm × 10 cm swatch of 70/30 polyester cotton-blended fabric to make a total patch of size 20 cm × 20 cm. This fabric patch was marked on the wrong side to differentiate between the exposed and non-exposed fabric sides. The prepared sterilized patch was stitched with gloved hands over the abdominal area (below belt) of the nurses' coat—one on the left and another on the right side using a sterile needle and thread at 8:00 am as shown in Fig. 1. The nurse would wear the coat and perform their normal duties. The patch was removed after the two shifts (twelve hours of exposure), and the patch on the right side was removed and taken immediately for plating. Then the coat with an exposed patch on the left side was folded and kept in a sterile polythene bag which was then home-laundered by the nurse. At home, it was soaked in water along with commercial detergent at room temperature for 30 min. This is followed by scrubbing, rinsing and finally drying in the open. After drying, the patch on the left side was removed from the laundered coat, bagged and taken to the microbiology laboratory for plating on BHI plates.

A total of 40 swatches were sampled in this study—10 nurses × 2 fabrics × 2 methods (before and after laundering).

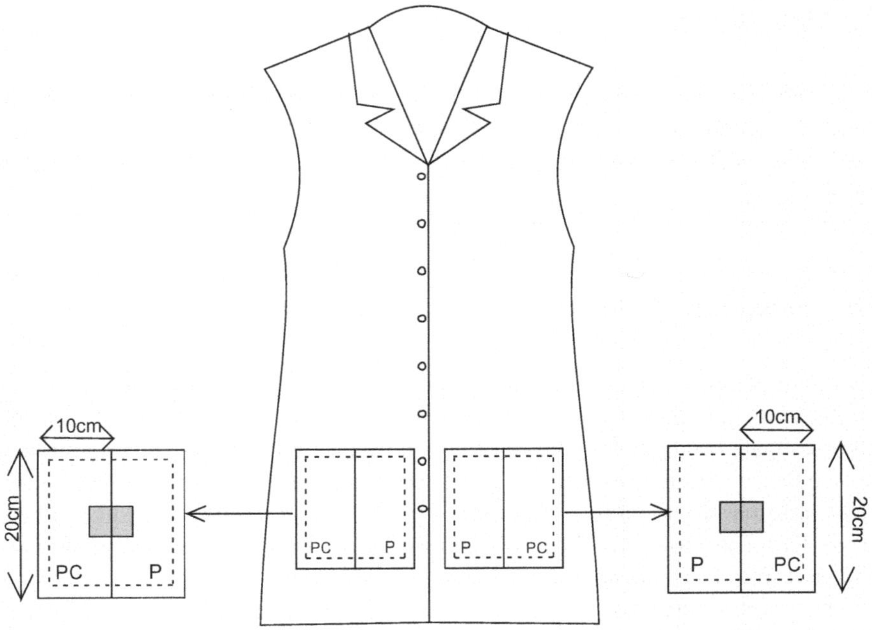

Fig. 1 White coat with test patch attached. P, polyester; PC, polyester cotton. Shaded portion indicates the location of the swatch cut for sampling

2.2 Determination of Bacterial Counts

In the microbiology laboratory, the collected patches of polyester and blend fabric were cut into swatches of 4 cm × 3 cm as marked, inside a biological safety cabinet. The cut samples were kept for five minutes in the non-nutritive medium phosphate-buffered saline (PBS, 1X) to moisten and maintain pH and osmolarity of the bacteria present on the sample, taken out and placed on the brain heart infusion (BHI) broth plates for 15 min with the exposed area facing the media. After 15 min of contact time, the swatches were removed and discarded. The inoculated plates were dried for 10 min and then incubated for 24 h at 37 °C. The colonies were counted manually the next day.

Control: Sterilized 100% polyester and 70/30 polyester cotton blend fabric patches were plated and incubated on BHI media, and the colonies were counted on the next day.

Table 1 Bacterial count (CFUs) on polyester and blend fabric

Ward	Polyester		Polyester cotton	
	BL	AL	BL	AL
Paediatrics	280	41	398	53
	287	38	374	42
Gynaecology	192	58	363	71
	184	50	347	66
Medicine	155	62	297	77
	164	40	278	53
Casualty	150	43	183	57
	134	52	187	69
Intensive care unit	255	57	387	63
	270	49	376	54
Total (CFUs)	**2071**	**490**	**3190**	**605**

(Before laundering: BL; after laundering: AL)

3 Results

There was no growth observed on the two control samples, indicating no contamination prior to sample collection and testing.

3.1 BHI Broth

Test patches were sampled for total bacterial contamination before and after laundering of coats that had been worn for two consecutive shifts. Total colony counts obtained on test samples in BHI broth, before and after laundering, are compiled in Table 1. The total colony count on polyester was around 2071 CFUs, and on polyester cotton was around 3190 CFUs. After the home-laundering process, bacterial colonies were reduced by nearly 76% (60–87%) on polyester and 81% (63–89%) on the blend. Mean colony counts on blend after laundering were 60.5 CFUs (42–77 CFUs) which was 23.4% higher than polyester with 49 CFUs (38–62 CFUs) as shown in Fig. 2.

4 Discussion

All white coats sampled in the study across different wards of the hospital were found to be contaminated even after laundering. The difference in a number of colonies from various coats could be because each nurse used a different method of laundering. The temperature of laundering, type of detergent or vigorousness of scrubbing with

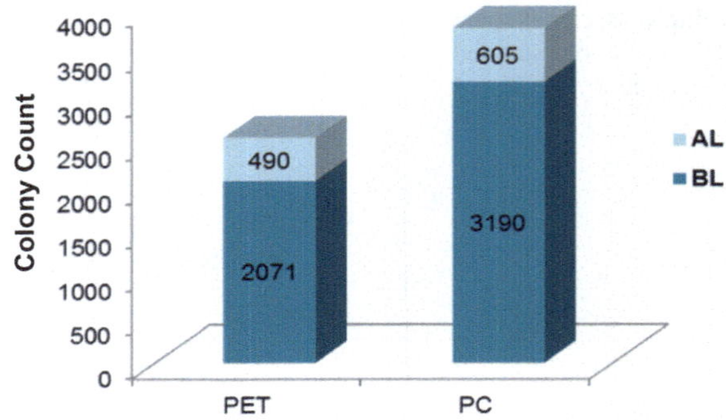

Fig. 2 Total bacterial count (CFUs) on polyester (PET) and blend fabric (PC) in BHI media (BL, before laundering; AL, after laundering)

hand or brush can have an effect on the degree of decontamination. However, it is worthwhile to note that variations in the method notwithstanding, about 11–13% microbes were always remaining after laundering. In some cases, the values were as high as 40%. This may also be attributed to the low washing temperature of less than 70 °C as per the recommendation of US CDC and also due to lack of chlorine bleach treatment on the laundered coats [1].

In another study when the bacterial contamination was compared on new, disposable, laundered and unlaundered hospital scrubs, it was documented that significantly higher bacteria counts were isolated from home-laundered scrubs and unwashed scrubs than from new, hospital-laundered and disposable scrubs [3]. Similar results were observed in this study.

Laundering of contaminated scrubs at home also increases the chances of contamination of the home environment. Inappropriate laundering temperature and lack of decontamination instructions may lead to contamination of an entire load of clothing [2, 4].

5 Conclusion

This research provides the latest insights on the degree of contamination of nurses' uniform in Indian hospital across different wards. It was found that nearly 20% of bacteria were retained on the nurses' coat after home laundering, indicating that the nurses, in fact, were carrying potentially harmful infections both into the hospital and their homes. The nurses should use a new set of laundered and sterilized uniform in each shift to avoid nosocomial infection and build-up of microbial contamination on the uniform from one shift to the next. To ensure complete decontamination policy needs to be included in determining the type of detergent, the temperature of wash and the method of drying and ironing. Methods for safe handling and transporting of soiled textiles based on hygienic storage should also be defined.

Acknowledgements We would like to extend our gratitude to the participants (nurses) involved in this study. We are thankful to Medical Superintendent of Government hospital, Director School of Natural Sciences, Shiv Nadar University and Head of Department Textile Technology, IIT Delhi.

References

1. Boyce, J.M.: Environmental contamination due to methicillin-resistant Staphylococcus aureus: possible infection control implications. Infect. Control. Hosp. Epidemiol. **18**, 622–627 (1997)
2. Belkin, N.L.: Home laundering of soiled surgical scrubs: surgical site infections and the home environment. Am. J. Infect. Control **29**(1), 58–64 (2001)
3. Nordstrom, J.M.: Comparison of bacteria on new, disposable, laundered, and unlaundered hospital scrubs. Am. J. Infect. Control **40**(6), 539–543 (2012)
4. Parish, C.: Washing work clothes at home risks spreading infection, conference told. Nurs. Standard. **20**(12), 12 (2005)
5. Perry, C.: Bacterial contamination of uniforms. J. Hosp. Infect. **48**(3), 238–241 (2001)
6. Steinberg, J.P.: The role of the hospital environment in the prevention of healthcare-associated infections by contact transmission. HERD: Health Environ. Res. Des. J. **7**(1 suppl), 46–73 (2013)
7. Wilson, J.A.: Uniform: an evidence review of the microbiological significance of uniforms and uniform policy in the prevention and control of healthcare-associated infections. Report to the Department of Health (England). J. Hosp. Infect. **66**(4), 301–307 (2007)

Part III
Smart Textiles

Hybrid Cover Yarn's Element Orientation and Its Impacts on Mechanical/Tensile Behavior of Conductive Yarns and Fabrics

Ali Asghar, Mohd Rozi Ahmad, Mohamad Faizul Yahya, Syed Zameer Ul Hassan and Muhammad Kashif

Abstract Of the different techniques to produce conductive yarns for e-textiles, multicomponent hybrid yarns with continuous metallic filaments are well known for its ease in processability, better durability, and conductivity. Continuous metallic wires in yarn's core have been researched vigorously, but since the metal wires are stiff, they produce fabrics with open interstices and make the fabric transparent towards higher-frequency electromagnetic waves. Moreover, this problem further elevates with an attempt to increase the amount of metallic content to improve the electromagnetic shielding capabilities. Continuous metallic filament, wound as spiral covers on textile core in form of hybrid cover yarns, is proven to have better control of the metallic component via altering the turns per meter of the coverings, which also improves the electromagnetic shielding capabilities of the ultimate fabrics. However, the consequence of this alternate orientation on the tensile and mechanical properties of the fabrics has never been studied. This study analyzes the mechanical properties of copper-covered polyester yarns. This orientation of hybrid cover yarns is far much superior in terms of the yarn's tensile properties with around fourfold increase in the tenacity values and around 30–70% reduction in modulus values. The fabrics prepared from the copper-covered yarn design required around 50–200% more force at around 200% more elongation to rupture, as compared to the conventional fabric design. Moreover, the fabric stiffness and abrasion properties in the copper-covered orientation also improve, but at the cost of increased static friction of the fabrics.

Keywords Conductive fabrics · Tensile properties · Mechanical properties · Electromagnetic shielding fabrics · E-textiles · Yarn design

---

A. Asghar · M. R. Ahmad (✉) · M. F. Yahya
Textile Research Group, Faculty of Applied Sciences, Universiti Teknologi MARA, Shah Alam, Selangor, Malaysia
e-mail: rozitex@salam.uitm.edu.my

A. Asghar · S. Z. U. Hassan · M. Kashif
Textile Engineering Department, Faculty of Engineering, Balochistan University of Information Technology, Engineering and Management Sciences, Quetta, Pakistan

A. Majumdar et al. (eds.), *Functional Textiles and Clothing*,
https://doi.org/10.1007/978-981-13-7721-1_7

1 Introduction

Electronic textiles are well recognized for their conductive characteristics and have involved many researchers recently. Smart fabrics with embedded sensors/computers, which can actively or passively interact with its user or the environment [1, 2] are one of the major application areas. These fabrics have been utilized as wearable health monitoring systems [3], flexible rechargeable solar batteries [4], wearable antennas [5], and in various other protective applications including electromagnetic interference (EMI) shielding materials [6] for industrial/device protection [7] and as safety workwear [8].

Multicomponent hybrid yarn with continuous metallic filament is well known for its ease in processability and electromagnetic shielding effectiveness [9]. Continuous metallic wires in yarn's core have been the attention of many researches in terms of its conductive and electromagnetic shielding capabilities, by producing yarns from techniques such as core-sheath and plied yarns [10, 11]. The stiff metal wires produce fabrics with open interstices and becomes transparent toward higher-frequency electromagnetic waves [12]. This issue further elevates with the increase in the amount of metallic content in an attempt to improve the electromagnetic shielding effectiveness (EMSE) [13].

Continuous metallic filament, wound as spiral covers on textile core, is proven to have efficient control of the metallic component by altering the turns per meter of the coverings [14], which also improves the EMSE [9] and thermal comfort properties [15]. However, the consequence of this alternate orientation on the mechanical properties of the fabrics has never been studied. This study analyzes the tensile/mechanical properties of copper-covered polyester yarns and fabrics and highlights the significant differences between both the orientations.

2 Experimental Methods

2.1 Materials

Copper wire was used as the conductive material to produce hybrid cover yarns (HCYs). The selection of copper was carried out due to its superior conductivity and better resilience [16], as compared to other low-cost conductive materials like stainless steel and aluminum. The copper wires were obtained from the open market manufactured by FE Magnet Wire (M) Sdn. Bhd, Malaysia, and TA Win Industries (M) Sdn. Bhd, Malaysia. Three diameters of copper wires were used (based on availability in the market) including 0.14, 0.11 and 0.08 mm. For the counter textile part, 100% polyester yarn was used, which was manufactured under the brand name Carnation® by Polytex Industries (M) Sdn. Bhd, Malaysia. The polyester yarns were dyed and had average Tex count of 31.

2.2 Sample Preparation and Testing

Twelve conductive hybrid cover yarns (two-component) from three genres of copper filaments (with respect to its size) were prepared using hollow spindle spinning technique. Three conventional (copper core) and nine proposed (copper cover) orientations of hybrid yarns were prepared, with 25, 20, and 15 turns per inch (TPI) of the covering component. These hybrid yarns were used as fillings in the fabrics by producing a 1/1-plain weave structure on an automatic rapier weaving machine, which was preloaded with 31 Tex polyester yarns as the warp component. Table 1 displays the yarn and fabric classifications.

The tensile properties of the resultant yarns and fabrics (weft direction only) were measured using ASTM D2256 and ASTM D5035, respectively, on a tensile testing machine. Martindale abrasion tester method ASTM D4966 was used to measure the

Table 1 Yarn and fabric sample classification

Fabric sample ID	Weave density (per 2.54 cm)	Warp yarns	Weft yarns (hybrid yarn ID)	Weft yarn's core component	Weft yarn's covering component	TPM/TPI (approximately) of covering component
FA_c	78 × 45	Polyester	A_c	0.14 mm copper	Polyester yarn	253/6
FA_{15}	78 × 45	Polyester	A_{15}	Polyester yarn	0.14 mm copper	603/15
FA_{20}	78 × 45	Polyester	A_{20}	Polyester yarn	0.14 mm copper	795/20
FA_{25}	78 × 45	Polyester	A_{25}	Polyester yarn	0.14 mm copper	984/25
FB_c	78 × 45	Polyester	B_c	0.11 mm copper	Polyester yarn	249/6
FB_{15}	78 × 45	Polyester	B_{15}	Polyester yarn	0.11 mm copper	601/15
FB_{20}	78 × 45	Polyester	B_{20}	Polyester yarn	0.11 mm copper	806/20
FB_{25}	78 × 45	Polyester	B_{25}	Polyester yarn	0.11 mm copper	970/25
FC_c	78 × 45	Polyester	C_c	0.08 mm copper	Polyester yarn	225/6
FC_{15}	78 × 45	Polyester	C_{15}	Polyester yarn	0.08 mm copper	590/15
FC_{20}	78 × 45	Polyester	C_{20}	Polyester yarn	0.08 mm copper	803/20
FC_{25}	78 × 45	Polyester	C_{25}	Polyester yarn	0.08 mm copper	965/25

abrasion properties of the resultant fabrics, while the static angle of surface drag was measured using Shirley friction tester and BS 3424-10 standard.

3 Results and Discussion

3.1 *Tensile Testing of Yarns*

Effect of Alignment Change on Tensile Properties of Yarns

Figure 1 displays the tensile properties of the hybrid yarns for the copper core alignment (A_c) and copper covering alignment with 25 TPI of cover (A_{25}). There was around twofold to sixfold increase in the peak tensile force for the copper cover alignment as compared to the copper core alignment. The maximum tensile force as sustained by the copper core yarns were 2.5, 1.5, and 3.7 N for A_c, B_c, and C_c, respectively, whereas, the maximum tensile force for the copper cover yarns were 11.6, 10.1, and 10 N for A_{25}, B_{25}, and C_{25}, respectively. Independent samples t test was conducted between the alignment factor (irrespective of copper size) and the peak force values. The results reveal significantly (t (28)=−24.019, $p < 0.001$) higher values of copper cover alignment *(M = 10.55, SD = 0.87)* as compared to the copper core alignment (*M* =2.54, *SD* =0.95).

For 0.14 mm copper yarns (*A*), an eightfold increase in the elongation value was recorded when the copper alignment was changed from core to cover. Similarly, a 17-fold increase in elongation was recorded for 0.11 mm (*B*) yarns, whereas the yarns with 0.08 mm copper (*C*) were recorded with 60% increase in elongation values when the copper core alignment was changed to copper cover alignment. The average copper core alignment values for elongation (*M* =9.46, *SD* =9.10) were significantly (t (20.38)=−11.77, p <0.001) lower as compared to the average elongation values for copper cover alignment (*M* =40.13, *SD* =4.45).

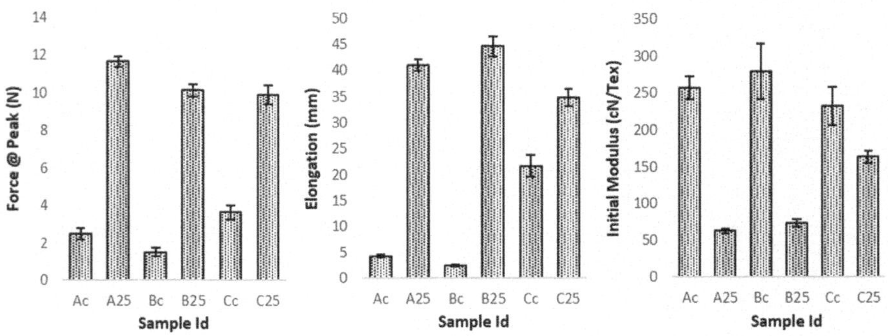

Fig. 1 Comparison of tensile properties between the alignments of copper in yarns

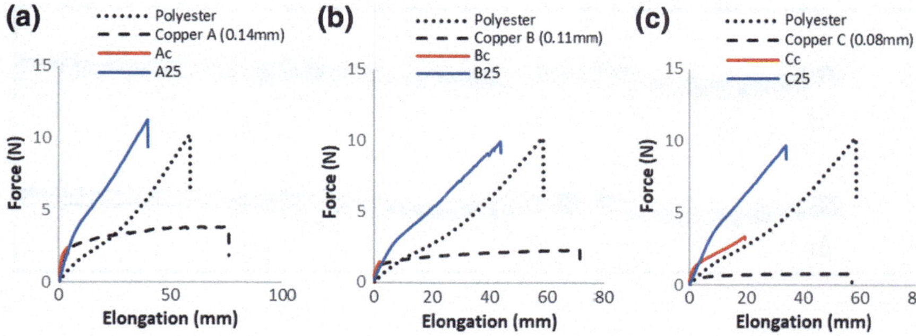

Fig. 2 Force–elongation curve comparison between copper core, copper cover, and components of HCY

Furthermore, the modulus values were reduced by 30–75% in the HCY structures when the copper core alignment was changed to copper cover alignment. The maximum reduction in initial modulus values was recorded in 0.14 mm (*A*) yarns, which was followed by 0.11 mm (*B*) and 0.08 mm (*C*) yarns in sequence. Irrespective of the copper size, the initial modulus values for copper yarns (M =100.05, SD =47.25) were significantly reduced (t (24.72)=10.57, p <0.001) as compared to the copper core yarns (M =256.12, SD =32.26).

Figure 2 is a comparison of force–elongation behavior of copper core and copper cover yarns with the force–elongation behavior of their components. In any hybrid yarn structure, the core component is mainly held responsible for the tensile characteristics [17], and the tensile properties of the core component supersede the tensile characteristics of any other component in that yarn [18]. In accordance with the literature, it is evident from Fig. 2a and b that the copper core hybrid yarns A_c and B_c mimic the force–elongation curve of their respective bare copper wires. However, since wrapped by polyester as the covering component and due to tension drafts during hollow spindle spinning [19], the core wires were not allowed to elongate as in the case of the bare wires. Moreover, the maximum force (N) as sustained by the copper core yarns was almost the same as the bare copper wires, for 0.14 and 0.11 mm copper.

On the other hand, the copper cover yarns A_{25} and B_{25} displayed the maximum force equivalent to the maximum force shown by the polyester yarn alone, which was around 3.7–5.8 times, respectively, higher as compared to copper core yarns. In addition, the A_{25} and B_{25} displayed 8–17 times, respectively, greater elastic elongations as compared to the copper core yarns. However, as evident from the initial Hookean region of the force–elongation curves, the yarns are much stiffer than the polyester component for both A_{25} and B_{25} yarns, but are approximately 75% less stiff than their respective copper core yarns.

The copper cover yarn C_{25} in Fig. 2c showed somewhat similar behavior as the copper cover yarns from other copper sizes. The maximum force, elongation, and the initial modulus values are far superior to its counter copper core yarn (C_c) and are much closer to the properties of bare polyester yarns, as so in the case

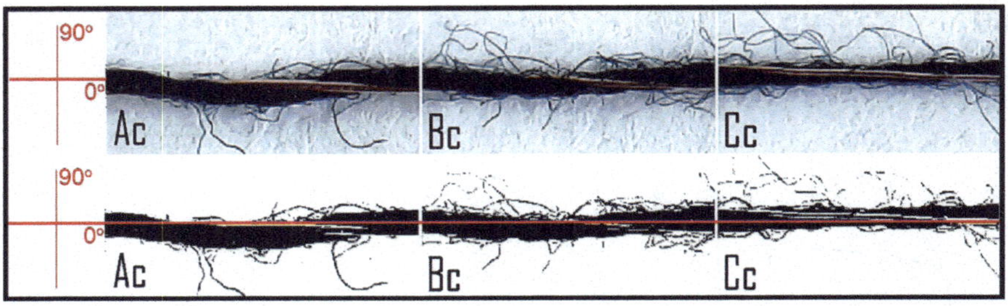

Fig. 3 Comparison of covering component's waviness in copper core yarns

of A_{25} and B_{25}. However, the copper core yarn C_c had a different behavior in the force–elongation curve as compared to other yarns of similar alignment. The peak force (*N*) and maximum elongation for this yarn were much superior to A_c and B_c; furthermore, it had a slightly lower initial modulus than A_c and B_c. These deviations of copper core yarn from class "*C*" are attributed to its copper positioning in the hybrid yarn structure as shown in Fig. 3. The images of copper core yarns, with one complete turn of the covering component, which were held at approximately one N force (images taken at the tensile testing machine), display the physical structure of HCYs. A horizontal line drawn at 0° (Fig. 3 threshold image, bottom) through the core of the yarns is supposedly the position of copper core inside the yarns. It can be approximately assessed that as the copper gets finer, the spiraling polyester cover changes its position, which is attributed to the fact that thin core cannot be held perfectly straight within the yarn's body and the covering component cannot follow a perfectly spiral arrangement on the core. The copper core belonging to class *C*, despite being held at similar core tension during yarn formation, changed its position relative to the cover yarn when the tension is released. Moreover, it can be observed that the spiraling arrangement of the polyester covering is more prominent where the core is thick; as we move from thick toward thin core (From A_c to C_c), the cover yarns become more and more straight. Hence, yarns belonging to copper core class *C* appear to have a plied yarn structure rather than a cover yarn structure. It is due to this plied structure of the yarn C_c that the yarns had greater maximum force and greater elongation as compared to A_c and B_c, since the tensile characteristics of polyester component were accompanied by the tensile characteristics of copper core.

Effect of Turns of Copper Covering on Tensile Properties of Yarns

The effects of the turns of copper coverings on the tensile properties are displayed in Fig. 4. There existed no statistically significant relationship (F (2, 42) = 0.42, p = 0.66) between the peak force values for 15 TPI (M = 10.54, SD = 0.85), 20 TPI (M = 10.80, SD = 0.87), and 25 TPI (M = 10.55, SD = 0.87) of copper cover yarns (one-way ANOVA).

The average elongation value for 25 TPI of copper cover yarn was 40.14 $\pm$ 4.45 mm, while the elongation values were 43.75 $\pm$ 3.0 mm for 20 TPI and 39.23 $\pm$ 2.9 mm for 15 TPI of copper coverings. One-way ANOVA results show significant

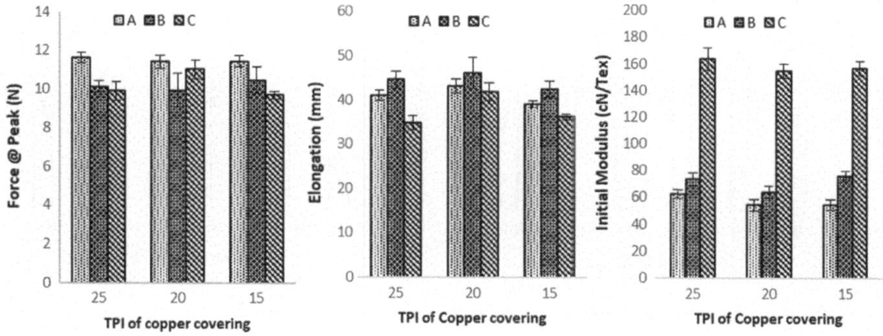

Fig. 4 Comparison of tensile properties between the TPI of copper coverings

differences ($F\,(2, 42) = 6.94, P = 0.002$) in the average values of elongations between 15, 20, and 25 TPI copper cover yarns. The multiple comparisons test revealed significantly higher values of elongation for 20 TPI yarns as compared to 15 TPI and 25 TPI yarns with *p values* of 0.003 and 0.02, respectively. Moreover, the average values of elongation for 0.11 mm (*B*) copper yarns were the highest as compared to other copper sizes in all the TPI classes.

Regardless of copper size, with only the number of turns as the governing factor, there exists no statistically significant relationship (F (2, 42) = 0.14, p = 0.87) between the average initial modulus values for 25 TPI (M = 100.05, SD = 47.25), 20 TPI (M = 91.02, SD = 47.05), and 15 TPI (M = 95.84, SD = 45.90) of copper coverings. However, the lowest initial modulus values were recorded for 0.14 mm (*A*) yarns.

3.2 Mechanical Properties of Fabrics

Effects of Alignment on Tensile Properties of Fabrics

The maximum tensile force (weftwise) as sustained by the fabrics is displayed in Fig. 5. The fabrics belonging to copper cover alignment showed the highest peak forces as compared to copper core yarn fabrics. The maximum force sustained by copper cover yarn fabric FA_{25} was 57% higher than its counter copper core yarn fabric, whereas the copper cover yarn fabrics FB_{25} and FC_{25} had almost 8 and 5% increased peak force, respectively, as compared to their respective copper core fabrics. Irrespective of copper sizes, the maximum force values as sustained by the copper cover fabrics (M = 461.92, SD = 25.42) were significantly higher (t (18.82) = −4.49, $p < 0.001$) as compared to the copper core fabrics (M = 386.02, SD = 60.33).

In terms of elongation of fabrics, the fabrics belonging to class A showed 75% higher elongation when the alignment of copper in weft yarns was changed from core

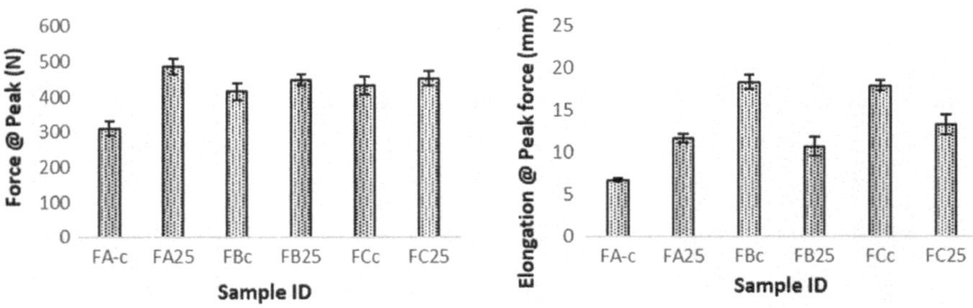

Fig. 5 Comparison of tensile properties of fabrics between alignments

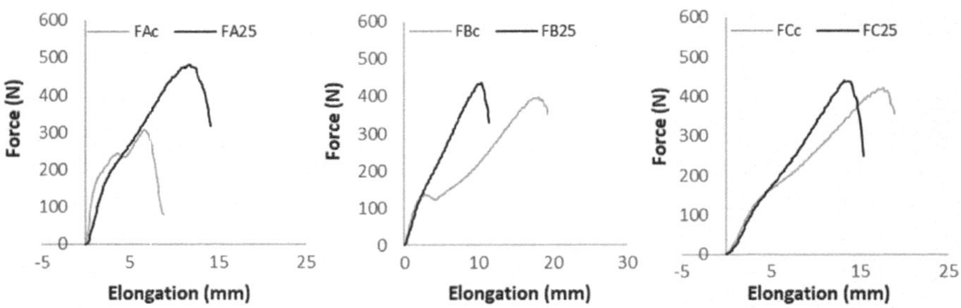

Fig. 6 Force–elongation comparison between copper core and copper cover fabrics

to cover; however, there was around 72% and 35% decrease in elongation values for fabric *B* and *C*, respectively. The average values for the copper core yarn fabrics ($M = 14.32$, $SD = 5.59$) had no significant differences ($t\,(15.86) = 1.63$, $p = 0.12$) over the copper cover yarn fabrics ($M = 11.90$, $SD = 1.44$).

The impacts of alignment on the tensile properties can be explained by the force–elongation curves of fabrics, which are displayed in Fig. 6. The force–elongation curves for copper core fabric FAc and FBc have negative slopes in the yield region of the curve. This negative slope during the tensile elongation displays the loss of force due to rupture of core component in the fabrics. To further explain this loss of force during the tensile extension, the change of force (delta *N*, difference between the succeeding and preceding value of *N*) was plotted against the elongation values for the copper core fabrics belonging to copper size *A* and *B* and is displayed in Fig. 7. The drop of force goes beyond the zero ΔN mark at around 3.7 mm elongation for fabric FAc and around 2.9 mm elongation for fabric FBc (dotted lines), which are approximately same as the maximum elongation for the component weft yarns (Fig. 1). From this point onwards, the fabrics may be considered ruptured as the conductive component of the fabrics is no longer intact. The values of force at these rupture points were 241.66 and 230.83 N for FAc and FBc fabrics, respectively. However, after this rupture point, the fabric continues to elongate more, since the second component of the hybrid yarns (polyester) takes over the copper core's position.

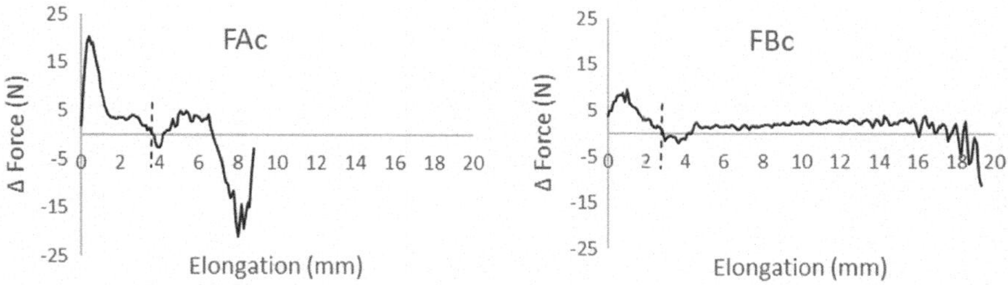

Fig. 7 Change of force against elongation for copper core fabrics *A* and *B*

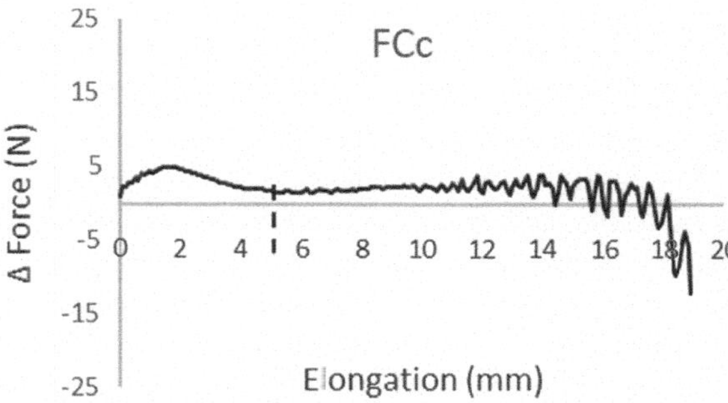

Fig. 8 Change of force against elongation for copper core fabric *C*

Additionally, there was no negative slope in the force–elongation curve for the copper core fabric *C*; moreover, the values of ΔN did not go beyond the zero ΔN mark. This different behavior of the copper core fabric Cc may again be attributed to the difference of weft yarn's design as explained in Fig. 3. The copper core yarn used for this fabric behaves more like a plied yarn structure with copper twisted with the polyester and not held straight in yarn's core. However, there is a maximum drop of force point (dotted line) at around 5.1 mm elongation (Fig. 8), which can again be attributed as the rupture point in fabric FCc with 252.56 N force.

Effects of Copper Covering's Number of Turns on Tensile Properties of Fabrics

The effects of the number of turns of the copper cover on the maximum force and elongation (in weft direction only) are displayed in Fig. 9. Irrespective of the copper size, the average values of the peak force for 25 TPI fabrics ($M = 461.92$, $SD = 25.42$), 20 TPI fabrics ($M = 462.65$, $SD = 39.06$), and 15 TPI fabrics ($M = 466.57$, $SD = 41.08$) had no significant differences (F (2, 36) = 0.117, $p = 0.9$); that is, the TPI of the covering component does not affect the tensile properties of the fabrics. Figure 6 also shows the effects of TPI of copper covering (weft yarns) on the weftwise elongation at peak force values for the fabrics. ANOVA analysis was conducted to check the influence of TPI of copper covering on the average elongation values of the fabrics. The effect of TPI of copper covering yielded an F ratio of F (2, 36) = 7.35, $p < 0.001$, indicating significant differences between 25 TPI ($M = 11.89$, $SD = 1.44$), 20 TPI ($M = 12.59$, $SD = 1.75$), and 15 TPI ($M = 13.15$, $SD = 1.18$) of copper cover

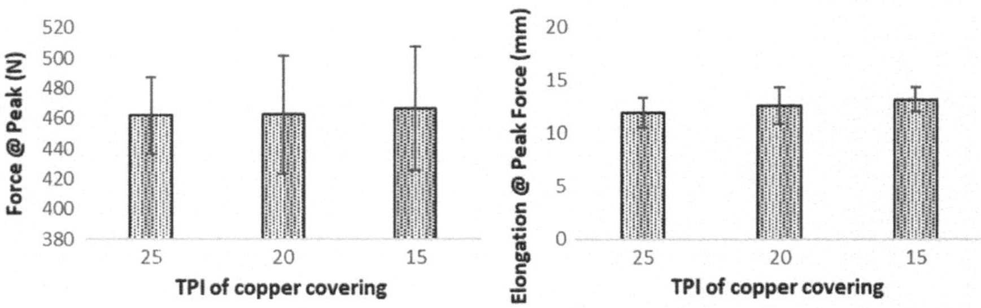

Fig. 9 Comparison of tensile properties of fabrics between TPIs of copper covering

on the elongation values of fabric. That is, as the TPI of copper covering increases, the elongation values for the fabrics decreases.

The tensile properties of the yarns mainly affect the tensile properties of the fabrics [20], and it mainly depends upon the core component's tensile behavior [18]; therefore, there was no significant change in the peak force values of the fabrics when the number of turns of the covering component was altered. Moreover, the copper covering on the weft yarns is like helical coil springs; upon tensile extension, these copper coverings behave like extension springs and starts winding up, hence causing reduction in its internal coil diameter [21]. The core component in copper cover yarns is fibrous (polyester yarn), and upon tensile extension, the interfiber friction plays an important role in the total extension of fabrics [22]. The reduction in the internal coil diameter of the copper covering upon tensile force causes an increase in the interfiber friction of the core yarn and hence making it less extensible. Therefore, greater the number of coils of covering, greater would be the interfiber friction and hence lower extension values. This might explain the significantly greater elongation values for fabrics with 15 TPI of copper covering as compared to 25 TPI of copper covering.

Stiffness/Initial Modulus

There are two main resources for fabric's tensile elongation, the de-crimping of the yarns under the test direction and the yarn's tensile elongation [23]. The de-crimping occurs in the initial Hookean zone of the curve, which is followed by the elastic deformation part until the yield point. This elastic or Young's module describes the fabric's resistance against tensile deformation. It can be observed from the slope of curves in Fig. 6 that the copper core fabrics have higher initial modulus as compared to the copper cover fabrics, and it is more prominent in fabrics belonging to copper size *A*. Table 2 summarizes the tensile properties of the fabrics along with the calculated modulus. There was approximately 128% increase in the initial modulus for the copper core fabric belonging to *A* class, as compared to its respective copper cover fabric. This higher modulus of the copper core fabrics explains its in-extensibility [23] and somewhat brittle nature [24], which causes the fabric to fracture at lower tensile load and lower tensile elongation as compared to the copper cover fabric. Similarly, fabrics belonging to copper class B behaved in the similar fashion, with the copper core fabrics showing approximately 38% more initial modulus values

Table 2 Comparison of tensile properties of copper core and copper cover fabrics

Sample ID	Force @ peak (N)	SD	Elongation peak (mm)	SD	Modulus (N/mm^2)	Yield strength (N/mm^2)
FAc	309.64	21.20	6.70	0.23	1141.00	14.83
FA25	486.00	23.31	11.70	0.51	501.22	14.94
FBc	415.84	23.68	18.32	0.88	524.54	10.76
FB25	447.69	14.55	10.68	1.10	379.84	17.91
FCc	432.59	24.80	17.92	0.64	304.02	13.91
FC25	452.07	20.20	13.30	1.20	316.23	15.84

as compared to copper cover fabrics, whereas, due to the yarn design difference as explained earlier, fabric *C* behaved otherwise.

Abrasion

Figure 10 displays the abrasive behavior of copper core and copper cover fabrics against ultra-fine silicon carbide (P3000) abrasive paper at different abrasive cycles. Generally, fabrics belonging to copper size *A* (both alignments) had the lowest reduction/loss of fibrous material as compared to other fabrics. If all the other parameters of fabrics remain consistent, thickness and weight per unit length play a decisive role in the abrasion of fabrics [25]. Accordingly, due to the use of heavier copper in the yarns of fabric *A* (0.14 mm), these fabrics were the heaviest and thickest as compared to the rest and showed the least percentage reduction ($M = 5.48$, $SD = 0.50$) in fabric weight at 250 rubbing cycles as compared to fabric FB ($M = 7.17$, $SD = 1.58$) and fabric FC ($M = 8.26$, $SD = 1.20$). In terms of copper alignment in the weft yarns of fabrics, the average reduction of weight after 250 rubbing cycles in copper core fabrics ($M = 7.92$, $SD = 1.62$) was significantly higher ($t\ (13.02) = 3.02$, $p = 0.01$) as compared to copper cover fabrics ($M = 6.03$, $SD = 0.96$); hence, the copper cover fabrics exhibit greater resistance against abrasion. Since the increase in density of yarns increases the abrasion resistance of fabrics [26] and so as the greater amount of metal in fabrics [27], the copper cover fabric belonging to copper size *A* offered the highest resistance with the percentage weight loss of 5.11%, while the weight loss in case of copper cover fabrics *B* and *C* was 5.78 and 7.18%, respectively. Moreover, since the copper cover yarns exhibit lower hairiness factors due to the coverage of fibrous core, which is more prominent in heavier copper covering yarns [9], therefore the resistance against abrasion was high.

Friction

Figure 11 displays the coefficient of static friction calculated from the static angle of surface drag [28] for the fabrics for both alignments. The average coefficient of static friction for copper core alignment fabrics ($M = 0.77$, $SD = 0.08$) was significantly lowered ($t\ (16) = -2.92$, $p = 0.01$) as compared to the copper cover alignment fabrics ($M = 0.87$, $SD = 0.05$). The main reason for the increased surface friction for the copper cover fabrics was the roughness created by the copper

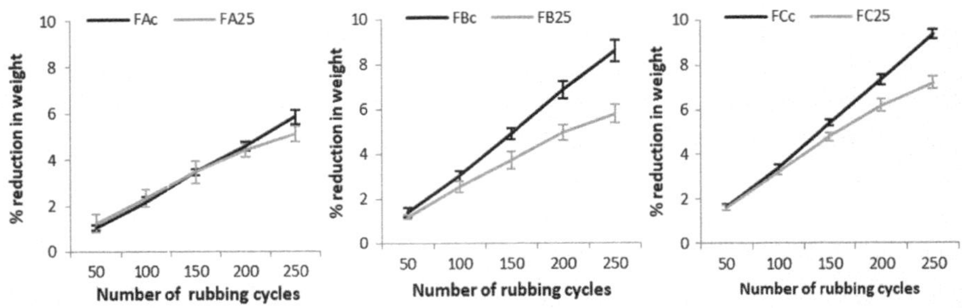

Fig. 10 Comparison of loss of weight in fabrics due to abrasion

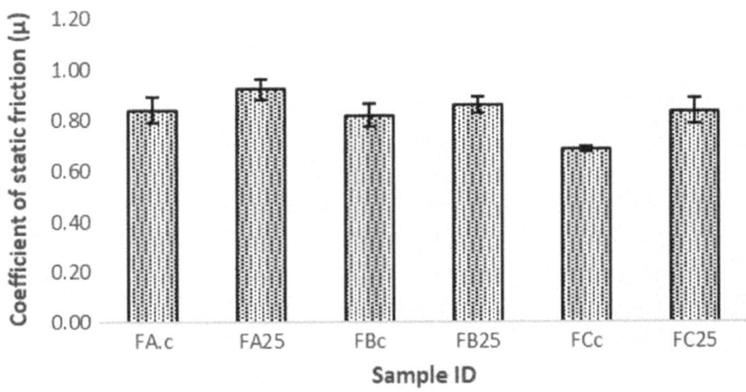

Fig. 11 Coefficient of static friction between alignments

spirals on the fabric face. These bulged copper spirals increase the point of contact (real area of contact) between the two frictional surfaces [29]. The greater this real area of contact, the greater would be the frictional resistance [30]. That is why, copper cover fabrics belonging to copper size *A* had the highest coefficient of friction ($\mu = 0.92$), which was followed by fabric *B* ($\mu = 0.86$) and then fabric *C* ($\mu = 0.84$). Furthermore, since the fabrics were tested in the weft direction, that is, warp yarns running perpendicular to the direction of fabric travel, therefore, increased linear density of the weft yarns caused greater crimping of the warp yarns. This greater crimp resulted in increasing the real area of contact between the two surfaces and hence resulted in greater friction.

4 Conclusion

The copper cover orientation of hybrid cover yarns is far much superior in terms of the yarn's tensile properties with around twofold to sixfold increase in the tensile strength values. The conventional copper core yarn design mimics the tensile properties of copper filament, since the core component is mainly responsible for the tensile behavior of the yarns. The elongation values for the yarns were also significantly improved. The copper coverings in the proposed yarns extend its spiral length (like conventional springs), to accommodate the increase in length of the core due to the axial forces, thus allowing the polyester core to retain its extension capabilities. Furthermore, the initial modulus values of the copper cover yarn designs were significantly reduced. Similarly, the fabrics prepared from the copper cover yarn design required around 80–100% more force to rupture than the conventional copper core design (considering the rupture point of copper core fabrics). This greater strength is attributed to the greater yarn's tensile behavior. The fabric with copper cover yarns could extend (elongation) around 200% more, before failure, as compared to the copper core fabric. Moreover, the fabric stiffness and abrasion properties in the copper cover yarn orientation also improved, but at the cost of increased static friction of the fabrics. Conclusively, the copper cover yarn design and the fabrics prepared thereof are better in retaining the tensile/mechanical properties of an otherwise normal textile structure. These fabrics can be utilized in applications where higher mechanical performances are required with concurrent or even better technical functionalities (such as EMSE).

Acknowledgements The authors would like to acknowledge the continuous support and technical guidelines provided by the Research Management Centre (RMC) of Universiti Teknologi MARA, Malaysia.

References

1. Ahmad, M.R., Hassan, M.R., Salleh, J., Ahmad, W.Y.W., Hassim, N.: Production of shape memory alloy core-sheath friction yarns. FIBRES Text. East. Eur. **3**(99), 68–72 (2013)
2. Stoppa, M., Chiolerio, A.: Wearable electronics and smart textiles: a critical review. Sensors (Switzerland) **14**(7), 11957–11992 (2014)
3. Custodio, V., Herrera, F.J., López, G., Moreno, J.I.: A review on architectures and communications technologies for wearable health-monitoring systems. Sensors (Switzerland) **12**(10), 13907–13946 (2012)
4. Lee, Y., et al.: Wearable textile battery rechargeable by solar energy. Nano Lett. **13**(11), 5753–5761 (2013)
5. Priya, A., Kumar, A., Chauhan, B.: A review of textile and cloth fabric wearable antennas. Int. J. Comput. Appl. **116**(17), 1–5 (2015)
6. Palanisamy, S., Tunakova, V., Militky, J.: Fiber-based structures for electromagnetic shielding—comparison of different materials and textile structures. Text. Res. J. (2017)
7. LessEMF, Shielding and conductive fabrics: Hightech and industrial conductive fabrics, 2014. [Online]. Available https://www.lessemf.com/fabric4.html. Accessed 14 Apr 2014
8. Yu, Z.C., et al.: Electromagnetic shielding, wicking, and drying characteristics of CSP/AN/SSW hybrid yarns-incorporated woven fabrics. J. Ind. Text. **46**(3), 950–967 (2016)
9. Asghar, A. et al.: An alternative approach to design conductive hybrid cover yarns for efficient electromagnetic shielding fabrics. J. Ind. Text. p. 152808371772192 (2017)
10. Tezel, S., Kavuşturan, Y., Vandenbosch, G.A., Volski, V.: Comparison of electromagnetic shielding effectiveness of conductive single jersey fabrics with coaxial transmission line and free space measurement techniques. Text. Res. J. **84**(5), 461–476 (2014)
11. Sekerden, F.: Effect of the constructions of metal fabrics on their electrical resistance. Fibres Text. East. Eur. **21**(6), 58–63 (2013)

12. Cheng, K.B., Cheng, T.W., Nadaraj, R.N., Giri Dev, V.R., Neelakandan, R.: Electromagnetic shielding effectiveness of the twill copper woven fabrics. J. Reinf. Plast. Compos. **25**(7), 699–709 (2006)
13. Perumalraj, R., Dasaradan, B.S.: Electromagnetic shielding effectiveness of doubled copper-cotton yarn woven materials. Fibers Text. East. Eur. **18**(3), 74–80 (2010)
14. Asghar, A., Ahmad, M.R., Yahya, M.F.: Effects of metal filament's alignment on tensile and electrical properties of conductive hybrid cover yarns. Fash. Text. **3**, 3 (2016)
15. Asghar, A., Ahmad, M.R., Yahya, M.F., Hassan, S.Z.U., Kashif, M.: Characterization based on the thermal capabilities of metallized fabrics equipped with hybrid conductive yarns for protective clothing. J. Text. Inst. **109**, 1–11 (2018)
16. Chung, D.D.L.: Materials for electromagnetic interference shielding. J. Mater. Eng. Perform. **9**, 350–354 (2000)
17. Vasile, S., Githaiga, J., Ciesielska-Wróbel, I.L.: Comparative analysis of the mechanical properties of hybrid yarns with superelastic shape memory alloys (SMA) wires embedded. Fibres Text. East. Eur. **89**(6), 41–46 (2011)
18. Ertekin, M., Kirtay, E.: Tensile properties of some technical core spun yarns developed for protective textiles. J. Text. Apparel/Tekst. ve Konfeksiyon **25**(2), 104–110 (2015)
19. Lawrence, C.A.: "Fancy yarn production", in fundamentals of spun yarn technology, pp. 491–494. CRC Press LLC, Boca Raton, Florida, USA (2003)
20. Grosberg, P.: The tensile properties of woven fabrics. In: Structural mechanics of fibres, yarns and fabrics, 1st edn, pp. 340–353, Wiley (1969)
21. Schmid, S.R., Hamrock, B.J., Jacobson, B.O.: "Springs", in fundamentals of machine elements, 3rd edn, pp. 502–505. CRC Press, Boca Raton, Florida, USA (2014)
22. Jinlian, H.U.: Tensile properties of woven fabrics. In: Structure and Mechanics of Woven Fabrics, 1st edn, pp. 91–93, Boca Raton, Florida, USA, CRC Press LLC (2004)
23. Zupin, Z., Dimitrovski, K.: Mechanical properties of fabrics from cotton and biodegradable yarns bamboo, SPF, PLA in weft. In: Dubrovski, P.D. (ed.) Woven Fabric Engineering, pp. 30–31. SCIYO, Rijeka, Croatia (2010)
24. Maleque, M.A., Salit, M.S.: Materials selection and design. In: Material selection and design. Springer Briefs in Materials, pp. 19–20, Singapore, Springer (2013)
25. Ozdil, N., Kayseri, G.O., Menguc, G.S.: Analysis of Abrasion Characteristics in Textiles. In: Adamiak, M. (ed.) Abrasion Resistance of Materials, pp. 124–125. InTech, Rijeka, Croatia (2012)
26. Bedeloglu, A.: Investigation of electrical, electromagnetic shielding, and usage properties of woven fabrics made from different hybrid yarns containing stainless steel wires. J. Text. Inst. **104**(12), 1359–1373 (2013)
27. Bedeloglu, A.: Electrical, electromagnetic shielding, and some physical properties of hybrid yarn-based knitted fabrics. J. Text. Inst. **104**(11), 1247–1257 (2013)
28. Mercier, A.A.; Coefficient of friction of fabrics, Bereau of Standards Journal of Research, 1930. [Online]. Available http://nvlpubs.nist.gov/nistpubs/jres/5/jresv5n2p243_A2b.pdf. Accessed 27 Jul 2017
29. Czichos, H.: Introduction to friction and wear. In: Friedrich, K. (ed.) Friction and Wear of Polymer Composites, 1st edn, pp. 5–6. Elsevier, Amsterdam (1986)
30. Ajayi, J.O.: Effects of fabric structure on frictional properties. Text. Res. J. **62**(2), 87–93 (1992)

Facile Metallization Technique of Textiles for Electronic Textile Applications

Md. Momtaz Islam, Musa Ali Reza, Dewan Murshed Ahmed, Md. Abdullah Al. Mamun and Hasan Shahariar

Abstract Decoration of insulating textile with electrically conductive nanoparticle has brought about a novel-application field for us. Nanoparticle-coated fabric can be used for different types of electronic purposes. In this work, cotton-based knitted and woven fabrics are coated with copper and silver nanoparticles simultaneously following a cheaper and easier dip and dry process with less chemical utilization. Cu and Ag nanoparticles are demonstrated using a highly reactive, cheaper, and available reducing agent. Our sample shows constant conductivity for about a few months in the open air. In addition, the change of resistance of the fabric with respect to bending is examined. Further, a simple LED circuit was integrated with the fabric to test the compatibility of the fabric for the application of wearable electronic textiles. This paper will provide a future direction to the researchers who would like to work with electronic textiles.

Keywords Metallization · Dip and dry method · Combined coating · Conductive textile

Md. Momtaz Islam · M. A. Reza · D. M. Ahmed · Md. A. Al. Mamun
Department of Fabric Engineering, Bangladesh University of Textiles, Dhaka, Bangladesh
e-mail: momtaztex@gmail.com

M. A. Reza
e-mail: mdmusaalireza17@gmail.com

D. M. Ahmed
e-mail: murshed@butex.edu.bd

Md. A. Al. Mamun
e-mail: almamuntex36@gmail.com

H. Shahariar (✉)
Fiber and Polymer Science Program, North Carolina State University, Raleigh, NC 27606, USA
e-mail: hshahar@ncsu.edu

A. Majumdar et al. (eds.), *Functional Textiles and Clothing*,
https://doi.org/10.1007/978-981-13-7721-1_8

1 Introduction

The end use of classical fabric is not limited to traditional dress wear. The property and functionality of the fabric can be enhanced by integrating nanomaterials in it [1–3]. The electrically conductive fabrics are used for organic photovoltaic (OPV) [4], wearable radio-frequency (RF) devices such as textile-based antennas and transmission line [5–8], solar cells [9], electromagnetic interference (EMI) shielding [10], wearable activity monitors [11], fiber-field-effect transistors [12], heatable textiles [13]. There are several methods to grow conductive nanoparticle. Among them, the polyol synthesis process is most frequently used [14]. The polyol synthesis process is very complicated, and the chemical utilization is also high. In this work, a conductive feature is added to the cotton-based woven and knitted fabric using both $CuSO_4$ and $AgNO_3$ by means of cyclewise layer-by-layer coating of nanoparticle. In this technique, there is no need to grow nanoparticles at the very first step. Conductive nanoparticles can be grown on the surface of the textile. $AgNO_3$ is a very pretty substance to grow conductive nanoparticles. But, it is too much costly. Therefore, $CuSO_4$ can be used as a cheap metallic salt to grow conductive nanoparticles on the textile material surface. But the problem of using copper nanoparticles is their inherent tendency to oxidize in ambient conditions. In the recent study, it has been mentioned that copper nanoparticles can resist oxidation under ambient conditions, if they are coated with a proper protective layer [15]. That is why copper and silver nanoparticles have been concurrently used. Silver nanoparticle acts as a protective layer on the copper nanoparticles. This technique is very cost-effective as well as there is no possibility to oxidize copper nanoparticles in ambient condition. We can grow nanoparticle by using sodium borohydride [16] and trisodium citrate [17], tricalcium phosphate [18] as reducing agent of silver salt. These chemicals are toxic and hazardous. Ascorbic acid works both as reducing and protecting agent, which makes the nanoparticle-growing process economical, non-toxic, and environment-friendly [19]. So, ascorbic acid has been chosen to grow both copper and silver nanoparticles on the surface of sample using dip and dry method. As it has been mentioned that it is a cheaper process and involves less chemical utilization, simply distilled water has been used as the solvent. A comparison is established among knitted and woven fabric in terms of conductivity. To find out the optimum concentration 1% (1 g/100 mL water), 3% (3 g/100 mL water), and 5% (5 g/100 mL water) chemical concentration were taken into consideration. A simple LED circuit was integrated with the fabric to test the compatibility of the fabric for the application of wearable electronic textiles. It is observed that the fabric sample is very suitable for current passing. Further, the resistance of the sample does not change due to bending up to 100 cycles. In the previous studies, it has been found that the silver-nanowire-decorated conductive fabric cannot be considered as waterproof and loses its conductivity after washing for a few cycles [13]. To overcome this challenge, a thin layer of silicon binder was applied onto the fabric surface which leads to make the fabric waterproof without the changing of conductivity at all.

2 Experimental Process

Hundred percent cotton-based knitted fabric and hundred percent cotton-based woven fabric were used for this experiment. All the glassware used in this experiment were cleaned with acetone and finally with distilled water. All the chemicals were purchased from Merck KGaA, Germany, and used without further purification. Ascorbic acid, $CuSO_4$, $AgNO_3$ were used for the synthesis of copper and silver nanoparticles. Only distilled water was used as the solvent to prepare ascorbic acid, $CuSO_4$ and $AgNO_3$ solution. 100 mL of ascorbic acid, $CuSO_4$, and $AgNO_3$ solutions were prepared, respectively, in three separate beakers. After that, a 3 × 3-inch pre-cleaned and mercerized cotton-based woven fabric was immersed into the ascorbic acid solution for 5 min which is shown in Fig. 1a. Then, the sample was dried at 80 °C for 10 min. After that, the dried sample was immersed into the $CuSO_4$ solution for about 5 min which is shown in Fig. 1b. Then, the sample was dried at 80 °C for 10 min. Now, one cycle has been completed. In this way, 3–5 cycles have been completed. Resistance of the sample was measured after each cycle. Figure 1c shows the copper-coated sample. The following few cycles were carried out using ascorbic acid and $AgNO_3$ solution until the resistance of the sample reached below 20 Ω per inch length. Figure 1d, e represent the silver-coating procedure and Fig. 1f represents the copper and silver combined coated sample. Knitted fabric was made conductive following combined coating process as it has been described for woven fabric. In the previous studies, it has been found that silver-nanowire-coated fabric loses its conductivity when undergoes a few washing cycles. It is because the water molecules weaken the continuous integration of the conductive network. As a result, sample loses its conductivity gradually. Since textile material is frequently subjected to water, it is important that fabric sample retains its conductivity when exposes to water. To meet the requirement, a very thin layer of silicon binder was used on the surface of the conductive fabric sample.

JIONSIL™ 1018 is a low modulus, one part, acid curing at room temperature silicon sealant. It is a general purpose silicon rubber sealant that offers good adhesion performance in interior and exterior applications. Before applying this silicon binder, it was diluted with mineral spirit and a fine layer of silicon binder was applied on the conductive fabric surface. After the application of silicon binder, the fabric sample became totally waterproof without the changing of conductivity at all.

3 Result and Discussion

3.1 Optimum Concentration

To find out optimum concentration 1, 3, and 5% chemical concentrations have been used, respectively, to make cotton-based woven fabric conductive. When 1% chemical concentration was used, it was required 15 cycles to obtain resistance of the

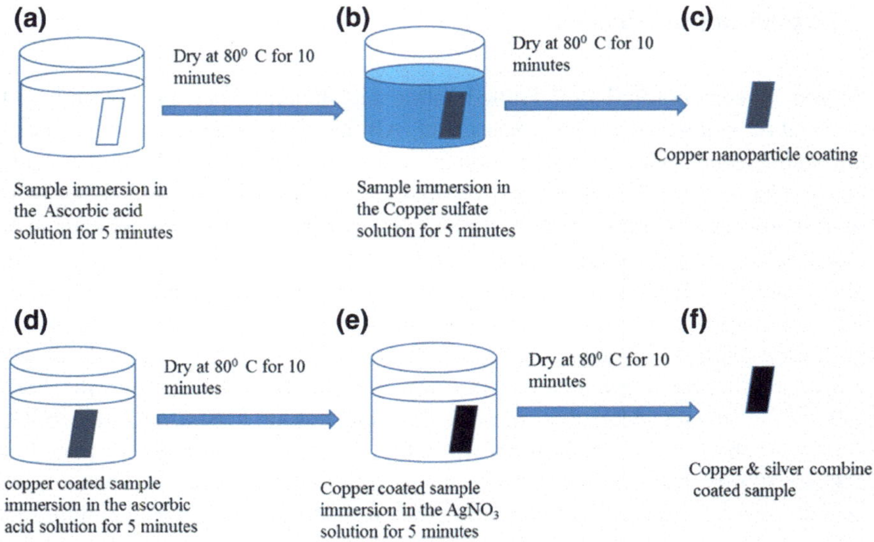

Fig. 1 Copper and silver combined coating procedure

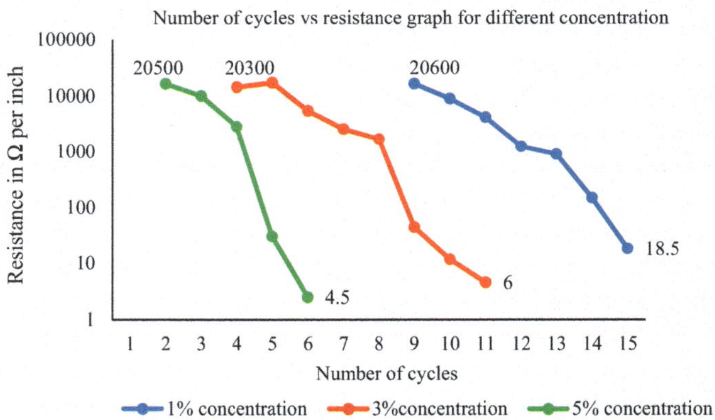

Fig. 2 Change in resistance with respect to number of cycles

fabric sample below 20 Ω per inch length. On the other hand, 3% chemical concentration required 11 cycles and 5% chemical concentration required 6 cycles to obtain resistance below 20 Ω per inch length. Though 5% chemical concentration required lowest number of cycle, the chemical consumption for 3% was lower than that of 5% concentration. Considering both the factors of the number of cycle and chemical consumption, 3% concentration can be regarded as the optimum concentration. Figure 2 shows the change in resistance of sample for different chemical concentrations.

3.2 Comparison Among Woven and Knitted Fabric

To compare among cotton-based woven and knitted fabric, a woven fabric sample and a knitted fabric sample each of similar size were taken. Single-jersey knitted fabric is not suitable for coating because of its inherent tendency to curling. Rib knitted fabric has a porous structure which is a major problem to build up continuous conductive network on the fabric surface. That is why interlock knitted fabric with higher GSM is suitable. We have selected interlock fabric with course per inch-60, wales per inch-35, and GSM-190 for the experiment. In case of woven fabric, tight structure gives better conductivity. For this reason, woven fabric with the specification: 1/1 plain structure, ends per inch-64, picks per inch-48, and GSM-110 was selected for the experiment. Both of the cotton-based woven and interlock knitted fabric samples were coated using 3% chemical concentration. It was observed that knitted fabric required 9 cycles and woven fabric required 11 cycles to reach the resistance below 5 Ω per inch length. But the knitted fabric is not suitable to withstand mechanical stress. Knitted fabric loses its conductivity when undergoes any mechanical deformation. Since, knitted fabric has enough elongation property, when it undergoes an external load, it shows the tendency to elongate. As a result, the continuous conductive network is broken and knitted sample fails to show constant conductivity. On the other hand, woven fabric sample shows constant conductivity because of its lowest elongation property. Figure 3 represents the comparison among the cotton-based woven and knitted sample in terms of resistance.

3.3 Bending Test

Textile materials experience different mechanical deformations and forces. Thus, it is necessary to test the change in resistance of the sample due to bending. For bending deformation, we folded the sample in the middle and unfolded again. The

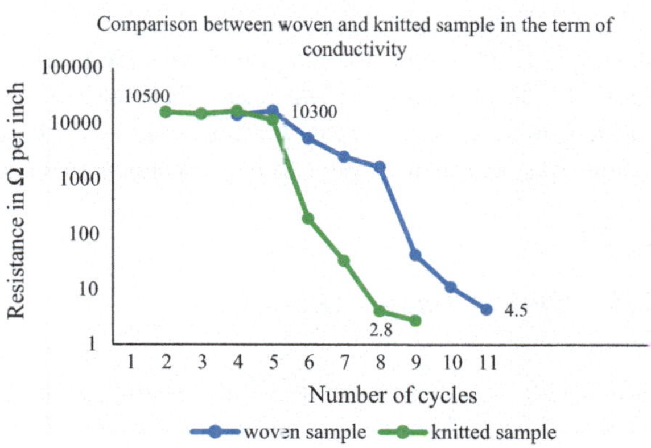

Fig. 3 Comparison among woven and knitted sample

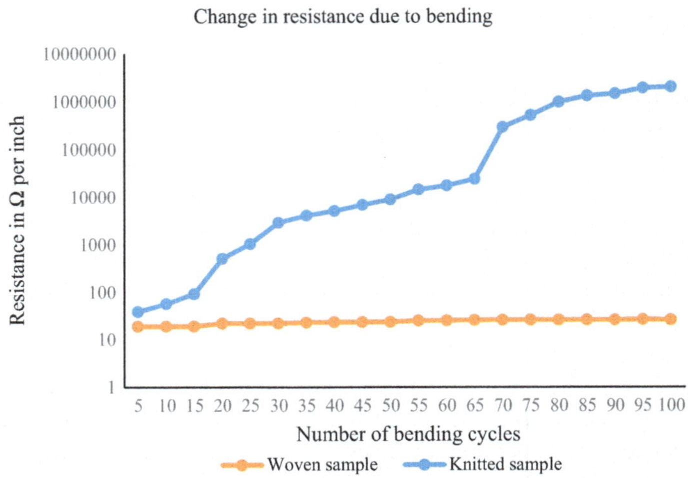

Fig. 4 Change in resistance with respect to bending cycles

same cycle has been repeated 100 times for woven and knitted sample. It is observed that woven sample does not change its resistance due to bending up to 100 cycles, whereas knitted sample starts to increase its resistance due to bending for a few cycles and after about 65 cycles its resistance increases significantly due to dimensional deformation which causes the breakage in the conductive network. Note that, the metallized fabrics (both woven and knit) were not coated with waterproof silicon coating, while the bending test was performed. The goal of this test was to analyze the particle-to-particle and particle-to-fiber adhesion of the metallized woven and knitted fabric without the addition of silicon coating under mechanical deformation. Figure 4 shows the change in resistance due to bending.

3.4 LED Demonstration

A simple LED circuit was integrated with the fabric to test the compatibility of the fabric for the application of wearable electronic textiles. It has been observed that the fabric sample is very suitable for current passing. Figure 5a shows the conductivity below 10 Ω per inch length and Fig. 5b shows LED demonstration.

3.5 Optical Image Analysis

When fabric sample undergoes dip coasting, a thin, uniform conductive layer of nanoparticle is deposited on the fabric surface. The conductive layer reduces the

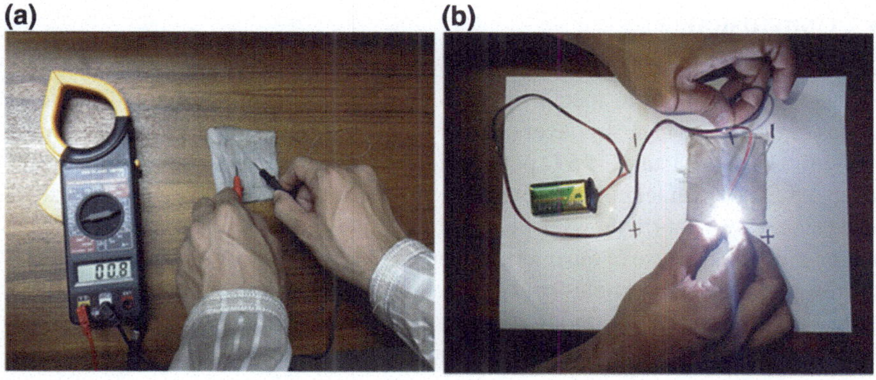

Fig. 5 Showing the conductivity of the fabric

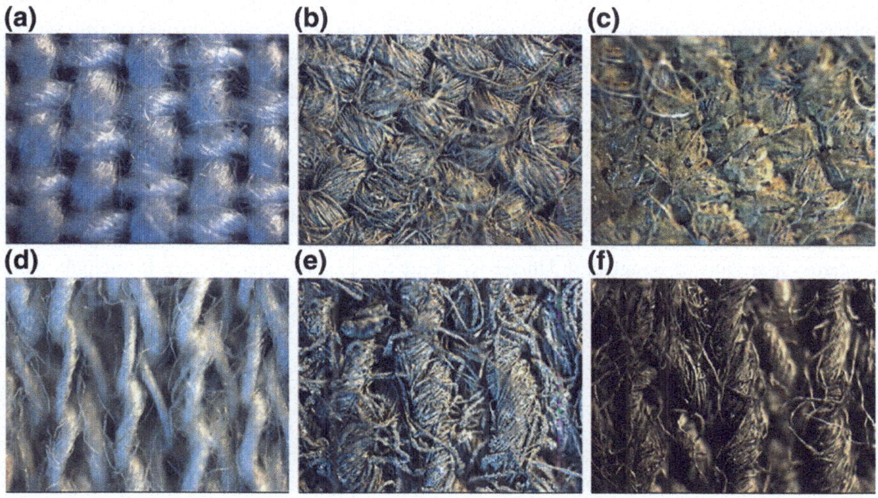

Fig. 6 Optical images of **a** native woven fabric, **b** woven fabric metallized with 3 weight % of $AgNO_3$ solution, **c** silicon binder coating of metallized woven fabric, **d** native knitted fabric, **e** knitted fabric coated with 3 weight % of $AgNO_3$ solution and **f** silicon binder coating of metallized knitted fabric

porosity of the fabric. It is shown from the optical images that the fibers of both knitted and woven fabrics have been properly coated with a metal layer. It is noteworthy that the dimensional stability of knitted fabric has significantly changed from its native condition after the metallization process. Figure 6e shows the untwisting phenomenon of the yarn due to the cyclic drying process of the knitted fabric during the metallization procedure. This explains the reason for higher resistance of the metallized knitted fabric and the lower stability of its electrical resistance during bending than that of woven fabric.

4 Conclusion

In this work, we tried to metallize the cotton-based woven and knitted fabric by means of cyclewise layer-by-layer coating of metal salts and reducing agent. Woven fabric shows the resistance 4.5 Ω per inch length, whereas knitted fabric shows resistance 2.8 Ω per inch length. However, in case of durability of maintaining electrical resistance, woven fabric shows better durability than knitted fabric. Woven fabric has a tight structure because of the interlacement between the yarns. As a result, it is easy to grow and fix metal particles between the interlacement points. On the other hand, knitted fabric shows poor durability due to dimensional instability. The resistance of woven fabric does not significantly increase due to bending up to 100 cycles, whereas there is a significant increase in the resistance of knitted fabric. Application of waterproof coating is desired to improve the durability of both metallized woven and knitted fabric. The detailed experimental procedures need to be carried on to understand the durability of the silver-coated textiles. As a continuation of this work, we are conducting washing, bending, and abrasion resistance test of the presented samples with varying the concentration of Ag salt and silicon binder. This study will help to develop a facile and durable integration of conductive (silver/copper) materials on textile platform.

Acknowledgements We acknowledge all the faculty members of Fabric Engineering department of Bangladesh University of Textiles, Dhaka, Bangladesh. We express special gratitude to all the laboratory assistants of Bangladesh University of Textiles, Dhaka, Bangladesh, for helping us to conduct the experiments.

References

1. Kim, I., Cho, G.: Polyurethane nanofiber strain sensors via in-situ polymerization of polypyrrole and application to monitoring joint flexion. Smart Mater. Struct. **27**, 075006 (2018). https://doi.org/10.1088/1361-665X/aac0b2
2. Harifi, T., Montazer, M.: Application of nanotechnology in sports clothing and flooring for enhanced sport activities, performance, efficiency and comfort: a review. J. Ind. Text. **46**, 1147–1169 (2017). https://doi.org/10.1177/1528083715601512
3. Commission, E.: Textiles and Clothing. (1998)
4. Steim, R., Chabrecek, P., Sonderegger, U., Kindle-Hasse, B., Siefert, W., Kroyer, T., Reinecke, P., Lanz, T., Geiger, T., Hany, R., Nüesch, F.: Laminated fabric as top electrode for organic photovoltaics. Appl. Phys. Lett. **106**, 51 (2015). https://doi.org/10.1063/1.4919940
5. Salvado, R., Loss, C., Gon, R., Pinho, P.: Textile materials for the design of wearable antennas: a survey. Sensors **12**, 15841–15857 (2012). https://doi.org/10.3390/s121115841
6. Shahariar, H., Soewardiman, H., Muchler, C.A., Adams, J.J., Jur, J.S.: Porous textile antenna designs for improved wearability. Smart Mater. Struct. **27**, 4 (2018). https://doi.org/10.1088/1361-665X/aaaf91
7. Shahariar, H., Soewardiman, H., Jur, J.S. (2017) Fabrication and packaging of flexible and breathable patch antennas on textiles. In: Conference Proceedings IEEE SOUTHEASTCON (2017). https://doi.org/10.1109/secon.2017.7925306
8. Chen, T., Qiu, L., Yang, Z., Peng, H.: Novel solar cells in a wire format. Chem. Soc. Rev. **42**, 5031 (2013). https://doi.org/10.1039/c3cs35465b

9. Das, A., Krishnasamy, J., Alagirusamy, R., Basu, A.: Electromagnetic interference shielding effectiveness of SS/PET hybrid yarn incorporated woven fabrics. Fibers Polym. **15**, 169–174 (2014). https://doi.org/10.1007/s12221-014-0169-0
10. Jung, S., Lee, J., Hyeon, T., Lee, M., Kim, D.H.: Fabric-based integrated energy devices for wearable activity monitors. Adv. Mater. **26**, 6329–6334 (2014). https://doi.org/10.1002/adma.201402439
11. Doganay, D., Coskun, S., Genlik, S.P.: Silver nanowire decorated heatable textiles. Nanotechnology. **27**, 1–8 (2016). https://doi.org/10.1088/0957-4484/27/43/435201
12. Hamedi, B.M., Herlogsson, L., Crispin, X., Marcilla, R.: Fiber-embedded electrolyte-gated field-effect transistors for e-textiles. 573–577 (2009). https://doi.org/10.1002/adma.200802681
13. Magdassi, S., Grouchko, M., Kamyshny, A.: Copper nanoparticles for printed electronics: routes towards achieving oxidation stability. Mater. (Basel). **3**, 4626–4638 (2010). https://doi.org/10.3390/ma3094626
14. Song, K.C., Lee, S.M., Park, T.S., Lee, B.S.: Preparation of colloidal silver nanoparticles by chemical reduction method. Korean J. Chem. Eng. **26**, 153–155 (2009). https://doi.org/10.1007/s11814-009-0024-y
15. Tarafdar, J.C., Raliya, R., Rathore, I.: Microbial synthesis of phosphorous nanoparticle from tri-calcium phosphate using aspergillus tubingensis TFR-5. J. Bionanoscience. **6**, 84–89 (2012). https://doi.org/10.1166/jbns.2012.1077
16. Umer, A., Naveed, S., Ramzan, N., Rafique, M.S.: Selection of a suitable method for the synthesis of copper nanoparticles. NANO **07**, 1230005 (2012). https://doi.org/10.1142/S1793292012300058
17. Shahariar, H., Jur, J.S.: Correlation of printing faults with the RF characteristics of coplanar waveguides (CPWs) printed on nonwoven textiles. Sens. Actuators, A Phys. **273**, 240–248 (2018). https://doi.org/10.1016/j.sna.2018.02.043
18. Hamedi, M., Forchheimer, R., Inganäs, O.: Towards woven logic from organic electronic fibres Nat. Mater. **6**, 357–362 (2007). https://doi.org/10.1038/nmat1884
19. Polte, J., Ahner, T.T., Delissen, F., Sokolov, S.: Mechanism of gold nanoparticle formation in the classical citrate synthesis method derived from coupled in situ XANES and SAXS evaluation. Jacs. **132**, 1296–1301 (2010). https://doi.org/10.1021/ja906506j

Development of Smart Textiles for Medical Care

Arindam Basu, Saurab Jain and V. S. Khoiwal

Abstract The advent of electronic sensors and other developments has helped the textile technologists to develop smart/intelligent textiles. In the present study, a smart textile garment has been developed which can sense the heartbeat, body temperature and geographical position of the wearer. The software, developed under this project, sends information to the caregivers and relatives whenever these values cross the threshold limits, which has been fixed based on practical data.

Keywords Smart textiles · Intelligent textiles · Global positioning system · Conductive fabrics

1 Introduction

Smart textiles or smart clothing term is very commonly used nowadays. In the 1990s, when electronic become very popular and product development peaked a number of companies came out with electronic textile products. Broadly, they were classified as wearable computers, wearable electronics, intelligent clothing, etc. In the beginning, most of these products were for fancy items such as MP3 player, cell phones, GPS, etc. Also some products were developed for health monitoring purpose [1–3]. People use terms like smart or intelligent textiles loosely but there are clear distinction between these two.

Smart textiles are textile materials which can sense the external condition or stimuli, react or use those information for taking next steps such as change of colour, activate the expected response, etc. The external condition may be mechanical, electrical, chemical and/or magnetic actions [4, 5]. Smart clothing consists of software or other mechanisms which can analyse the senses and send the information to desired source [6]. Intelligent clothing can analyse the stimuli and make internal modification in pre-defined manner [3]. Most of the smart textiles are equipped with sensors, hard or soft, data analysers, communicators to decision-making units, etc. Often electri-

A. Basu (✉) · S. Jain · V. S. Khoiwal
Northern India Textile Research Association, Sector-23, Raj Nagar, Ghaziabad, UP, India
e-mail: drabasu@nitratextile.org

A. Majumdar et al. (eds.), *Functional Textiles and Clothing*,
https://doi.org/10.1007/978-981-13-7721-1_9

cally conductive textiles with specific electrical conductivity are used in smart and intelligent textiles [7]. These can be fibres, yarns, fabrics and final products.

Though a number of products were developed by different companies most of them could not become commercially successful due to its high price, difficult maintainability and not so important functions. The smart textiles used for medical and health purpose could survive as they deal with very important purpose. In India, textiles and apparel industry plays a very important role, and due to the presence of highly developed companies, it is possible to produce high-value technical textile products. Now many Indian companies successfully have entered into technical textiles sector and many others in the process of entering in the business of this segment.

Social and economic conditions force many old-age people to stay alone and away from their children either in their home or old-age home. People at advanced age are more vulnerable for diseases such as high/low blood pressure, heart problem and diabetes. Though looked after by caregivers (doctors, nurses), they cannot accompany these old age people all the time. The aim of this work is to provide affordable smart garments to these old age people which will measure their body temperature, blood pressure continuously. Whenever the values cross threshold limit, the message will be sent to caregivers, relatives (total five in number) immediately. Timely action can save lives of the people in most of the cases.

Same clothing can be used by security forces engaged in duties in international borders or terrorist infested areas where it is important to reach to the soldier immediately.

2 Experimental

Smart textiles can be made from any textile fibres such as cotton, polyester, nylon or viscose rayon. It is preferred to have sensors, conductive fibres/yarns integrated to the textile materials. In this present study, health monitoring such as blood pressure, heartbeat sensors and conductive material need to be accommodated in the garment. The conductive textile yarns can be produced by mixing metallic fibre during spinning of textile yarn, in situ conductivity generation or using metallic yarn along with standard textile yarns.

2.1 Development of Conductive Fabric

In the context of the present research, conductive threads are arranged in a strictly organized way in fabrics to form an intelligent functionality. Modified circular knitting technology has been used to incorporate electrically conductive yarn into fabric to convert into the same as smart fabric. The fabric is prepared by flat weft knitting method. A conductive steel filament yarn has been inserted in course direction of fabric. It is connected with the temperature sensors on the body and the signal acquisition electronics (Fig. 1).

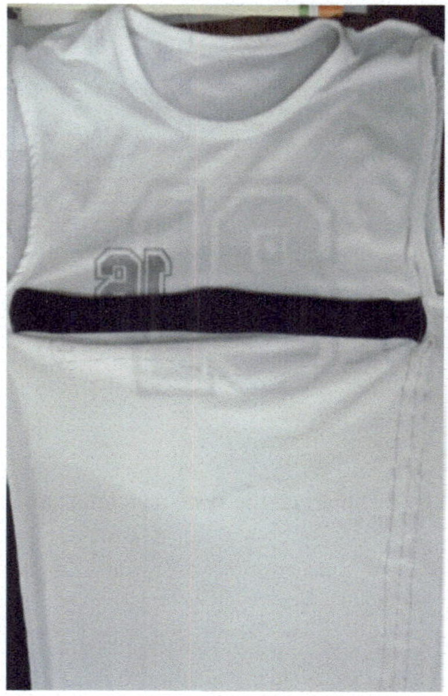

Fig. 1 Vest with heartbeat sensor and temperature sensor

2.2 Preparation of Garment with Incorporation of HeartBeat Sensor, Temperature Sensor and GPS

The main textile material in this application is an inner vest, in which heartbeat and temperature sensor have been incorporated. This system is an implementation of a multichannel wearable wireless textile-based sensor that monitors heartbeat and body temperature.

The developed smart textile with sensor cum software fixed with garments will continuously measure the heartbeats and body temperature of the wearer. When the reading crosses the threshold limit, the message will be sent to the caregiver. The message will be sent directly to the mobile phone of the caregiver. Presently, developed system can be connected to predetermined five people. The message will send to one to five in sequences. The conductive part of fabric will supply power to the system. Figure 2 shows the architecture of the system.

The device consists of the following three tiers

1. WBAN (Wireless body area network) consist of heartbeat sensor which senses the heartbeat of the body.
2. Data Aggregation, collect the data provided by the sensors.
3. Communication, GSM network is used for communication.

Figure 3 shows the flowchart of communication part. When an alarm is being triggered, i.e. the reading has crossed the threshold limit, SMS will be sent to five different predetermined people or places using GSM network. The SMS will contain

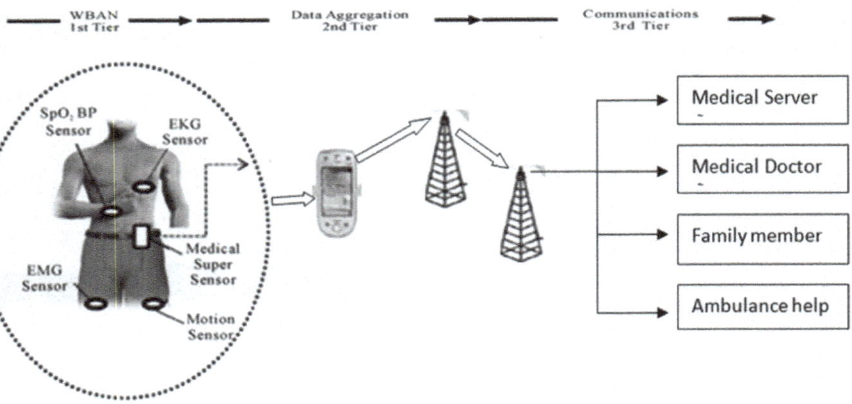

Fig. 2 Smart textile system architecture

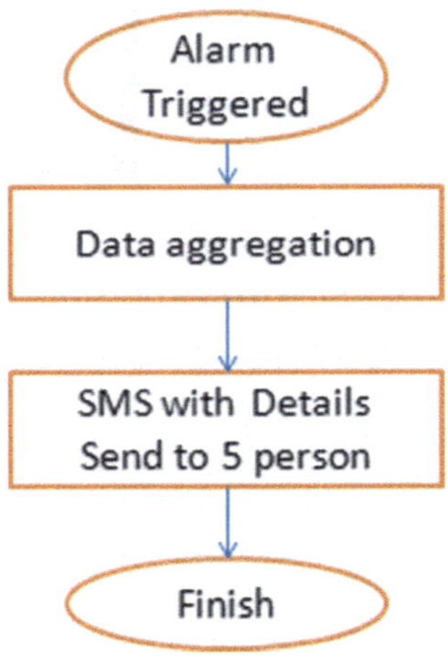

Fig. 3 Flow chart of communication

the details of heartbeat rate, body temperature and location of the concerned person who is wearing the smart textile.

Figure 4 shows the working of the electronics. Heart rate and temperature sensor continuously sensing the heartbeat and temperature of the body and send the data to the controller. Controller continuously compares these values with the set threshold values. GPS system continuously tracks the position of the concerned person wearing the device. When the heartbeat or temperature crosses the threshold, control system gathers the details of heartbeat, temperature of body and position. GSM module has been activated and details have been sent to five persons using the GSM network. Upper and lower heartbeat rate has set to 105 and 50 per minute, respectively, while upper and lower threshold value of temperature is set to 40 and 32 °C.

Fig. 4 Electronic assembly

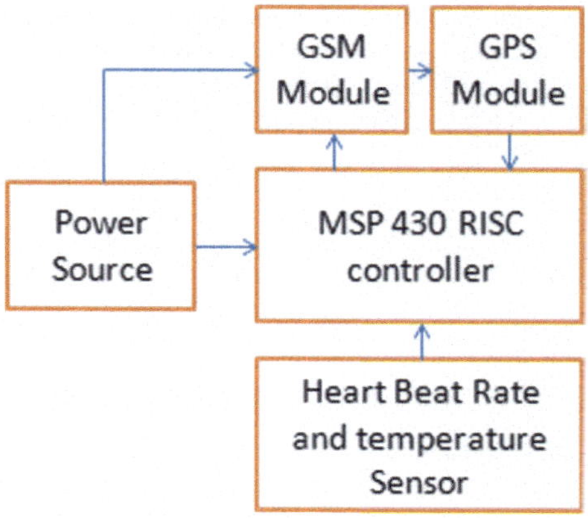

3 Results and Discussion

In this development, we have chosen a sensitive sensor, for heartbeat monitoring, after consultation with cardiologists. The sensor readily available in the market for using during jogging, exercise have been avoided considering their accuracy. For cases, where a wrong reading can cause death, it was advised by the cardiologists to use very sensitive sensors. To avoid complications during washing, the garment has been made in such a manner that the sensors and the battery can be removed before washing and can be put back to their original place easily. Considering that it has to be worn next to skin, finishes such as antimicrobial, dust proof finish are applied to the garment. Initial simulation trials have shown that the performance satisfactory. Broad-based and multi-user trials are in progress.

Acknowledgements The authors are thankful to the Ministry of Textiles, Government of India for kindly sponsoring the project. They are thankful to other colleagues from NITRA for helping us in performing the project activity.

References

1. Marculescu, D.: E-textiles: towards computational clothing. IEEE Pervasive Comput. **2**(1), 89–95 (2003)
2. Kirstein, I., Cottet, D., Grtyb, J., Troster, G.: Wearable computing systems—electronic textiles. Woodhead Publishing Ltd., England (2003)
3. Anderson, K., Seyan, A.M.: The road to true wearable electronics. In: Proceedings of TI & 2nd World Textile Conference, The Textile Institute, Cairo, Egypt, Mar 2002
4. Lam Po Tang, S., Stylios, G.K.: An overview of smart technologies for clothing design & engineering. Int. J. Cloth Sci. Tech. **18**, 108–129 (2006)
5. Cadenius, E., Gartvall, K.: Next generation of active underwear—implementing smart textiles & technologies. M.Sc. thesis, KTH Industrial Engineering & Management, Sweden (2015)

6. Clim 8 TM: What's the difference between "smart" and "intelligent clothing? https://www.myclim8.com/2016/12/08/whats-the-difference-between-smart-and-intelligent-clothing/. Accessed 23 Nov 2016
7. Knittel, D., Schollmeyer, E.: Electrically high conductive textiles. Synth. Met. **159**, 1433–1437 (2009)

Photoluminescent Printed Fabrics: An Innovative Solution to Natural Nightlight

Richa Sharma and Nilanjana Bairagi

Abstract Photoluminescent pigments are rare earth-based luminous materials activated by divalent europium. A photoluminescent pigment emits a bright phosphorous shade on excitation by daylight, incandescent, fluorescent, or ultraviolet light. It exhibits a high initial brightness and a long afterglow. Limited literature is available to guide textile and fashion designers on how these pigments can be used to create novel illuminated surface patterns. Patterns may be created of different intensity of luminescence for different application and products ranging from safety to fashion. Therefore, this research aims to systematically study the properties and design potential of photoluminescent blue pigments on textiles, printed by screen-printing method. The objective of this research is to study the effect of the concentration of photoluminescent pigments of different particle size on the emission of photoluminescence (luminosity and decay). The results are also correlated with the visual perception of photoluminescence by users in a nighttime environment.

Keywords Photoluminescent pigment · Scoptic light · Natural nightlight · Circadian cycle

1 Introduction

Photoluminescent (PL) pigments are rare earth-based luminous materials activated by divalent europium. These pigments possess remarkable characteristics such as extremely narrow emission bands and high internal quantum efficiencies. PL pigment emits a bright phosphorous shade on excitation by daylight, incandescent, fluorescent, or ultraviolet light. It exhibits a high initial brightness and a long afterglow. The effect of afterglow is dependent on the pigment concentration, surface area, and

R. Sharma (✉)
Department of Textile Design, National Institute of Fashion Technology Bengaluru, Bengaluru 560102, Karnataka, India
e-mail: richa.sharma@nift.ac.in

N. Bairagi
Department of Knitwear Design, National Institute of Fashion Technology Bengaluru, Bengaluru 560102, Karnataka, India

A. Majumdar et al. (eds.), *Functional Textiles and Clothing*,
https://doi.org/10.1007/978-981-13-7721-1_10

amount of radiant energy absorbed [1]. Although extensive research projects have explored ways of creating light emitting fabric displays using LEDs, electroluminescent wires, optical fibers, and embedded photoluminescent pigment in polymer fibers, limited experimental research has focused on the ways of designing a novel illuminated surface pattern using photoluminescent pigments in textile printing and fashion design [2].

There is a lack of detailed experimental study on the effect of the pigment concentration, light sources, substrate, and particle size and on the intensity of luminescence by the pigments. Limited literature is available to guide textile and fashion designers on how these pigments can be used to create novel illuminated surface patterns of different intensity of luminescence for different application and products ranging from safety to fashion.

Light is a powerful modulator of cognition [3–6] through its long-term effects on circadian rhythm and direct effects on brain function. Many sleep specialists suggest that this widespread exposure to blue light, long after the sun has set, is a major contributor to the modern epidemic of insomnia [7, 8]. Therefore, alternative sources of natural nightlight imitating moonlight or starlight, which does not retard the circadian cycle of humans, may be an area to be explored using PL printed textiles.

Therefore, the aim of the research is to study the effect of luminescence (luminosity and decay) emitted by different concentrations of photoluminescence pigments of different particle size, printed on textiles. The results are also correlated with the visual perception of photoluminescence by users in a nighttime environment.

2 Experimental Details

The luminosity of PL blue pigment was tested using a three-pronged approach. Initially, the absolute lux values were ascertained using an illuminometer (lux meter). As the lux values were below 1 lx, the second approach of time-resolved photoluminescence (TR-PL) spectroscopy was carried out to study the decay and intensity with higher precision. Additionally, visual perception study by user trials was undertaken to validate the above data in real-life environment.

2.1 Materials

The study was carried out on 100% ready to dye cotton fabric with both EPI and PPI of 68, thread count of 10 s Ne for both warp and weft and GSM of 290. The cotton fabric was printed with blue photoluminescent pigments of different particle sizes. The details and the given sample codes are presented in Table 1. The PL pigments were procured from supplier Jash Marketing Services, India, with the chemical composition, $Sr_4Al_{14}O_{25}$: Eu^{2+}, Dy^{3+} with a yellowish appearance exhibiting an aqua glow emission at λ max of 490 nm.

Table 1 Pigment and concentration matrix of blue PL pigments

S. No.	Pigment sample code	Particle size (μm)	Sample code for printed fabrics for variable concentration (%)				
			C5 (20%)	C4 (10%)	C3 (5%)	C2 (2%)	C1 (1%)
1	B4	50–60	B4C5	B4C4	B4C3	B4C2	B4C1
2	B3	45–55	B3C5	B3C4	B3C3	B3C2	B3C1
3	B2	25–35	B2C5	B2C4	B2C3	B2C2	B2C1
4	B1	10–15	B1C5	B1C4	B1C3	B1C2	B1C1

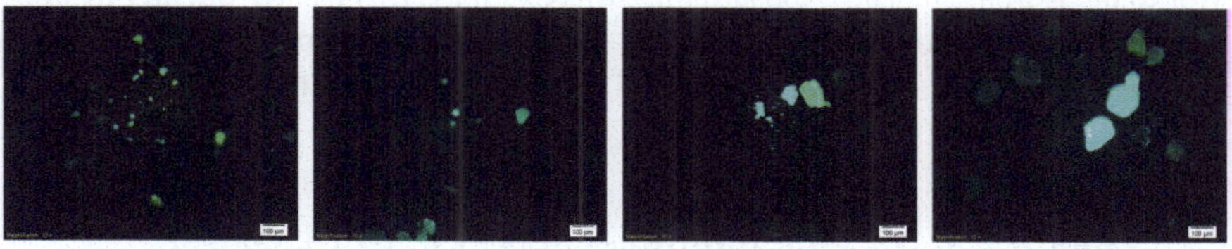

Fig. 1 Confocal images of photoluminescent pigment in variable particle size B1–B4 left to right

To study the effect of an increase in particle size, four pigments of blue series B1, B2, B3, and B4 have been chosen of variable particle size. The particle size was analyzed using confocal images as shown in Fig. 1. The particle size analysis was not successful as the pigment was not able to be stabilized in suspension.

2.2 *Printing with PL Pigments*

The fabric was industrially screen-printed (mesh size 200) with 1, 2, 5, 10, and 20% concentration of pigment in the print recipe. The curing was done at 120 °C. As presented in Table 1, the samples were coded for variable pigment size and concentration. A basic printing recipe consisting of pigment, commercial binder, fixer, and urea was used for printing.

The proposed usage of end products is for upholstery, drapes, and soft furnishings to facilitate interiors of home spaces to augment nightlight, and thus, the pigment concentration is taken on the higher range from 1 to 20%. The viscosity of the print paste was stable even at 20% pigment concentration.

2.3 *Evaluation of Intensity of Luminosity of PL Blue Pigment Using Illuminometer*

Illuminometer is a device used to measure lux of PL emitted by the samples as suggested in the literature [9]. The lux meter of brand HTC with a measurement range of 0–200,000 lx and resolution 0.01 + 3% lux was used to evaluate the intensity of the printed samples. The lux meter consists of a photon detector, which is connected to a selenium chip. The specimen is placed on a photon detector, which absorbs the photons emitted by the specimen.

Initially, the sample was kept in a dark room for two hours for the complete discharge of luminescence in line with similar trials conducted by Gulrajani et al. [9]. When the samples were measured after 2 h, without exposure to any excitation energy, no readings were exhibited by the lux meter. The sample was thus exposed to both UV and D65 lamp (closest to natural light) for a period of 1 h in a color cabinet prior to measurement. The exposure time was extended to 5 h subsequently to study the effect of exposure time on luminosity, but no appreciable increase in luminosity was recorded. Thus, for all further studies, exposure time of 1 h was maintained.

2.4 Time-Resolved Photoluminescence (TR-PL) Spectral Studies

Time-resolved photoluminescence spectral studies were conducted using high-resolution spectrophotometer QE Pro with scientific graphing, data analysis, and image processing computer tool Igor Pro. PL spectra, time scans, and lifetime study were undertaken to observe the irradiation from the PL pigments. Additionally, SEM and confocal imaging were carried out to observe the aggregation of the PL pigments on the printed samples.

The PL spectra for all the fabric samples have been recorded at 480 nm LED using single point light source of 220 V. This is the excitation wavelength as referred in the literature [10]. The experimental setup was a direct probe connected using fiber optics without any filters (so as to capture emission without any loss) to the fiber optic connected to spectrophotometer QE Pro equipped with Igor Pro, data management software to obtain spatial intensity maps and PL intensity time scans. The time-resolved photoluminescence (TR-PL) decay profiles for the PL pigments in the respective samples were observed at an integration time of 2 s/4 s sleep time over 60 data points and total time of 360 s. Thus, the decay was recorded for 360 s over 60 data points. As the PL intensity was very low, very high integration time of 2000 ms or 2 s was used. Due to high integration time and thus limitation of the system, the total no. of data points was limited to 60.

2.5 Visual Perception Study

The main objective of the study was to ascertain the visual perception of the samples printed with blue (B series 1–4) in scoptic light (dark room) environment with respect to change in concentration and overall perception (PL irrespective of the particle size).

A user-centric approach was taken to study the visual perception by users in real-life environment through rating of the luminosity of the samples. These observations were correlated with the experimental data.

The pilot study was conducted with ten participants. Eight boards were created with five samples of variable concentration for each board and randomly arranged. The color of the board chosen was black so that it does not interfere with the PL. The boards were kept in a room with both indirect sunlight and CFL and were exposed to about 5 h before the experiment was conducted. The perception rating was taken on a scale of 5 using the questionnaire with 5 being the best and zero being the least score. Scoptic lighting was maintained inside the room while recording data.

3 Research Findings

3.1 *Effect of Pigment Concentration on the Luminosity of PL Pigment Printed Textiles*

The luminosity at variable concentration of the blue PL pigment has been tabulated in the graph, Fig. 2. The *X*-axis depicts the concentration, and the *Y*-axis depicts the lux readings of the luminosity. Figure 2 depicts the increase in luminosity at respective particle size. B4 depicts the largest particle size of 50–60 μm, and B1 depicts the smallest particle size of 10–15 μm. It was observed that irrespective of the size of the pigment PL increases as the concentration of pigment increases in the print. This is seemingly the obvious outcome, and similar studies have been reported in the literature as well [11]. At 10 and 20% concentration, the PL intensity was the highest, with the average value being 0.05–0.15 lx. At lower concentrations of 1 and 2%, the intensity ranged from 0 to 0.02 lx. Particle size of about 10–15 μm gave the best results in PL, especially above 5% concentration.

The spectral image in Fig. 3 shows the decay of the PL printed textiles over a period of 360 s. The *X*-axis depicts the wavelength (nm), and the *Y*-axis depicts the time in sec. The color depicts the decay intensity. It is observed that pigment blue of wavelength 500 nm in the first 50 sec exhibits maximum intensity depicted by color blue which after 200 sec changes to yellow. We can clearly observe that the PL does not become nil, but reduces in intensity significantly. Figure 3 shows the spectral image of particle blue series B4 of the size 50–60 μm at three concentrations 5, 10, and 20%. Similar imaging is seen in other pigments as well. The spectral scans give the visual evidence of our previous findings with lux meter and thus evidently display the increase in PL with an increase in pigment concentration in the print paste.

TR-PL study for blue pigment B4 (50–60 μm) is presented in Fig. 4a, b. Figure 4a shows the intensity of PL versus wavelength, and Fig. 4b shows the exponential decay of PL intensity versus time. It can be observed that with the increase in the concentration of the pigment, the PL intensity increases at 500 nm for pigment B4. At 20% concentration, the intensity is almost two times of 10% pigment concentra-

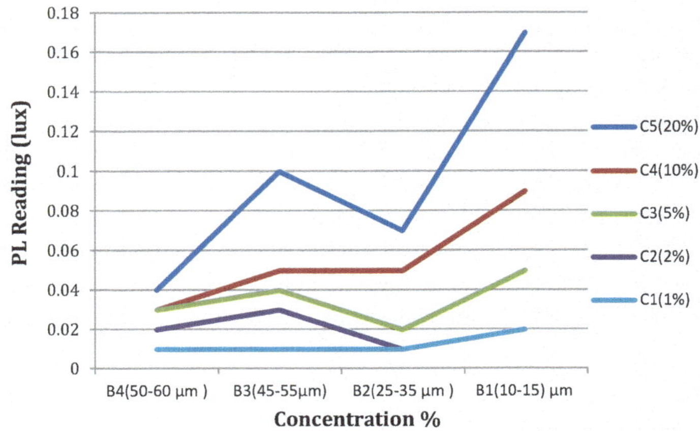

Fig. 2 Lux readings of PL pigment: blue

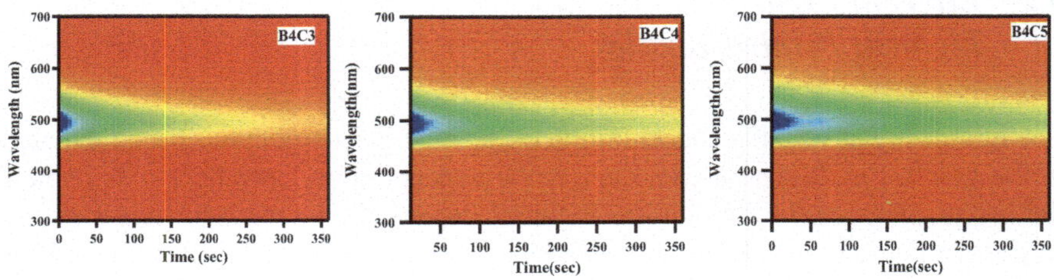

Fig. 3 Spectral image of B4C3, B4C4, and B4C5

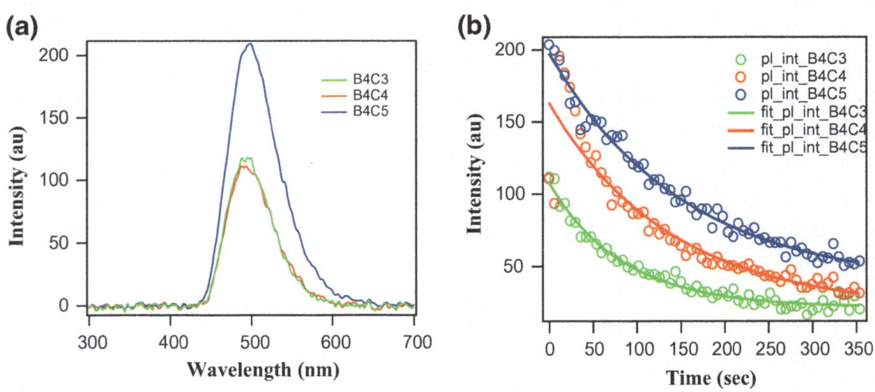

Fig. 4 **a**, **b** TR-PL study of blue pigment B4 with variable concentration

tion, and therefore, there is a direct correlation of pigment intensity with pigment concentration.

On the other hand, the TR-PL decay for pigment B4 as shown in Fig. 4b is an exponential curve, where luminosity does not become zero till the end but reduces exponentially in intensity. It can be correlated with the spectral scans, which show the initial high intensity in blue region. The luminosity curve plotted at variable concentration C3 (5%), C4 (10%), and C5 (20%) follows a similar path only the intensity increases with increase in concentration. Also, the intensity runs parallel to the time curve after the peak. This was correlated with user trials wherein even after 60 min the luminescence is perceptible by human eye. The study by Yan et al. [1] demonstrated the use of decay curves and spectral curves for their luminosity study.

The PL decay curve was again verified using a time-correlated photon counting (TCPC) of blue pigment B4 of particle size 50–60 μm. It was found that the path of the curve remains similar to the TR-PL study. Similar results have been plotted for blue pigments B2, B3, and B4 with particle size being constant and the variable component being the pigment concentration from C3 (5%), C4 (10%), and C5 (20%).

3.2 *Effect of Pigment Particle Size on the Luminosity of PL Pigment Printed Textiles*

To study the effect of pigment particle size on the luminosity, the studies were carried out with a constant pigment concentration of 20% (Fig. 5). The TR-PL decay follows an exponential curve similar to earlier results, where luminosity is highest at the start and gradually reaches a consistent level where it is sustained for longer period of time.

It was observed that the blue pigment B1, smallest particle size, was exhibiting the highest intensity. This may be attributed to the packing density and the structure of the PL particles on the printed samples. The surface-to-volume ratio, i.e., packing in a given volume, can be observed in the confocal microscopic image in Fig. 6. It is observed that the color of glow for blue pigments was bluish green. The confocal images clearly explain that the packing volume plays an important role in the lumi-

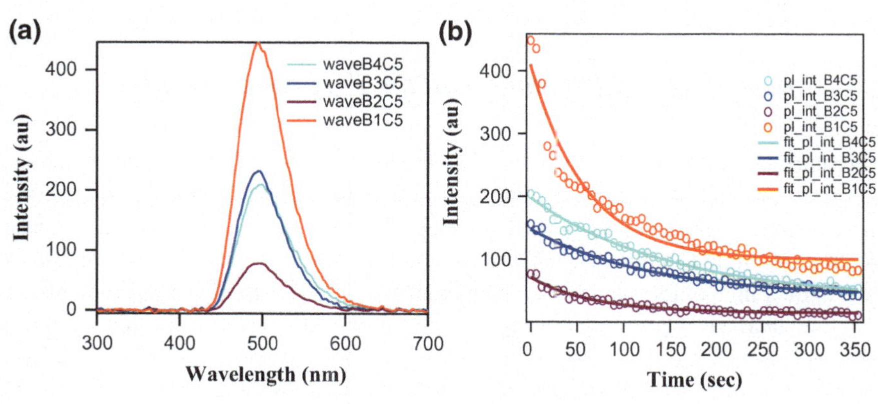

Fig. 5 **a, b** TR-PL of blue pigment blue with variable particle size at C5 (20%) concentration

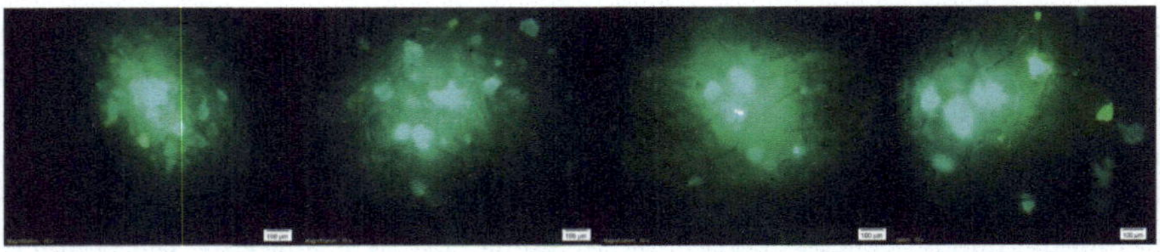

Fig. 6 Confocal images of blue pigment in increasing pigment size from left to right B1–B4 at constant exposure

Table 2 Time constant values of blue PL pigment with variable particle size at C5 (20%) concentration

S. No.	Pigment	Swatch code	Particle size (μm)	Tau (s)
1	B4	B4C5	50–60	151.43
2	B3	B3C5	45–55	129.8
3	B2	B2C5	25–35	68.75
4	B1	B1C5	10–15	59.1

nosity of the PL printed swatches, and lower particle size shall have the best packing volume and thus the best results [12]. Also stated by Hom, "the bigger particles by their orientation scatter the emitted light in undesired direction that reduces the emission intensity." Therefore, smaller particle size may perform better in luminous intensity.

The decay profile is single exponential curve with the experimental lifetimes (Tau) measured from the decay curves for all the swatches under investigation. Contrary to the luminosity readings, the time constant, Tau(τ) value as shown in Table 2, increases with increase in particle size, which means that the glow is sustained for longer time period as the size of the particle increases or

$$\text{Tau}(\tau) \propto \text{pigment particle size}$$

3.3 *Effect of Visual Perception on the PL Printed Textiles*

Literature survey indicates that the visual effectiveness at low adaptation states, i.e., mesopic and scoptic vision changes in a nonlinear fashion [13]. As per the recommended light levels [14], the lux captured in the samples ranges from overcast skies, starry night to a full moon. Thus, we find that the luminosity of the PL pigments falls in the range of mesopic and scoptic lighting [15, 16]. Visual perception with the user data analysis is presented in Fig. 7. The study clearly establishes that swatches with as low as C1 (1%) concentration with recorded luminosity of (0.01–0.02 lx) were also visually perceptible in spite of low concentrations as in the case of pigment B1, B2, and B4. At C3 (5%) and C2 (2%) level of concentration, it has been observed

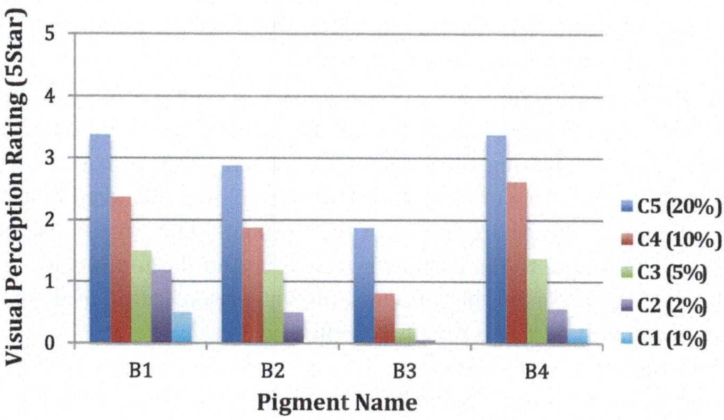

Fig. 7 Visual perception rating with change in concentration at variable particle size

that visible perception rating is low at 0.5–2.5 on a 5-point scale. At C4 (10%) and C5 (20%) concentration, the PL perception results were very good with the average value being 3–4 on a 5-point scale. PL pigment at C1 (1%) concentration has very low intensity and sometimes zero and thus can be ignored. As the concentration of pigment increases in the print, PL increases very evidently irrespective of the pigment. The visual effectiveness of pigment B1 and B4 of particle size (50–60 μm) and (10–15 μm), respectively, has the best rating.

Even though the lux values of the printed swatches is below 1 lx which is equal to starry night as per DIN67510 standard, it was observed that the user was able to perceive the PL even at lower concentration in low light conditions. Thus, intensity of luminescence on printed textiles may provide alternate light source so as to improve the impact of light cognition [3, 4, 7, 8] in comparison with conventional light sources.

4 Conclusion

It may be concluded that both pigment size and concentration play a significant role in the intensity of luminescence exhibited by the printed textiles. Increase in concentration of the pigment in the print paste increases the intensity of PL irrespective of the pigment size used as seen evidently in all three approaches, i.e., illuminometer, time-resolved spectroscopy, and visual perception study. At 10 and 20% concentration, the PL perception results were very good with the average value being 3–4 on a 5-point scale. The time constant (Tau) value increases with increase in particle size, which means that the glow is sustained for longer time period as the size of the particle increases. The visual effectiveness of PL blue pigment of particle size (50–60) and (10–15) μm, respectively, has the best performance rating. This is explained better with respect to surface-to-volume ratio, i.e., higher packing density decides the spectral intensity, and thus, we find that B1 (10–15) μm, though smallest in size, gives better performance when compared to rest.

Intensity of PL exhibits high initial brightness and a long afterglow and follows an exponential decay curve as seen in the TR-PL scans. The PL gradually reaches a consistent level, never touching zero, where it is sustained for considerable length of time. By increasing the exposure time/excitation time, it is observed that there is no appreciable change in the performance of the pigment. The time constant remains at the given range specific to the particle size. Even after the luminosity failed to be measured by the high precision instruments, the visual perception trials indicate that the human eye can still perceive the photoluminescence.

The non-visual effects of light on the circadian rhythm of humans have already been reported in the literature, and PL pigment printed textiles may provide the optimum luminosity to facilitate the circadian cycle. Thus, aqua glow from these blue PL printed textiles may play a key role in the design of lighting systems to optimize cognitive performance. Phosphorescent light emitted by PL pigments may be used as prints on textiles as alternative source of light.

The lux values of the printed samples were below 1 lx, which is equivalent to starry night as per DIN67510 standard. Even at low lux, the luminosity of the PL printed samples is visually perceptible by all the users. Therefore, PL printed fabrics may be used as an alternative for ambient lighting especially in areas inside homes for nighttime navigation. Thus, aqua glow from these PL printed textiles may play a key role in the design of lighting systems to optimize cognitive performance especially to facilitate easy and safe navigation in dark environment, especially for safety and preventive measures in edge definition and path illumination.

Acknowledgements I would like to acknowledge the support of Dr. G. Vijaya Prakash and Dr. Mohini Gupta, Nanophotonics Laboratory, Department of Physics, Indian Institute of Technology-Delhi, Hauz Khas, New Delhi, 110016, India, for providing laboratory and support for time-resolved photoluminescence spectral studies.

References

1. Yan, Y., Zhu, Y., Guo, X., Ge, M.: The effects of inorganic pigments on the luminescent properties of colored luminous fiber. Text. Res. J. **84**, 785–792 (2014). https://doi.org/10.1177/0040517513507361
2. Kooroshnia, M.: Designing a two-phase glow-in-the-dark pattern on textiles. In: Shapeshifting: A Conference on Transformative Paradigms of Fashion and Textile Design, pp. 1–16 (2014)
3. Vandewalle, G., Archer, S.N., Wuillaume, C., et al.: Effects of light on cognitive brain responses depend on circadian phase and sleep homeostasis. J. Biol. Rhythms **26**(3), 249–259 (2011). https://doi.org/10.1177/0748730411401736
4. Vandewalle, G., Maquet, P., Dijk, D.J.: Light as a modulator of cognitive brain function. Trends Cogn. Sci. (2009). https://doi.org/10.1016/j.tics.2009.07.004
5. Riemersma-van der Lek, R.F., Swaab, D.F., Twisk, J., Hol, E.M., Hoogendijk, W.J., Van Someren, E.J.: Effect of bright light and melatonin on cognitive and noncognitive function in elderly residents of group care facilities. **299**, 2642 (2008). Available from: http://jama.jamanetwork.com/article.aspx?doi=10.1001/jama.299.22.2642

6. Monteoliva, J.M., Rodriguez, R.G., Pattini, A.E., Ison, M.S.: Daylighting and cognition: experimental studies on working memory and attention in clerical and educational contexts. In: Laboratory of Human Environment and Housing, INCIHUSA—CONICET, Mendoza, Argentina 2 Group of Evolutionary and Educational Psychology, Argentina. Experiencing Light 2012, At Eindhoven, Netherlands (2012). http://2012.experiencinglight.nl/doc/29.pdf. Viewed on 8 Sept 2016
7. Cajochen, C., Frey, S., Anders, D., et al.: Evening exposure to a light-emitting diodes (LED)-backlit computer screen affects circadian physiology and cognitive performance. J. Appl. Physiol. **110**, 1432–1438 (2011). https://doi.org/10.1152/japplphysiol.00165.2011
8. Schlangen, L.: Circle of Light the Effect of Light on Our Sleep/Wake Cycle Principal Scientist at Philips Executive Summary (2013). https://www.lighting.philips.com/b-dam/b2b-li/en_AA/Experience/Topics/Education/Lighting_Academy/how-to-beat/Daily-sleep-wake-cycles-whitepaper-FINAL.pdf
9. Gulrajani, M.L., Agarwal, A., Bajaj, A., Gupta, A., Lohia, C.G.P.: Self-illuminated safety jackets. Fiber to fashion e-magazine (2008). http://www.fibre2fashion.com/industry-article/12/1187/self-illuminatedsafety-%0Djackets2.asp. Viewed on 31 Jan 2016
10. Gfroerer, T.H.: Photoluminescence in analysis of surfaces and interfaces. In: Encyclopedia of Analytical Chemistry, pp. 9209–9231 (2000). https://doi.org/10.1002/9780470027318.a2510
11. Zhang, X., Zhang, X., Zhang, J., et al.: Size manipulated photoluminescence and phosphorescence in CaTiO3:Pr3+ nanoparticles. J. Phys. Chem. C **111**(49), 18044–18048 (2007). https://doi.org/10.1021/jp0761281
12. Hom, N.L.: Preparation and Properties of Long Persistent Sr4Al14O25 Phosphors Activated by Rare Earth Metal Ions, p. 145 (2010)
13. Zele, A.J., Cao, D.: Vision under mesopic and scotopic illumination. Frontiers Psychol. **6** (2015). https://doi.org/10.3389/fpsyg.2015.00594
14. Department of Public Works B of SL: Design Standards and Guidelines (2014). http://bsl.lacity.org/downloads/business/BSLDesignStandardsAndGuidelines0507Web.pdf
15. Stockman, A., Sharpe, L.T.: Into the twilight zone: the complexities of mesopic vision and luminous efficiency. Ophthalmic Physiol. Opt. **26**(3), 225–239 (2006). https://doi.org/10.1111/j.1475-1313.2006.00325.x
16. Barbur, J.L., Stockman, A.: Photopic, Mesopic and Scotopic Vision and Changes in Visual Performance, vol. 3 (2010)

Development and Characterization of Metal Woven Electric Heating Fabrics

N. Muthukumar, G. Thilagavathi, T. Kannaian and S. Periyasamy

Abstract The textile industry focuses on providing value-added functional textile materials to cater to the demands of the consumer world. Recently, a new field of research has emerged, which combines the benefits of the textiles with the world of electronics and technology yielding products and devices referred to as electro-active textiles or smart textiles. The various textile technologies such as weaving, knitting, sewing, and printing have been used to incorporate the electrically conducting elements with textile materials. In this work, a metal woven electric heating fabric was developed and characterized for heat generating property. The electric heating fabric was developed for weaving fine copper wire as weft and acrylic yarn as warp. The temperature–electrical resistance relationship of the developed fabric was determined. The main idea of the work is to design and develop electric heating fabric which can be used with a minimum power supply in order to obtain a wearable system.

Keywords Copper wire · Electro textiles · Electrical resistance · Heating fabric · Smart textiles

1 Introduction

Textiles, being the second most important needs of the human kind, it is the product which would not be replaced by any others though the technology is tremendously producing various consumer products. In the present scenario of increasing fashion, the textiles could be regarded as the materials that make man more than 50% than the usual perception of man and textiles being 50% each. Apart from the protection requirement of the body, there has been also the interest on various aspects such as the comfort, clothing care aspects, durability, and other stability conditions

N. Muthukumar (✉) · G. Thilagavathi · S. Periyasamy
Department of Textile Technology, PSG College of Technology, Coimbatore, India
e-mail: nmk.textile@psgtech.ac.in

T. Kannaian
Department of Electronics, PSG College of Arts and Science, Coimbatore, India

A. Majumdar et al. (eds.), *Functional Textiles and Clothing*,
https://doi.org/10.1007/978-981-13-7721-1_11

from the customers, nowadays value addition to textiles. In addition to the general requirements on textiles as just normal apparel, textiles are also expected to perform some of the human life supporting/enhancing/defending systems. Such textiles are recognized as either under functional textiles or even as smart textiles if they are responsive. Development of such textile structures uses various interdisciplinary products such as electronic gadgets, particularly the recent interests of flexible electronics, which may be used for various vital body symptoms continuous monitoring and the likes. The applications of functional and smart textiles are many and are increasing [1]. Following are the focused discussion on electro-active textiles, one of the functional/smart textiles, as the emphasis of the work and the article falls in this category.

One of the growing research areas of electro-active textiles is the electric heating garments. The electric heating garments can be used for military and other outdoor applications [2]. There is a lot of research work going on in the field of electrically heating garments in the past few years. The electric heating fabrics can be developed by introducing electrically conductive textile-based materials in the fabric. Materials like silver-coated and copper-plated yarns, carbon-based fibers, and tiny inox cables are textile-based conductive materials and used as a heating elements for electric heating fabric development. Weaving, knitting, sewing, and finishing are the different technologies available to incorporate the heating elements into the textile materials [3]. A power supply, electrical heating element, temperature sensor, and a user interface are the components required to design an electric heating fabric. In electric heating fabric, the function of the heating element is the energy converstion that is from electrical energy into thermal energy. The temperature sensor is used to measure the temperature of the human skin and atmosphere. The heating energy of the electric heating system was provided by the power supply. The electric heating system of heating fabric can be adjusted by the wearer by using the user interface. The electric heating fabric should be designed in such a way that the incorporation of these components into the clothing does not affect the comfort and user-friendliness of the wearer [4].

The working concept of electric heating fabric is shown in Fig. 1. The electric heating fabric is kept in the middle layer of the three-layer clothing. The total heat (Q_{tot}) given by the electric heating garment is calculated by

$$Q_{tot} = H_{sk} + H_{env} + H_{abs} \quad (1)$$

where H_{sk} is heat loss of the human body through the heating fabric and outer clothing to the atmosphere, H_{env} is heat loss of the electric heating fabric by radiation/convection to the atmosphere, and H_{abs} is heat absorbed by the heating fabric from heating element [5].

Numerous different techniques for creating electric heating fabrics have been studied by many research groups. Wang et al. investigated the heating performance of three-layer electric heating garment using a thermal manikin. The better heating performance of the garment was observed when the heating fabric was in the middle

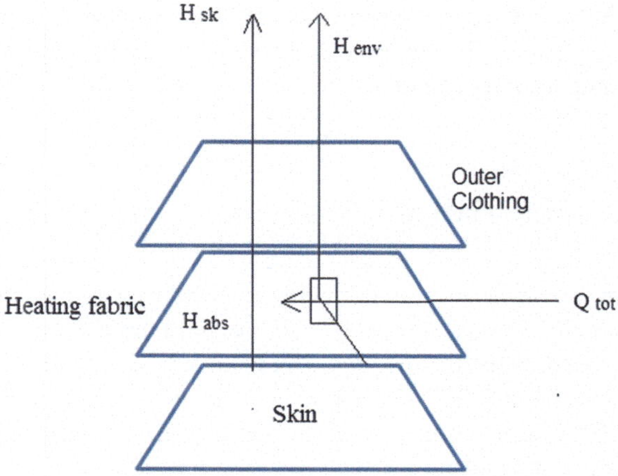

Fig. 1 Heating transfer mechanism of a heating fabric

layer of the garment compared to the other two [6]. Kanyan et al. developed electric heating pads using steel fabric and investigated the heat-generating properties. It was found that the heat-generating property of pads vary for a particular period of time with respect to the number of plies used to construct the heating pad. They concluded that the size of the heating pad, number of plies used or amount of conductive yarn used to construct the heating pad, and power supply used are the important parameters influencing the heat-generating properties of the heating pad [7]. Hao et al. investigated the heating behavior of plain woven electric heating fabrics made from silver filaments and silver-coated yarns. The results showed that the fabrics made from silver filaments had better heating performance compared to fabrics made from silver-coated yarns [8]. Poboroniuc et al. manufactured electric heating fabrics using different types of conductive yarns by knitting technology and investigated heat-generating properties of the developed fabrics. They found that the developed heating fabrics generate heat up to 20 °C with electric power of 1.7 W [9]. Mey et al. designed an electric heating pad with the size of 8 cm × 45 cm using silver-coated nylon yarn in a shirt by knitting. It was found that the temperature of the shirt incorporated with the heating pad increased up to 12–14 °C from the room temperature for 5 W power supply. They also demonstrated that a portable battery can supply this amount of power for 8–10 h [10]. In our previous work, we have fabricated an electric heating fabric by stitching silver-coated nylon yarn over the polyester fabric. In this, silver-coated nylon yarn acts as a heating element. It was found that the temperature of the developed electric heating fabric was increased up to 16 °C from the room temperature for 9 V power supply. It was also found that 0.92 W required for 1 °C rise in temperature [11].

It was concluded from previous studies that, the electrically conductive fabrics made from woven structure have better performance for heating applications compared to the knitted structure due to its structural properties. Hence, in this study, weaving technology was chosen for electric heating fabric development. We have developed a metal woven electric heating fabric using copper wire. The heating fabric was designed in such a way that it can be operated with a minimum power supply and also user-friendly.

2 Materials and Methods

In this work, fine copper wire has been used as a heating element. Copper has a low specific electrical resistivity. It has good thermal conductivity and low heat capacity compared to other heating wire materials. Due to its high ductility, the risk of fracture is very small [12]. The copper wire having a diameter of 150 μm with linear resistance of 1.8 Ω/m was used as weft. The acrylic yarn having 40^s Ne (143 μm) was used as warp. Acrylic yarn was chosen as warp because of its good thermal resistance value [13]. Acrylic yarn and copper wire were sourced from in and around Coimbatore, Tamilnadu, India.

2.1 Fabrication of Electric Heating Fabrics

The copper woven electric heating fabric was developed on a desk loom by introducing copper wire in the weft direction and acrylic yarn as warp in a plain weave pattern. Because of stable conformation, dense structure, smooth surface, and less shrinkage, the plain weave was selected in this study. The developed fabric had a size of 8 cm × 8 cm and is shown in Fig. 2. The developed fabric had EPI of 60, PPI of 56, and thickness of 0.6 mm. The developed copper woven fabric had GSM of 170.

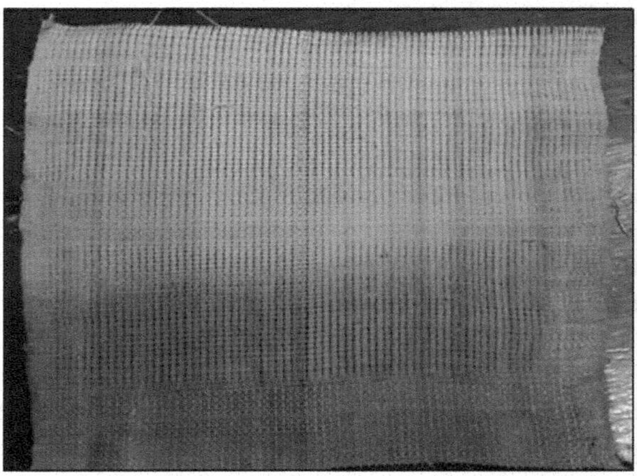

Fig. 2 Image of the copper woven electric heating fabric

2.2 *Characterization of Heat-Generating Property*

As per joules law, if the current I flows through a resistance R, then voltage V and power absorbed P are determined by using Eqs. 2 and 3.

$$U = R \times I \tag{2}$$

$$P = V \times I \tag{3}$$

In an electrical resistance heating, this power is converted into heat. In case of copper woven heating fabric, the resistance consists of thin metal wires. The resistance offered to the current flow by the internal friction of the electrons in the lattice generates heat in the developed fabric.

Figure 3 shows the experimental setup used to study the heating behavior of the developed electric heating fabric. It has a power supply (9 V battery), AC and DC multimeter (voltage, resistance, etc.), and a temperature sensor. The temperature of developed metal woven fabric can be raised by employing the power supply using 9 V battery and the heating behavior of the developed fabric was observed in point of temperature and electrical characteristics. A carton was used to keep the fabric during characterization in order to measure the actual surface temperature of the electric heating fabric.

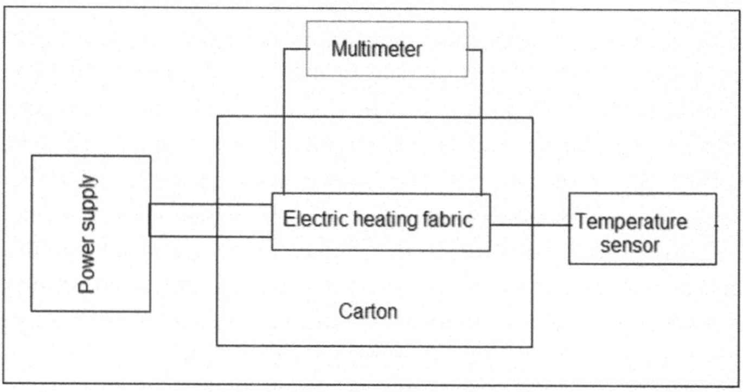

Fig. 3 Experimental setup for heating fabric characterization

3 Results and Discussion

3.1 *Heat-Generating Property of Copper Woven Fabric*

In this research work, to study the heating behavior of the copper woven fabric, a 9 V battery was used as a power supply. During experiment, the copper woven fabric provided approximately 20 °C heating within 5–6 min period and temperature of the fabric remained the same for the rest of the period. The maximum temperature rise of the developed copper woven fabric was 40 °C. The current flow in a wire is because of the free electrons movement and as they move they tend to also colloid with each other which would increase with the increase in their flow rate driven by the potential difference. Such collisions would result in heat generation and would depend on the square of the flow of current and hence the relationship is given as following

$$Q = I^2RT \tag{4}$$

where,

Q Amount of heat,
I Electric current,
R Amount of electric resistance in conductor, and
T Time.

Figure 4 shows the relationship between recorded temperature and electrical resistance of the fabric for the applied voltage of 9 V. It was observed that there is an increase in temperature with an increase in electrical resistance. The surface resistance of copper woven fabric increased from the initial resistance of 8–35 Ω/square. It was also observed that the amount of heat generated is proportional to the electrical resistance of the wire. This may be due to there is no change in the current in the circuit and also current flow. The positive coefficient value of a variable in the equation indicated that the increase in electrical resistance value increases the temperature of fabric. The coefficient of determination (R^2) of the linear regression curve is 0.986 indicating its goodness of fit. The relationship between the temperature (T) of the metal wire to its electrical resistance can be expressed as

$$R_{\mathrm{T}} = R_{\mathrm{ref}}(1 + \alpha(T - T_{\mathrm{ref}})) \tag{5}$$

where α is the temperature coefficient of resistance and it is resistance change factor per °C of temperature change [14]. A positive coefficient for a material means that its temperature increases with an increase in resistance. The electrical resistance of copper wire is dependent upon collision processes within the wire. Hence, it is expected that there is an increase in electrical resistance with an increase in temperature because of more collisions. The temperature coefficient of resistance of the developed copper woven fabric was 2.5×10^{-3}.

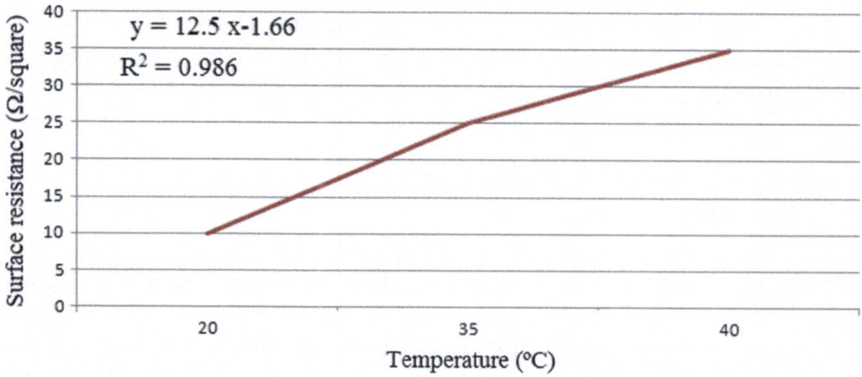

Fig. 4 Relation between fabric temperature and electrical resistance

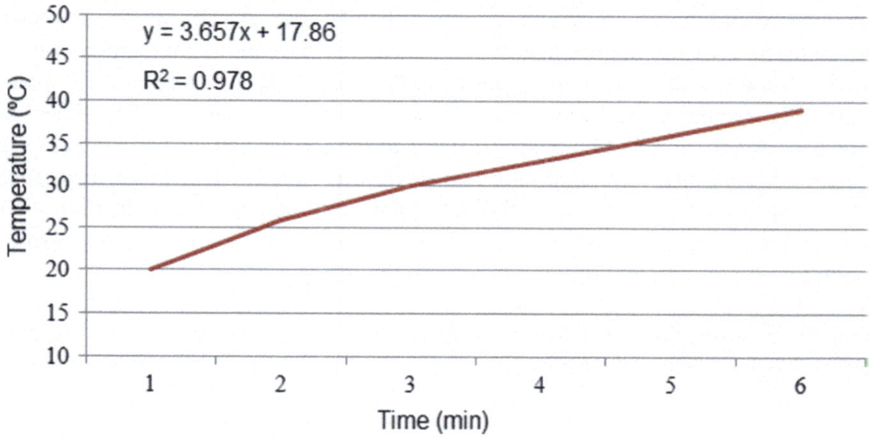

Fig. 5 Relation between fabric temperature and time

Figure 5 shows the relationship between the temperature rise of copper woven fabric and time. It was found that the temperature of the developed copper woven fabric increases with increase in time. It was also observed that the amount of heat generated is proportional to time. This may be due to there is no change in electrical resistance and current flow. The positive coefficient value of variable in the equation indicated that the increase in time increases the temperature of the copper woven fabric. The coefficient of determination (R^2) of the linear regression curve is 0.978 indicating its goodness of fit.

In this study, a battery having an output voltage of 9 V was used. The capacity of the battery was 2400 mAh which means that 2.4 A can be delivered for one hour. Such a battery can supply 21.64 W of power for one hour. From the experimental trails, it was found that the power consumed by the fabric to rise 1 °C temperature was 0.5 W; hence, 10 W is required to give a temperature rise from 20 to 40 °C. The developed fabric can be used for heat-generating application for approximately 2 h with 9 V battery.

4 Conclusion

In this work, a metal woven electric heating fabric was developed by weaving fine copper wire as weft and acrylic yarn as warp. It was found that the temperature of the developed copper woven electric heating fabric rises during the power application. The increase in temperature of the copper woven fabric is proportional to the increase in electrical resistance value and time. The temperature–resistance relationship demonstrated a linear trend with a goodness of fit (R^2 value) of over 98%. The developed metal woven fabric can be used with a 9 V battery. The developed metal woven fabric provides approximately 20 °C heating above the room temperature.

The developed electric heating fabric can be used domestic or in medical treatments. For example, localized heating of the human body can help a patient for quicker recovery. Some of the major features of the electro-active textile developed in this work are structure simplicity and flexibility which would make them find a wide application particularly in the field of warmth demanding applications. To provide the required power supply to the fabric, an adjustable regulator can be incorporated.

Acknowledgements The authors thank M/s PSG TECHS COE INDUTECH, Coimbatore for providing facilities to carry out this research.

References

1. Rossi, D., Carpi, F., Lorussi, F.: Electroactive fabrics and wearable biomonitoring devices. AUTEX Res. J. **3**, 180–185 (2003)
2. Van Langehove, L.: Smart Textiles for Medicine and Healthcare: Materials, Systems and Applications. CRC Press, New York (2007)
3. Husain, M.D., Kennon, R.: Preliminary investigations into the development of textile based temperature sensor for healthcare applications. Fibers **1**, 2–10 (2013)
4. Kanyan, O., Ender, B., Ozge, S.: Implementation of steel-based fabric panels in a heated garment design. Text. Res. J. **79**, 1427–1437 (2009)
5. Wang, F., Gao, C.: Protective Clothing—Managing Thermal Stress. Wood Head Publishing, United Kingdom (2014)
6. Wang, F., Gao, C., Kuklane, K., Holmér, I.: A review of technology of personal heating garments. Int. J. Occup. Saf. Ergon. **16**, 387–404 (2010)
7. Ozan, K., Ender, B.: Heating behaviors of metallic textile structures. Int. J. Cloth Sci. Technol. **21**, 127–136 (2009)
8. Hao, L., Yi, Z., Li, C., Li, X., Yuxiu, W., Yan, G.: Development of flexible heating fabric based on conductive filaments. Measurement **45**, 1855–1865 (2012)

9. Poboroniuc, M.S., Curteza, A., Cretu, V., Macovei, L.: Designing wearable textile structures with embedded conductive yarns and testing their heating properties. In: Proceedings of International Conference and Exposition on Electrical and Power Engineering, pp. 778–783 IEEE, Romania (2014)
10. Mey, G.D., Özçelik, M., Schwarz, A., Gürsoy, N.C.: Designing of conductive yarn knitted thermal comfortable shirt using battery operated heating system. Tekstil ve Konfeksiyon **24**(1), 26–29 (2014)
11. Thilagavathi, G., Muthukumar, N., Kannaian, T.: Development and characterization of electric heating fabric based on silver coated nylon yarn. J. Text. Eng. Fashion Technol. **1**(6), 1–3 (2017)
12. Lou, C.W., Lin, T.A., Chen, A.P., Lin, J.H.: Stainless steel/polyester woven fabrics and copper/polyester woven fabrics: Manufacturing techniques and electromagnetic shielding effectiveness. J. Ind. Text. **46**(1), 214–236 (2016)
13. Akbar, A.R., Kamruzzaman, M., Xu, W., Gull, S., Ahmed, W., Khalid, J.: Development of weft knitted heating pads on V-bed hand flat knitting machine by using conductive yarns. Am. J. Poly. Sci. Eng. **4**(1), 1–9 (2016)
14. Ali Hamdani, S.T., Potluri, P., Fernando, A.: Thermo-mechanical behavior of textile heating fabric based on silver coated polymeric yarn. Materials **6**, 1072–1089 (2013)

Part IV
Textile Chemical Processing

Mechanical Properties of the Silk Degummed with Citric Acid and Ultrasound

Ruzica Brunsek, Ivana Schwarz and Mirta Than

Abstract Silk, called the "Queen of Fibers," is a continuous protein fiber produced by silkworm to form its cocoon. Raw silk fiber obtained by unwinding the cocoons of the silkworm Bombyx mori requires a great deal of handling and processing, which makes it one of the most expensive fibers. Because of the presence of sericin, raw silk is rough, hard, brittle, rugged, and without luster. The traditional method for silk degumming is with Marseilles soap, which is very expensive. Today, with new ecological and economic requirements, there are new methods for silk degumming, such as degumming with enzyme or with polycarboxylic acid, degumming with alkalis or with water. Recently, acid agents with the purpose to enhance the physical properties of silk replace soap. In general, the action of alkali is more aggressive than the action of organic acid. Citric acid, compared to the other organic acids, is cheaper, proven lack of toxicity, and ready availability. The aim of this paper is to investigate the influence of citric acid and ultrasound on silk degumming. For this purpose, silk fibers were degummed by the classic procedure, and with different concentrations of citric acid (15 and 30%), with and without ultrasound. The efficiency of the degumming was determined by measuring degumming ration and mechanical properties (fineness, tenacity, and elongation at break) of untreated and treated silk yarn.

Keywords Silk degumming · Citric acid · Ultrasound · Properties

1 Introduction

Silk is one of the oldest fibers known to man called the "Queen of Fibers" for its luster, luxury appeal, comfort, elegance, sensuousness, and glamour. Silk is a natural animal fiber, mainly made of protein fibroin. The greatest significance is the silk produced by the mulberry silk moth Bombyx mori and a few others in the same genus, which is cultivated for obtaining silk [1, 2].

R. Brunsek · I. Schwarz (✉) · M. Than
Faculty of Textile Technology, University of Zagreb, Prilaz Baruna Filipovica 28a, 10000 Zagreb, Croatia
e-mail: ivana.schwarz@ttf.hr

A. Majumdar et al. (eds.), *Functional Textiles and Clothing*,
https://doi.org/10.1007/978-981-13-7721-1_12

Raw silk fiber obtained by unwinding the cocoons of the silkworm Bombyx mori requires a great deal of handling and processing, which makes it one of the most expensive fibers. Because of the presence of sericin, raw silk is rough, hard, brittle, rugged, and without luster. Raw silk is composed mainly of sericin, fibroin, water, and mineral salts. The proportion of fibroin in raw silk is 70–82% by weight, and it is insoluble in hot water. The sericin, which covers fiber in raw silk, is a compound of proteins with gum-resilient properties and acts as a gum binder to maintain the structural integrity of the cocoon. In addition, the sericin is more water-soluble than fibroin, and it is easily removable from the filaments through various processes without considerable damage to the filaments [3].

Silk processing from cocoons to the finished clothing materials consists of a series of steps (reeling, weaving, degumming, dyeing or printing, and finishing), but two most important processes of silk are degumming and dyeing. The sericin, natural gum, protects the fibers from mechanical injury and its removing improves the softness, color, and gloss of silk. To remove the sericin is necessary to conduct a process called degumming.

After removing sericin, the silk fibroin fibers get good brightness with excellent elasticity (45%) and tenacity reducing (30%). Properties of silk such as handling, luster, and rubbing behavior primarily depend on the quantity of sericin remained on the silk fibroin which shows that the degumming process is thus very important [4–7].

The traditional method for silk degumming is with Marseilles soap, which is very expensive. Today, with new ecological and economic requirements, there are new methods for silk degumming, such as degumming with enzyme or with polycarboxylic acid, degumming with alkalis or with water [8]. Recently, acid agents with the purpose to enhance the physical properties of silk replace soap. In general, the action of alkali is more aggressive than the action of organic acid. Citric acid, compared to the other organic acids, is cheaper, proven lack of toxicity, and ready availability [8–11].

Ultrasound has been used in processes of textile industry, especially in the processes of wet finishing which reduces processing time, saving energy and chemical substance, and increase the quality of the product [12, 13]. Ultrasound, in the degumming process, relieves the removal of the substances existing on the raw silk like dirt and sericin [14].

Therefore, the aim of this paper is to investigate the influence of citric acid and ultrasound on silk degumming. For this purpose, silk fibers were degummed by the classic procedure, and with different concentrations of citric acid (15 and 30%), with and without ultrasound. The efficiency of the degumming was determined by measuring degumming ration and mechanical properties (fineness, tenacity, and elongation at break) of untreated and treated silk yarn.

2 Experimental

2.1 *Testing Samples*

Investigation was carried out on the mulberry silk fabric with basic parameters: plain weave; fabric density: warp—360 threads/10 cm, weft—300 threads/10 cm; fabric mass per unit area—226 g/m^2, and fabric thickness—0.33 mm.

2.2 *Traditional Degumming*

Silk fabric was traditional degummed at 80 °C for 2.5 h in the degumming solution containing 15% Marseilles soap, 1.5% sodium carbonate (Na_2CO_3), 0.05% nonionic detergent, and liquor ratio was 1:30. After degumming, the sample was treated at 80 °C for 5 min with 0.2% Na_2CO_3 to remove soap, which remains on the surface of the sample. Then, degummed samples were washed first with cold and then with warm water and finally dried immediately at 80 °C for 1 h, and then kept at room temperature for 48 h.

2.3 *Citric Acid Degumming*

Silk fabric was degummed in the degumming bath containing citric acid solutions at the concentration of 15 and 30%, respectively, and 0.2% nonionic detergent SVN at 80 °C for 2.5 h, and liquor ratio was 1:20 in laboratory apparatus with mechanical agitation.

2.4 *Ultrasound Degumming*

Silk fabric was degummed with and without ultrasound employing ultrasound unit tt. Elmasonic with ultrasound frequency 80 kHz and different media (water, 15% Marseilles soap, 15 and 30% citric acid) at constant temperature (80 °C) for 2.5 h.

Due to the analysis of the effectiveness of the different degumming processes, raw silk fabric was referred as untreated control sample, and silk fabric treated in water containing without soap and citric acid at the same treatment condition with and without ultrasound was referred as "Blank" sample.

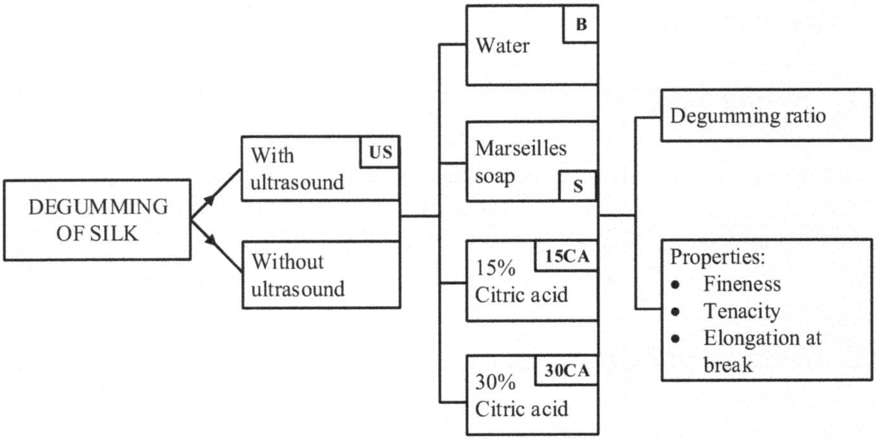

Fig. 1 Protocol of investigation of silk degumming

2.5 *Determining the Treatment Efficiency*

Determining of degumming ratio. Degumming ratio, i.e., the amount of sericin removed by different degumming treatment, was calculated from the weight loss of silk fiber before and after degumming treatment.

Determining of textile technology properties. Fineness (ISO 1973:1995), tenacity, and elongation at break (ISO 5079:1995) determined on tensile testers Vibroscop and Vibrodyn 400. Measurements of samples of tested properties were performed on conditioned samples.

The protocol of investigation was carried out as it is shown in Fig. 1.

3 Results and Discussion

A large number of measurements ($n = 50$) of the silk yarn properties are performed. Therefore, statistical indicators are shown: arithmetic mean ($\bar{X}$), standard deviation (σ), variation coefficient (CV), and practical limit of error (p_{gg}). The degumming ratio was calculated comparing silk fabric degumming in hot water with samples degumming with Marseilles soap, 15 and 30% citric acid solution with and without ultrasound. Differences (%) are expressed as percentages and are shown in Fig. 2.

The difference in loss of mass was established between the blank sample and samples treated with Marseilles soap, 15 and 30% citric acid solution. The degumming ratio of silk fiber treated with hot water, soap solution, 15 and 30% citric acid solution was 3.47, 18.56, 9.36, and 12.69%, respectively.

As expected, almost all of the sericin was removed after degumming with soap solution. Initially, the degumming ratio of silk fibers increased gradually with raising

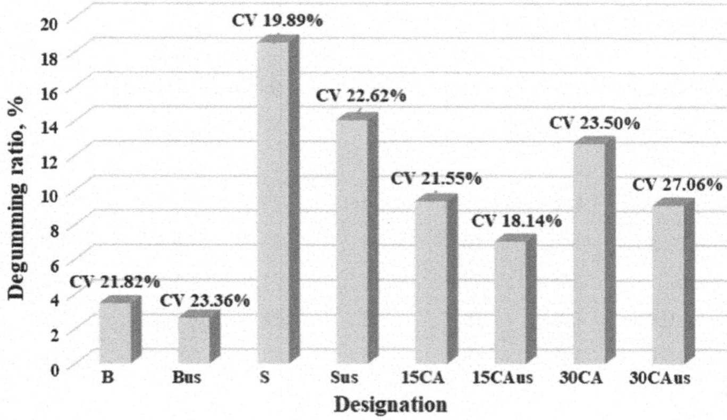

Fig. 2 Degumming ratio of degummed silk

Table 1 Fineness of silk yarn—warp and weft

Designation	$\bar{X}$ (dtex)	σ (dtex)	CV (%)	p_{gg} (%)	$\bar{X}$ (dtex)	σ (dtex)	CV (%)	p_{gg} (%)
	Without ultrasound				With ultrasound			
Warp								
N	196.6	45.9	23.4	6.5	196.6	45.9	23.4	6.5
B	186.5	54.1	29.0	8.0	220.7	87.9	39.8	11.0
S	184.2	49.8	27.1	7.5	251.7	116.3	46.2	12.8
15CA	194.2	57.9	29.4	8.1	222.2	77.6	34.9	9.7
30CA	193.3	41.7	21.6	5.9	225.6	96.5	42.8	11.9
Weft								
N	215.9	58.9	27.3	7.6	215.9	58.9	27.3	7.6
B	206.7	47.1	19.9	5.5	251.8	73.7	29.3	8.1
S	188.4	57.7	30.6	8.5	219.3	70.8	37.0	10.3
15CA	205.9	46.6	22.6	6.3	222.5	65.1	29.3	8.1
30CA	198.0	46.5	23.5	6.5	227.2	78.4	34.5	9.6
n	50							

the concentration of citric acid and the amount of residual sericin, which remains, on the raw silk fiber decreased with increasing citric acid concentration. After degumming with ultrasound, degumming ratio decreased, and amount of residual sericin on the raw silk increased in all treatments carried out.

Test results of fineness of silk yarn (Table 1) indicate a little change in silk yarn fineness, degummed without ultrasound, in relation to the untreated sample. The yarn becomes slightly finer. Comparing the degumming processes with and without ultrasound, results show that fineness of yarn degummed with ultrasound little bit increased which is in accordance with the results for degumming ratio.

Table 2 Tenacity of silk yarn—warp and weft

Designation	$\bar{X}$ (cN/tex)	σ (cN/tex)	CV (%)	p_{gg} (%)	$\bar{X}$ (cN/tex)	σ (cN/tex)	CV (%)	p_{gg} (%)
	Without ultrasound				With ultrasound			
Warp								
N	27.1	4.9	17.9	4.9	27.1	4.9	17.9	4.9
B	22.4	3.9	15.0	4.2	23.7	6.4	27.0	7.5
S	6.2	2.9	47.9	13.3	21.8	5.5	25.3	7.0
15CA	26.6	4.4	19.7	5.5	28.9	7.4	31.5	8.7
30CA	15.7	2.8	18.1	5.0	27.5	5.9	24.9	6.9
Weft								
N	21.6	3.7	16.9	4.7	21.6	3.7	16.9	4.7
B	20.4	3.8	16.4	4.6	20.7	4.8	23.2	6.4
S	9.4	3.9	41.6	11.5	11.6	4.2	35.9	9.9
15CA	20.9	3.6	16.6	4.6	24.9	5.9	23.8	6.6
30CA	16.2	3.5	21.8	6.1	24.2	4.8	19.9	5.5
n	50							

Morphological and structural changes have occurred after exposure of fibers to the ultrasound treatment. The fiber structure is loose, and cross-linking with citric acid is much better. Therefore, the cross-link distribution is uniform, and as a result, fiber fineness increased.

The analysis of the tenacity results, given in Table 2, of silk yarn after degumming without ultrasound shows significant variations in relation to the untreated sample. As expected, traditional degummed samples in the soap show a significant decrease in tenacity. It can be concluded that samples degummed with 15% citric acid have the lowest tenacity reduction, 1.8%, while samples degummed with 30% citric acid have tenacity reduction for 6.6%. The reason for tenacity reduction is the process of cross-linking that is performed in an acid medium (pH 2.5), which aggressively acts on fiber structure and damages fibers.

The results of ultrasound degumming indicate a positive impact of ultrasound on the mechanical properties of the silk yarns tested. During the ultrasound treatment, it occurred changes in the fiber structure and fiber tenacity increased. Because of ultrasound vibrations, the segment of macromolecules come close each other and forming new area with better-arranged structure and reduction of internal stresses.

Samples degummed with ultrasound loose less of their tenacity, while the loss of tenacity of samples degummed without ultrasound treatment is higher. By analyzing the obtained results, it is evident that tenacity of the yarn degummed with 15% citric acid and with ultrasound is increased for 6.2%, and samples degummed with ultrasound and 30% CA slightly increased for 1.5%.

In addition, the results of tenacity (Table 2) are in accordance with change of fineness (Table 1) and degumming ratio (Fig. 2) of silk yarn. It can be observed that

Table 3 Elongation at break of silk yarn—warp and weft

Designation	$\bar{X}$ (%)	σ (%)	CV (%)	p_{gg} (%)	$\bar{X}$ (%)	σ (%)	CV (%)	p_{gg} (%)
	Without ultrasound				With ultrasound			
Warp								
N	14.3	1.6	11.2	3.1	14.3	1.6	11.2	3.1
B	14.2	1.4	10.0	2.8	12.9	2.7	20.6	5.7
S	5.0	1.3	25.1	6.9	11.4	2.4	20.6	5.7
15CA	11.3	1.7	15.2	4.2	10.8	2.1	19.8	5.5
30CA	7.8	0.9	12.6	3.5	12.2	1.8	14.5	4.0
Weft								
N	17.2	2.2	12.6	3.5	17.2	2.2	12.6	3.5
B	15.0	1.7	11.1	3.1	13.3	2.1	15.6	4.3
S	5.2	1.5	29.1	8.1	5.7	1.5	25.7	7.1
15CA	10.9	1.5	14.1	3.9	13.1	2.6	19.8	5.5
30CA	7.8	1.1	14.4	4.0	12.4	1.5	12.4	3.4
n	50							

performed degumming silk with citric acid and with ultrasound have influenced fiber tenacity.

For all the samples, the elongation is continuously, but not significantly reduced what is shown in Table 3. The test results of the samples degummed with ultrasound have the lowest reduction of elongation at break in comparing with samples degummed without ultrasound.

4 Conclusion

The raw silk fabric was degummed with different concentration of citric acid and ultrasound to investigate their influence on silk degumming and mechanical properties. Based on the results of the investigations, which have been carried out under the same degumming conditions, with and without ultrasound, a positive effect of ultrasound treatment on the fineness of silk yarn cannot be confirmed with certainty.

However, a positive effect of ultrasound treatment during degumming on the tenacity of silk yarn should be pointed out. Samples degummed with ultrasound loose less of their tenacity, while the loss of tenacity of samples degummed without ultrasound treatment is larger. The results obtained by conducted treatments have shown that samples degummed with 15% citric acid, with or without ultrasound, have the lowest tenacity reduction.

It may be concluded that the use of ultrasound and citric acid, as an environmentally and economically acceptable degumming process in comparison with the traditional method, results in a successful removal of sericin, with significant savings of time and energy.

Acknowledgements The financial support was provided within the framework support scientific research 2017, "Natural Fibres Agrotextile" (TP 12/17), financed by the University of Zagreb.

References

1. Murugesh Babu, K.: Silk-Processing, Properties and Applications, 1st edn. Woodhead Publishing (2013)
2. Basu, A.: Advances in Silk Science and Technology, 1st edn. Woodhead Publishing (2015)
3. Niyaz, M.M., et al.: Degradation of sericin (degumming) of Persian silk by ultrasound and enzymes as a cleaner and environmentally friendly process. J. Clean. Prod. **18**(2), 146–151 (2010)
4. Freddi, G., Allera, G., Candiani, G.: Degumming of silk fabrics with tartaric acid. J. Soc. Dyers Colour. **112**(7–8), 191–195 (1996)
5. Freddi, G., Mossotti, R., Innocenti, R.: Degumming of silk fabric with several proteases. J. Biotechnol. **106**(1), 101–112 (2003)
6. Gulrajani, M.L., Chatterjee, A.: Degumming of silk with oxalic acid. Indian J. Fibre Text. Res. **17**(1), 39–44 (1992)
7. Gulrajani, M.L., Sethi, S., Guptha, S.: Some studies in degumming of silk with organic acids. J. Soc. Dyers Colour. **108**(2), 79–86 (1992)
8. Majibur, R.K., et al.: Physical properties and dyeability of silk fibres degummed with citric acid. Biores. Technol. **101**(21), 8439–8445 (2010)
9. Uddin, K., Hossain, S.: A comparative study on silk dyeing with acid dye and reactive dye. Int. J. Eng. Technol. **10**(6), 22–27 (2010)
10. Teli, M.D., Rane, V.M.: Comparative study of the degumming of mulberry, muga, tasar and ericream silk. Fibres Text. Eastern Eur. **19**(2), 10–14 (2011)
11. Šurina, R., Andrassy, M.: Effect of preswelling and ultrasound treatment on the properties of flax fibres crosslinked with polycarboxylic acids. Text. Res. J. **83**(1), 66–75 (2013)
12. Andrassy, M., Šurina, R.: Effect of ultrasound treatment on resistance of flax yarn to microorganisms. In: Book of Proceedings of the 3rd International Textile, Clothing & Design Conference—Magic World of Textiles, pp. 22–27. A&C Design, Zagreb (2006)
13. Andrassy, M., Pezelj, E., Čunko, R.: Single-bath bleaching and fiber bundle size reduction of flax using ultrasound. AATCC Rev. **5**(7), 20–23 (2005)
14. Yuksek, M., et al.: Effect of degumming performed with different type natural soaps and through ultrasonic method on the properties of silk fibers. Adv. Environ. Biol. **6**(2), 801–808 (2012)

Photocatalytic Decolorization of Rhodamine B Dye Solution Using TiO_2 Coated Cotton Fabric

Anu Mishra and Bhupendra Singh Butola

Abstract TiO_2 is known for its photocatalytic activity. It has ability to degrade a number of organic pollutants. In the current study, TiO_2 has been prepared in situ on cotton fabric, using its precursor through a modified sol-gel method. The in situ prepared TiO_2 on fabric was further given heat treatments via different routes. The crystalline form, surface morphology and Ti content of the samples coated with TiO_2 have been evaluated by XRD, FESEM and ICP-MS. The UV absorption behavior of the samples was studied using UV-visible spectroscopy. The photocatalytic degradation of Rhodamine B dye was conducted under UV light, using TiO_2 coated cotton fabric prepared via different routes. It has been found that the sample treated with TiO_2 via pad-dry-solvothermal route shows the highest rate of decolorization in comparison with samples prepared via pad-dry-cure and pad-dry-hydrothermal route. The mechanism of dye decolorization using TiO_2 coated fabric has also been discussed.

Keywords Photocatalyst · TiO_2 · Dye decolorization · Sol-gel · Rhodamine B

1 Introduction

Coloration of textile material requires a huge amount of water, in addition to dyes and chemical auxiliaries. About 15% of the total dye used in the coloration of textiles remains in un-reacted form. This eventually goes into wastewater and becomes a cause of high effluent load. This effluent load can be reduced either by removal or by degradation of the un-used dyes from the wastewater [1, 2]. Various chemical, electrochemical and biological processes have been used for the reduction of effluent from the wastewater. However, these processes suffer from low degradation efficiency. Sometimes, these processes require further addition of chemicals and thus become the cause of generation of secondary pollutants.

A. Mishra (✉)
Indian Institute of Carpet Technology, Chauri Road, Bhadohi, India
e-mail: mishraanu79@gmail.com

B. S. Butola
Indian Institute of Technology Delhi, New Delhi, India

A. Majumdar et al. (eds.), *Functional Textiles and Clothing*,
https://doi.org/10.1007/978-981-13-7721-1_13

In recent years, photocatalysis has emerged as a simple, economic and efficient process for destruction of various dyes from the aqueous medium. Photocatalysis is an advanced oxidation process (AOP) by which degradation or even complete mineralization of many complex organic species takes place [3].

Rhodamine B (RhB) dye is one of the organic dyes which are widely used in coloring paper, laser printing and dyeing of textile materials [4]. Rhodamine B belongs to triphenylmethane family and is among the most notorious pollutant dyes used in textile and paper industry. The chemical structure of RhB consists of four *N*-ethyl groups at either side of the xanthene ring [5–7]. The traces of dye can persist for a long period in the aquatic environment due to its resistivity toward chemical and biological attacks. This makes the removal or degradation of this hazardous dye a great challenge [8, 9].

Among the transition metal oxides, TiO_2 has been extensively used as photocatalyst due to its strong oxidizing power, non-toxicity, low cost, chemical stability and high photocatalytic activity [10, 11]. It has been utilized in various applications like sterilization, sanitation, air purification and water treatment [12, 13]. TiO_2 has also been used by many researchers for the decolorization of Rhodamine B dye solution. But in most of the studies, the powder form of TiO_2 is used directly for decolorization of dye wastewater in the presence of UV light. However, repeated use of TiO_2 powder and its recovery from the dye solution remains a challenge. Even the centrifugation process does not ensure a complete recovery of TiO_2 particles from the aqueous bath [14].

On providing a suitable template like cotton fabric, the deposited photocatalyst on it can be re-used. However, there are problems associated with the deposition of TiO_2 particles on cotton fabric. TiO_2 in its particle form does not have any affinity with the cotton substrate. Application of binders to hold the TiO_2 particles with fabric shields the effectiveness of the photocatalyst. These problems have further been resolved by using low-temperature sol-gel method of preparation of TiO_2 from its precursor and its subsequent coating on textile substrate.

The present study demonstrates three different process routes for application of TiO_2 on cotton fabric. The role of the process routes (pad-dry-cure, pad-dry-hydrothermal and pad-dry-solvothermal) on the development of TiO_2 morphology on fabric and its effect on rate of dye decolorization have also been studied in details.

2 Experimental Section

2.1 Materials

Cotton fabric with ends per inch = 124, picks per inch = 64 and gsm = 150 was supplied by Vardhman fabrics, Budhani. Titanium Tetra Isopropoxide (TTIP) was used as a precursor for synthesis of TiO_2 sol. TTIP was purchased from Spectrochem,

Mumbai. Analytical grade glacial acetic acid and ethyl alcohol were used in sol-gel preparation. Deionized water was used, whenever required.

2.2 *Preparation of TiO_2 Sol*

Preparation of 1% v/v TTIP sol was done in 50:50 ethanol/DI water solvent system. One mL of pure TTIP was stirred with 10 mL of pure ethanol for 10 min. One mL of glacial acetic acid was stirred with 10 mL of pure ethanol separately for 5 min and then added in the 1st mixture dropwise. 30 mL of pure ethanol was mixed separately with 50 mL DI water and this aqueous ethanol was added dropwise to acidified TTIP/ethanol solution with continuous stirring. Finally, the prepared sol was stirred for another 4 h.

2.3 *Application of TiO_2 on Cotton Fabric*

15 × 15 cm^2 (approx. 3 g wt) samples of untreated cotton fabric were taken. Each sample was dipped in prepared sol of TTIP, followed by sonication for a period of 10 min. Further, each sonicated sample was padded at a pressure of 2.75 bar and dried at 80 °C. The padding pressure of 2.75 bar ensures a uniform percentage expression of 100% in all the samples. The heat treatment to each fabric sample was subsequently given by curing (sample A), hydrothermal (sample B) or solvothermal method (sample C).

Curing was conducted at 120 °C in hot air for a period of 3 min. Hydrothermal treatment was given by transferring the sample in IR dyeing cylinder containing 100 mL of DI water. It was treated for a period of 3 h at 120 °C. Solvothermal treatment was given by transferring the sample in IR dyeing cylinder containing 100 mL of TiO_2 sol. It was also treated for a period of 3 h at 120 °C.

2.4 *Characterization*

The crystallographic phase of the prepared samples was investigated in a Panalytical X'Pert X-ray diffractometer (XRD) using Cu-K radiation. The surface deposition of TiO_2 on cotton fabric was studied using scanning electron microscope (SEM), model ZEISS EVO 50. Energy dispersive X-ray (EDX) analysis was carried out in EDX system Model QuanTax 200 to analyze the elemental composition.

The total amount of Ti in treated samples was determined by high-resolution inductively coupled plasma mass spectrometry (ICP-MS, make: Agilent 7900). A known weight of sample was digested in a microwave-assisted acidic digester at 280 °C for about 30 min. Spectral interferences arising from the plasma gases or

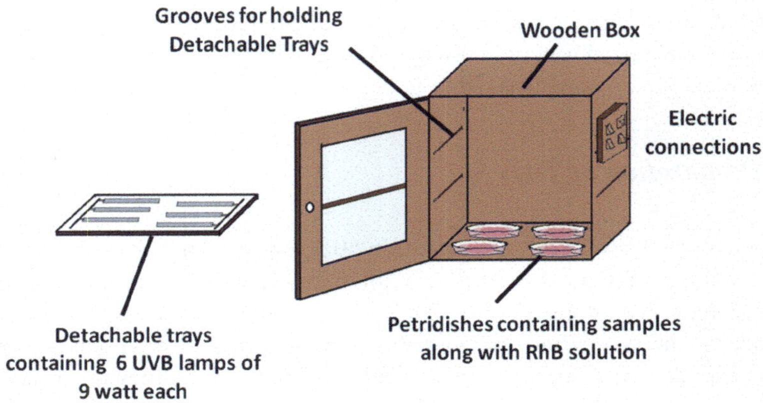

Fig. 1 Experimental setup used for dye decolorization

the major components of the samples were identified and removed. Eventually, the amount of Titanium present in fabric was determined in terms of microgram of Ti per gram of the fabric (ppm) [15].

ISO 105 C10: 2006—B was adopted to evaluate the performance of the treated samples after standard wash. The treated samples were washed in laundro-meter at 50 °C temperature using 5 gpl of soap. The material to liquor ratio was kept as 50:1 and the washing cycle was completed in 45 min.

UV-visible spectra of samples were recorded on a UV-VIS-NIR Photo-spectrometer, model LAMBDAL6020087 keeping air as a reference.

2.5 Setup Used for Decolorization of Rhodamine B Dye Solution

The photocatalytic activities of untreated and TiO_2 treated cotton fabrics were evaluated for decolorization of Rhodamine B dye under UVB light. Figure 1 shows the schematic diagram of setup used for dye decolorization studies. It consists of detachable trays containing UVB tube lights. UVB light of very narrow waveband emission with a peak at 311 nm is supplied with the help of 6 UVB lamps of 9 W each (make—Philips PL-S 9W/01). Dye solution of Rhodamine B with an initial concentration of 20 mg/L was used in the entire study. All the fabric samples were cut in a circular disk shape of diameter 7.5 cm. The individual samples were dipped in the Petri dishes containing 100 mL of prepared dye solution. The vertical distance between the detachable trays containing light source and the dye solutions was kept 45 cm.

3 Results and Discussion

3.1 XRD Analysis

The XRD scans of untreated fabric and fabric impregnated with TiO_2 precursor and subsequently prepared via different routes were analyzed. As can be seen from Fig. 2, the untreated fabric has only characteristic peaks of cellulose I at $2\theta = 16.5°$, $22.5°$ and $34.6°$. Among other samples, samples B and C show a distinct hump at around $25.5°$, indicative of anatase form of TiO_2. However, sample A in which TiO_2 is applied via Pad-dry-cure route, crystalline peak of TiO_2 could not be seen. In case of sample A, the curing time given was only 3 min. It is expected that in such a short duration of thermal treatment, transformation of amorphous phase to crystalline phase could not occur in sample A. Therefore, no crystalline phase of TiO_2 could be observed in sample A. On the other hand, sufficient time of 3 h and temperature of 120 °C has been able to develop crystalline peak of TiO_2 in samples B and C.

3.2 SEM of TiO_2 Coated Cotton Samples

The SEM images of TiO_2 coated cotton samples are shown in Fig. 3. It can be seen that PDC method results in formation of a layer of TiO_2 on fabric surface. In PDH method, the texture of TiO_2 turns to be flat-granular, whereas in case of PDS higher concentration of these granules can be seen. Therefore, it can be inferred that the texture of TiO_2 deposited on cotton surface can be altered depending upon the method used for application of heat treatment in the samples.

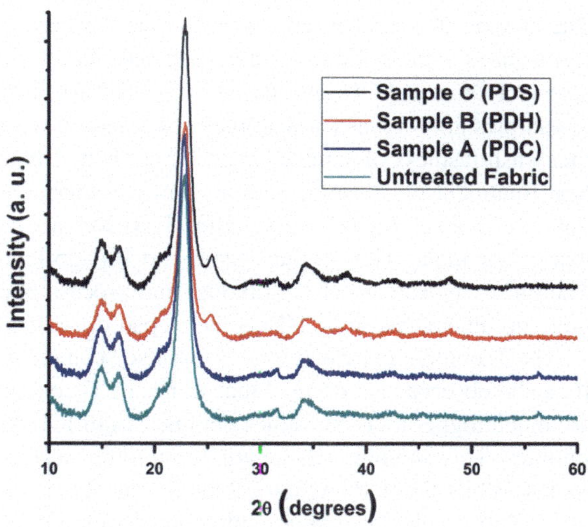

Fig. 2 XRD patterns of untreated and TiO_2 coated cotton fabrics prepared by various routes

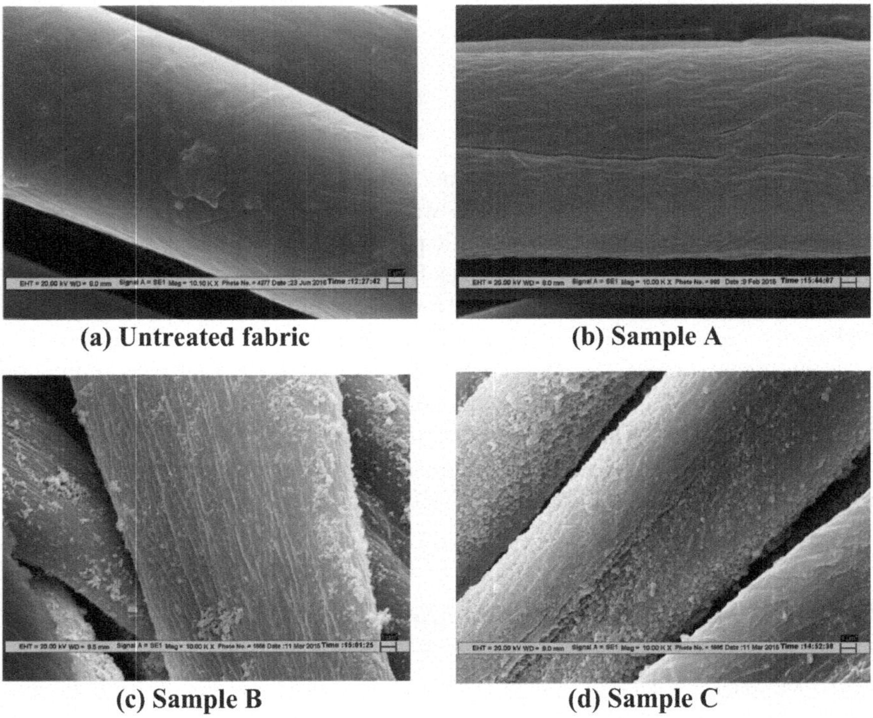

Fig. 3 SEM of (i) untreated fabric (ii) sample A (PDC), (iii) sample B (PDH) and (iv) sample C (PDS)

3.3 Measurement of Titania Content in Fabric

The content of Ti in treated cotton fabrics has been measured using ICP-MS. The results are shown in Table 1, which suggests that the add-on level of Ti is similar in case of PDC and PDH and higher with PDS. Moreover, comparing the Ti content of samples after 3 washes with their 1st standard wash samples, it can be observed that wash fastness of samples B and C are better than that of sample A. Prolonged heat treatment given to the samples in hydrothermal or solvothermal process help not only in development of crystalline TiO_2 coating, it also ensures the deposition of nano-particulates TiO_2 of the size around 100–250 nm on the cotton fabric surface. Therefore, hydrothermal or solvothermal process may be considered to provide a more durable deposition of TiO_2 than a pad-dry-cure process.

The Ti content in treated fabrics has also been measured through EDX technique. It can be observed from Fig. 4 that values of Ti content obtained by EDX technique are much higher than the values obtained from ICP-MS method for the same set of samples. This anomaly can be attributed to the surface deposition of TiO_2 by these routes. While the ICP method calculates the values on cotton fabric weight, in case of EDX, it is calculated taking into consideration only the surface layer deposition.

Table 1 Ti content values (using ICP-MS technique) in samples prepared by different routes (after 1st and 3rd wash)

Sample code	Process	Ti content in ppm (after 1st wash)	Ti content in ppm (after 3rd wash)	% reduction in Ti content
UNT	–	–	–	–
A	PDC	530	424	20.0
B	PDH	536	521	2.8
C	PDS	748	720	3.7

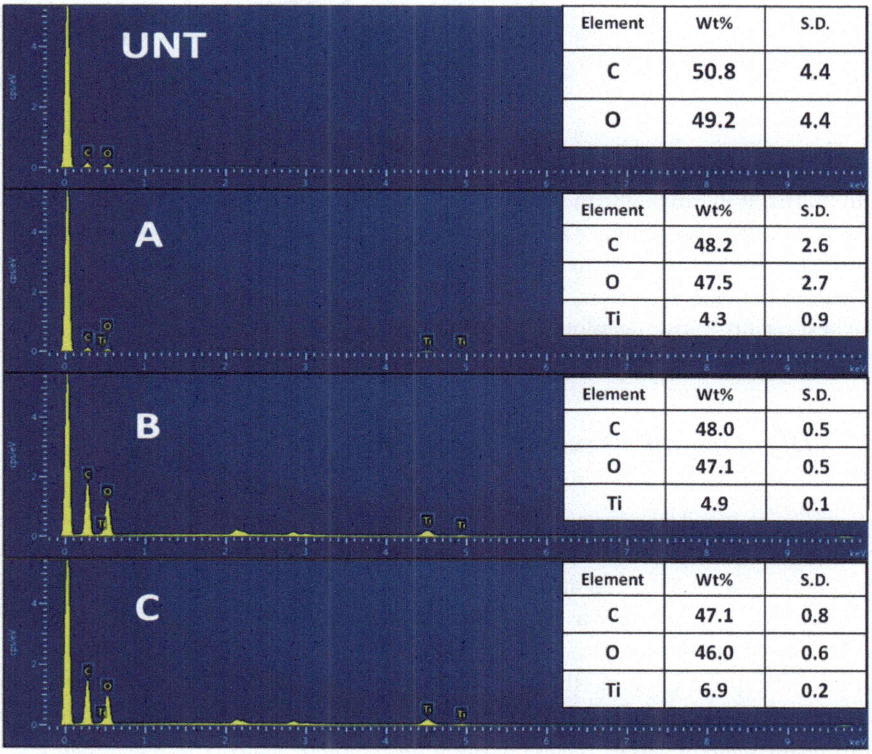

Fig. 4 EDX spectra of TiO_2 coated cotton fabrics prepared by various methods. (UNT) Untreated. **a** PDC, **b** PDH and **c** PDS

3.4 UV Absorption Studies

UV absorption spectra of all the samples were done to understand the absorption behavior of the fabric samples in UV light. The UV-visible absorption spectra of different samples are shown in Fig. 5. The untreated cotton has almost no ability to absorb UV light. On the other hand, the TiO_2 treated fabric samples exhibit good absorption of UV light, which confirms that TiO_2 is well deposited on samples A, B

Fig. 5 UV absorbance curve of untreated and TiO_2 coated fabrics prepared by various routes

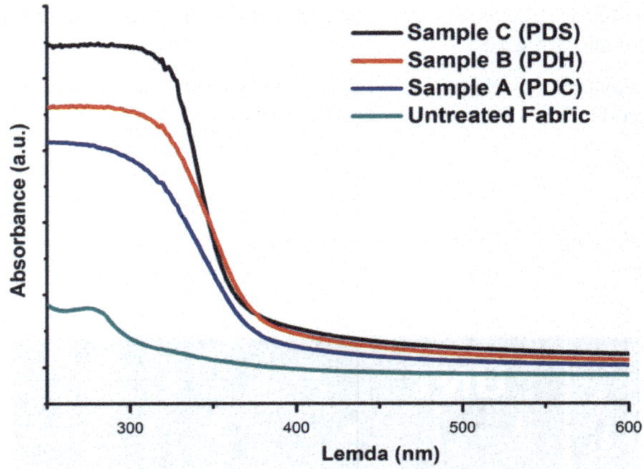

and C [16]. With respect to the TiO_2 coated fabric sample A, sample B and sample C show a comparatively higher absorption of UV light. Higher absorption of UV light facilitates the TiO_2 coating to exhibit better photocatalytic activity due to the ease in generation of charge carriers. Therefore, this could be one of the reasons for getting the best results of dye decolorization with sample C.

3.5 Decolorization Study of Rhodamine B Dye Solution

The photo-induced decolorization of dye solution was measured through UV-visible absorption spectroscopy. The characteristic wavelength maxima of Rhodamine B dye solution were observed at 553 nm. Therefore, change in absorbance of the dye solution at 553.6 nm peak was used to calculate the decolorization % of dye solution.

During the photocatalytic degradation of Rhodamine dye solution, the characteristic absorption band of RhB at 553.6 nm decreases. Later, in addition to this a hypsochromic shift of absorption maxima also occurs from 553.6 to 534 nm. This shift in wavelength can be regarded as the formation of many *N*-de-ethylated intermediates of RhB dye molecule in the process of its photocatalytic degradation [3].

The percentage decolorization of Rhodamine B solution in the presence of cotton fabric as a function of UV light exposure is represented in Fig. 6.

Another experiment was separately performed to study the effect of exposure of UV light directly on Rhodamine B solution of the same concentration in the absence of fabric. However, a direct exposure of light in dye solution in the absence of TiO_2 catalyst did not show any change in absorbance peak of dye solution. This confirms that aqueous solutions of RhB dye are stable in UV radiation.

Under UV light, the dye solution shows a significant decrease in the absorbance value in the presence of TiO_2 coated cotton fabrics. The absorbance peak of Rhodamine B dye solution at wavelength value of 553.6 nm has been made the basis of

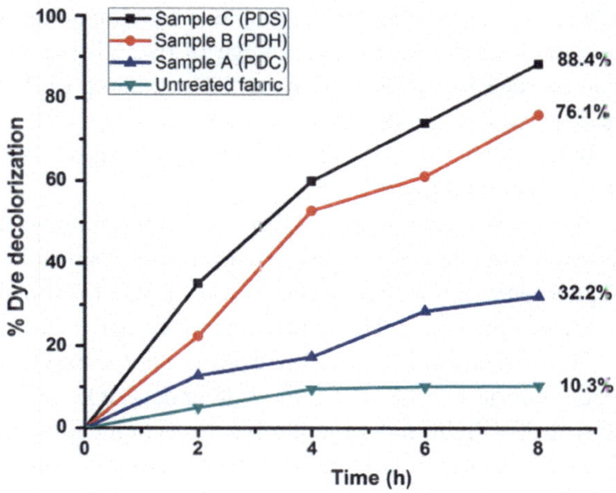

Fig. 6 Dye decolorization study of Untreated and TiO_2 coated samples under UV light

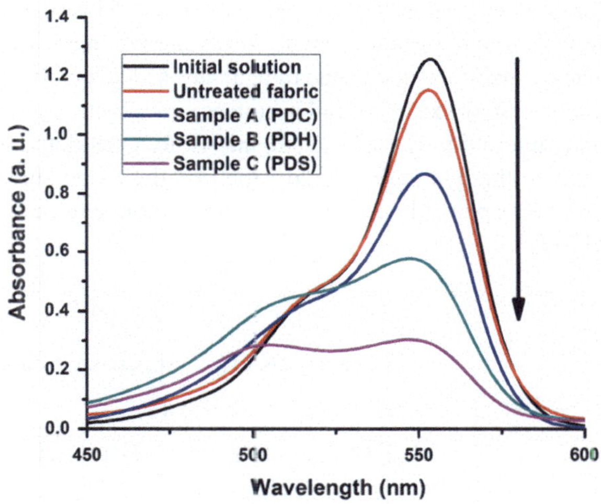

Fig. 7 Dye decolorization study of Untreated and TiO_2 coated samples under UV light

calculation of total dye decolorization in the entire study. Figure 7 represents continuous curves between absorbance and wavelength for all the samples. The ultimate decrease in absorbance value of the dye solutions after a complete cycle of exposure of 8 h under UV light has been recorded. Using the calibration curve, it has been found that in comparison with untreated fabric, the TiO_2 coated fabric sample prepared via PDC route (sample A) showed a total decolorization of 32.2% after a period of 8 h. Further, in comparison with untreated fabric, samples B and C showed a total decolorization of 76.1 and 88.4%, respectively.

Among the TiO_2 coated samples, sample A showed the lowest percentage of overall dye decolorization. As shown in Table 1, the Ti content in samples A and B is comparable to each other. Therefore, Ti content cannot be considered as a cause for this effect. The other reasons like the difference in the XRD pattern and

UV absorbance behavior of sample A in comparison with samples B and C can be assigned for the lowest percentage of overall dye decolorization in sample A. As can be seen from the XRD pattern, the coating of TiO_2 is amorphous in sample A and hence the pad-dry-cure treatment ends up with a low degree of photocatalytic activity. Also, the UV light absorption of sample A was the lowest among all the TiO_2 coated fabrics (Fig. 5).

In contrast, the coating of TiO_2 has been stabilized effectively in samples B and C, where sufficient crystallinity in TiO_2 coating is developed by prolonged exposure to higher temperature. The other obvious reason for showing the highest decolorization % by sample C may be the highest percentage of titania deposited in sample C.

The overall mechanism involved in photocatalytic degradation of Rhodamine B dye solution can be understood as under-TiO_2 nano-particles have a tendency to absorb UV light and generate reactive chemical species. These reactive chemical species take part in mineralization of various organic species. After the exposure of UV light, the electron of the valence band of titanium dioxide present in TiO_2 coated fabric gets excited to the conduction band. This leaves behind a positively charged hole (h^+) in the valence band. The excited e^- reacts with the atmospheric oxygen to convert itself to superoxide radical anion. The positively charged hole can react with water or hydroxyl ions and produces super hydroxyl radical. Super hydroxyl ($HO_2\cdot$) and superoxide $O_2{}^{\circ(-)}$ radical anions are considered as the reactive species, which oxidize the dye molecule adsorbed on the oxide surface. These generated radicals are the responsible active species for complete mineralization of the hazardous dyes [17–20].

$$TiO_2 + h\nu \rightarrow TiO_2\,(e^- + h^+)$$

$$e^- + O_2 \rightarrow O_2^{\circ(-)} \text{ (superoxide radical anion)}$$

$$h^+ + O_2^{\circ(-)} \rightarrow HO_2(\circ)$$

$$HO_2(\circ) + HO_2(\circ) \rightarrow H_2O_2 + O_2$$

$$H_2O_2 \rightarrow 2HO(\circ)$$

$$\text{Dye} + O_2^{\circ(-)} \rightarrow \text{Degradation of Dye}$$

$$\text{Dye} + HO(\circ) \rightarrow \text{Degradation of Dye}$$

3.6 Reusability of Samples for Dye Decolorization

The repeated use of the dye decolorization ability of TiO_2 coated samples is essential for its practical viability. Therefore, the evaluation of the reusability of the samples was carried out. At the end of 1st decolorization cycle, the fabric samples were washed with deionized water and dried at 50 °C before their use for the next cycle. The results of the three consecutive cycles of dye decolorization with all the fabric samples are shown in Fig. 8. In comparison with the decolorization of 76.1 and 88.4% in 1st cycle,

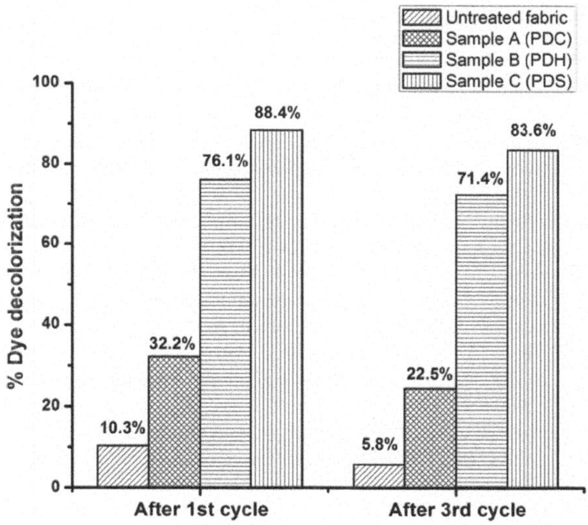

Fig. 8 Percentage dye decolorization of fabric samples after 1st and 3rd cycle

samples B and C showed 71.4 and 83.6% decolorization in the 3rd repeated cycle. However, there was a huge reduction in decolorization ability of sample A from 32.2% in the 1st cycle to 22.5% in the 3rd cycle. This reduction in decolorization ability of sample A may be assigned to the regular loss in titania content on the fabric after each cycle. In contrast, such loss in decolorization efficiency was not shown by samples B and C probably due to the more stable TiO_2 coating on these samples.

4 Conclusions

The present study demonstrates coating of TiO_2 on cotton fabric using its precursor. This has been done using three different process routes, i.e., pad-dry-cure, pad-dry-hydrothermal and pad-dry-solvothermal. The morphological structures of the coatings confirmed that the process routes play a critical role in deposition behavior of TiO_2 on fabric surface. The extent of titania deposition depends upon the process route followed, which in turn determines the extent of dye decolorization. The overall dye decolorization % also depends on the UV light absorption behavior of the prepared samples. The results showed that a uniform coating of TiO_2 on cotton fabric prepared by hydrothermal or solvothermal route gives better photocatalytic decolorization of Rhodamine B dye solution.

References

1. Vinu, R., Akki, S.U., Madras, G.: Investigation of dye functional group on the photocatalytic degradation of dyes by nano-TiO_2. J. Hazard. Mater. **176**, 765–773 (2010)
2. Chatterjee, S., Tyagi, A.K., Ayyub, P.: Efficient photocatalytic degradation of rhodamine B dye by aligned arrays of self-assembled hydrogen titanate nanotubes. J. Nanomater. **2014**, 7 p (2014)
3. Aarthi, T., Madras, G.: Photocatalytic degradation of rhodamine dyes with nano-TiO_2. Ind. Eng. Chem. Res. **46**(1), 7–14 (2007)
4. Abdel-Messih, M.F., Ahmed, M.A., El-Sayed, A.S.: Photocatalytic decolorization of Rhodamine B dye using novel mesoporous SnO_2–TiO_2 nano mixed oxides prepared by sol–gel method. J. Photochem. Photobiol. A **260**, 1–8 (2013)
5. Yu, K., Yang, S., He, H., Sun, C., Gu, C., Ju, Y.: Visible light-driven photocatalytic degradation of Rhodamine B over $NaBiO_3$: Pathways and mechanism. J. Phys. Chem. A **113**, 10024–10032 (2009)
6. Ayed, L., Chaieb, K., Cheref, A., Bakhrouf, A.: Biodegradation of triphenylmethane dye Malachite Green by *Sphingomonas paucimobilis*. World J. Microbiol. Biotechnol. **25**(4), 705–711 (2009)
7. Li, L., Dai, W., Yu, P., Zhao, J., Qu, Y.: Decolorisation of synthetic dyes by crude laccase from *Rigidoporus lignosus*. J. Chem. Technol. Biotechnol. **84**(3), 399–404 (2009)
8. Mishra, G., Parida, K.M., Singh, S.K.: Solar light driven Rhodamine B degradation over highly active β-SiC–TiO_2 nanocomposite. RSC Adv. **4**(25), 12918–12928 (2014)
9. Chen, Y., Li, Y., Zhu, A., Huang, Y., Liu, Z., Yan, K.: Degradation of aqueous Rhodamine B by plasma generated along the water surface and its enhancement using nanocrystalline Fe-, Mn-, and Ce-doped TiO_2 films. Environ. Sci. Pollut. Res. **21**(16), 9948–9958 (2014)
10. Di Paola, A., Cufalo, G., Addamo, M., Bellardita, M., Campostrini, R., Ischia, M., Ceccato, R., Palmisano, L.: Photocatalytic activity of nanocrystalline TiO_2 (brookite, rutile and brookite-based) powders prepared by thermohydrolysis of $TiCl_4$ in aqueous chloride solutions. Colloids Surf. A **317**(1), 366–376 (2008)
11. Chen, X., Mao, S.S.: Titanium dioxide nanomaterials: synthesis, properties, modifications, and applications. Chem. Rev. **107**(7), 2891–2959 (2007)
12. Fujishima, A., Rao, T.N., Tryk, D.A.: Titanium dioxide photocatalysis. J. Photochem. Photobiol. C **1**(1), 1–21 (2000)
13. Winkler, J.: Nano-scaled titanium dioxide–properties and use in coatings with special functionality. In: Macromolecular Symposia, pp. 317–324. Wiley-VCH, Verlag (2002)
14. Zhang, Q., Joo, J.B., Lu, Z., Dahl, M., Oliveira, D.Q., Ye, M., Yin, Y.: Self-assembly and photocatalysis of mesoporous TiO_2 nanocrystal clusters. Nano Res. **4**(1), 103–114 (2011)
15. Pranaitytė, B., Padarauskas, A., Naujalis, E.: Application of ICP-MS for the determination of trace metals in textiles. Chemija **18**, 16–19 (2007)
16. Koziej, D., Fischer, F., Kranzlin, N., Caseri, W.R., Niederberger, M.: Nonaqueous TiO_2 nanoparticle synthesis: a versatile basis for the fabrication of self-supporting, transparent and UV-absorbing composite films. ACS Appl. Mater. Interfaces **1**(5), 1097–1104 (2009)
17. Akpan, U.G., Hameed, B.H.: Parameters affecting the photocatalytic degradation of dyes using TiO_2-based photocatalysts: a review. J. Hazard. Mater. **170**, 520–529 (2009)
18. Mahmoodi, N.M., Arami, M., Limaee, N.Y., Tabrizi, N.S.: Kinetics of heterogeneous photocatalytic degradation of reactive dyes in an immobilized TiO_2 photocatalytic reactor. J. Colloid Interface Sci. **295**, 159–164 (2006)
19. Hasnat, M.A., Siddiquey, I.A., Nuruddin, A.: Comparative photocatalytic studies of degradation of a cationic and an anionic dye. Dyes Pigm. **66**(3), 185–188 (2005)
20. Laid, N., Bouanimba, N., Zouaghi, R., Sehili, T.: Comparative study on photocatalytic decolorization of an anionic and a cationic dye using different TiO_2 photocatalysts. Desalin. Water Treat. **57**(41), 19357–19373 (2016)

Chemical Modification of Indian Yak Fibre for Development of Jute/Yak Fibres Blended Warm Textile

Kartick K. Samanta and A. N. Roy

Abstract Yak wool, an important speciality protein fibre, is mostly available in China, Mongolia and India. The yak fibre is commonly categorized into down or fine fibre, guard or coarser fibre and middle type fibre based on their fibre fineness. A large quantity of coarser yak fibre, which is quite stiffer and thicker, remains underutilized including its application in textile. In our work, the coarser grade black colour yak (fibre) wool of 9.3 tex was suitably chemically modified for blending with jute (ligno-cellulosic) fibre to develop blended textile with jute to yak fibres blend ratio of 25:75. After chemical modification, the coefficient of friction between the yak fibre and the metal was found to enhance to 0.368 from 0.280 in the unmodified hair sample, which is closer to the coefficient of friction of jute fibre. This facilitates in developing jute and yak fibres blended yarn with higher percentage of yak fibre content in the yarn. Indeed after modification, the natural black colour of yak fibre turned into a colour alike to ligno-cellulosic jute fibre with a marginal decrease in breaking load (12%) and linear density (16%). Scanning electron microscope (SEM) and EDX elemental analysis showed no detrimental changes in surface morphology and chemical composition of the yak wool after requisite chemical modification. Different fabrics were developed from the jute and yak wool blended un-dyed and dyed yarns. The jute–yak fibres (50:50) blended woven fabric of 385 g/m^2 areal density showed a thermal insulation value of 1.4 (Tog).

Keywords Yak fibre · Jute fibre · EDX of hair · Jute–yak fibres textile

1 Introduction

Yak (*Bos grunniens*), an important domesticated bovid animal, plays an important role in the Indian economy of the tribal population living in the foothills of Himalayas [1]. The products and services provided by the yak animal are milk, meat, leather and

K. K. Samanta · A. N. Roy (✉)
ICAR-National Institute of Natural Fibre Engineering and Technology, 12 Regent Park, Kolkata 700040, India
e-mail: a_n_royin@yahoo.co.in

A. Majumdar et al. (eds.), *Functional Textiles and Clothing*,
https://doi.org/10.1007/978-981-13-7721-1_14

wool that are used for clothing, blankets, bags, implements, rugs and tents [1]. As far as the fibre (hair) part is concerned, it is a specialty animal fibre. The total population yak animal is approximately 14.5 million, and the yield of wool is approximately 410,000 tons per year; out of which 10,000 tons is down wool [2–5]. The yak fleece processes a large quantum of guard/coarse hair, which is quite coarser and stiffer in nature [5, 6]. A yak animal generates approximately 100 g of fine hair fibre (diameter of 16–20 μ) annually that comes with few natural colours, out of those white yak hair is the valuable one [7]. Yak herds are mainly available in the mountainous areas of China, Mongolia, India, Bhutan, Afghanistan, Kyrgyzstan, Russia and Nepal on the Central Asian Plateau. Yaks' coarser hair fibre as well as the fine fibre is seasonal character in nature that is gradually shed and replaced with the newer one, ensuring a requisite thermal insulation to the animal during the cold season [3]. The yak coat is consisting of three different types of fibres, which are greatly varied in properties throughout the seasons [7]. The coarse yak hair fibre (70–90 μ) contributes in forming the outer coat of guard hair fibre that also represents the appearance of the yak; the fibre is mostly used in tent making by the nomads. Down hair (fibre) with a length of 35–50 mm and approximately diameter of 16–20 μ is mostly used for textile application; such hair fibre is generally shed during the late spring or early summer [8]. On the other hand, the middle type yak wool fibre with a diameter of 20–50 μ shows properties in between the finer and the coarser fibre and also used for ropes and tents production. Fabric made of yak down fibre has good lustre and high degree of heat insulation [9]. Naturally occurring hair fibres, such as human and animal hairs consist of α-keratin protein in a filamentous structure [10, 11]. Fine yak fibre has reasonably good crimp and tensile strength (9.18 cN/tex) [4]. The yak animals as well as the hairs are available in four different colours, namely black, white, brown and blue; out of which black colour hair fibre has the major share [7]. Black colour yak hair fibre is some extent alike to human hair that is mainly consists of 65–95% keratotic protein, natural pigments, lipids, trace elements and water [12, 13]. Yak wool fibre was used for making tents, ropes, clothing and blankets by the nomads in the Trans-Himalayan region. In recent time, the yak wool fibre has been utilized to produce premium textile by well-known companies/brands likes, Louis Vuitton, British Heritage Brand Inc., Eileen Fisher and Dunhill [7]. Since the mid-twentieth century, emphasis has been given for encouraging the use of yak wool in the garment industry due to its exotic attribute and favourable performance. In all such above applications, mainly the down or finer fibres have been utilized. On the other hand, due to the higher diameter and rigidity, the coarser yak fibre is typically carded and then spun in worsted yarn production system [13]. Multiple plied guard hair threads are then subsequently braided or weaved into ropes, halters, belts, bags and rugs.

Jute, a ligno-cellulosic vegetable bast fibre, is presently being used as a biodegradable packaging material for food grains and sugar. The traditional usage of jute fibre, i.e. meant for packaging application, has undergone a substantial change in the last few decades. Some of the recent technological advancements and possible applications in this direction are: (i) jute geo-textile, (ii) jute agro-textile, (iii) jute composite in substitute of wood/plastic, (iv) handmade paper and (v) ornamental and decorative items, like shopping bags, jute shoes, jewellery and lifestyle products, either from

100% jute or in suitable blending with other fibres, such as cotton, viscose, ramie, flax, silk, wool, polyester and acrylic [14–16]. The inherent golden colour of jute fibre adds an elegant attribute and aesthetic appeal to the jute products. Jute fibre has high initial modulus along with low extensibility, and it is produced from the plant, genus *Corchorus* [17]. In the recent time, various high valued jute blended textile products, meant for apparel applications have been developed viz., use of jute in a sari, warm garment and jackets, knitted items, and many more after suitable physical and/or chemical intervention of jute or in the fibre processing system. Almost 85% of the world's jute cultivation is concentrated in the Ganges Delta. In the past, several research and developments on jute fibre have been carried out for its commercial application as apparel textile. However, such whole jute products have realized some degree of success owing to its limitation of itchiness, prickling effect, higher drape and higher bending rigidity.

Blending of jute fibre with yak wool fibre will help to exploit the positive attributes of both the individual fibres to produce finer and stronger yarns with a warm feel effect. Moreover, the jute fibre has the potential for colouration in different shades to produce a blended yarn with the decorative look, which can help in the production of fabrics suitable for warm garments with a premium price. On the other, the stiffness of jute yarns can be masked by blending with softer yak fibre to produce a fabric with good handle property. At present, India produces around 150–250 tonnes of guard (coarser) and 20–25 tonnes of down (fine) yak fibres. In our institute, an effort was made to develop jute–yak fibres blended warm textile to produce high-value outer garments that could be used in the cold climatic location. Coarser grade yak fibre mostly does not utilize for high-value products, because of its higher diameter and stiffness along with a high slippery surface that makes it difficulty in developing fine textile yarn, either from the 100% yak hair fibres or in suitable blending with other natural fibre. Also, there is no report on blending of yak wool with ligno-cellulosic jute fibre to produce fashionable textile to the best of our knowledge. It was found that after suitable chemical modification of black colour coarser yak fibre, it was possible to produce jute–yak fibres blended yarn with a higher percentage of yak fibre content in the blended yarn.

2 Experimental

2.1 *Materials and Modification of Yak Fibres*

Coarser yak fibre with natural black colour was received from ICAR-National Research Centre (NRC) on Yak, Dirang, Arunachal Pradesh, and the jute fibre was procured from the local market of Kolkata. For the purpose of production of jute–yak fibres blended yarn as well as textile, yak hair fibre was initially scoured and then successively bleached with an oxidative bleaching agent as per the standard recipe

Table 1 Yak hair fibre subjected to different dosages of chemical treatment in addition to the standard recipe [18]

Yak hair fibre	NaOH addition (g/l)	H_2O_2 addition (o.w.f.) (%)	Time (min)
Sample 1	0.75	11	60
Sample 2	2.3	27	110

[18]. Additionally, following dosages of chemicals were given to produce modified yak fibres with different degree of coefficient of friction and colour (Table 1).

2.2 *Evaluation of Physical Properties*

Instron Tensile Tester (Model 5567) was used to measure the physical properties, such as breaking tenacity, elongation percentage and initial modulus of yak hair fibres. Initial modulus and tenacity were measured using the data of stress–strain curve and fibre linear density. The reported data are the average of sixty measurements. Stress–strain curve is plotted for different yak fibres using one of such measurement. The gravimetric method was used to measure the fibre fineness in terms of 'tex'. Photographs of the different yak fibres were taken using Canon 5D Mark III SLR camera. Moisture regain of hair fibre samples was calculated by gravimetric method. Hair fibre to metal coefficient friction was measured using a friction tester that works on inclined plane principle as reported in literature [19]. Different colour parameters viz., L, a^*, b^*, K/S, C, H, R, G and B values were measured using computerized colour matching system (Model Spectrascan–5100). Colour depth of the different fibre samples was measured by measuring the colour strength (K/S) obtained from the reflectance data following the Kubelka–Munk equation. Surface micrograph of the different yak fibres and jute fibre was studied using SEM, Philips XL-30. The X-ray diffractometer (Model Philips PW 1877) was used to estimate the degree of crystallinity of the hair fibre samples.

2.3 *Evaluation of Chemical Property*

Field emission gun (FEG) scanning electron microscope, Model: JSM-7600F with energy dispersive X-ray (EDX) attachment, was used to measure the elemental composition of the different hair fibre samples.

3 Results and Discussion

3.1 Physical Properties of Fibre

As yak hair is a protein fibre, it possesses lots of surface impurities, such as vegetable matter, fatty matter, dust and dirt on its surface. It was observed that there was a 4.1% weight reduction due to scouring process of yak hair due to the removal of above surface impurities. Likewise presence of around 2.5–3.1% fat and dirt mixture has also been reported in literature for similar fibre [4].

Moisture regain of the untreated and chemical modified yak hair fibres was measured by gravimetric method after drying the samples at 110 °C. It was observed that an untreated yak fibre showed 15.6% moisture regain, and it improved a little (16.5%) after chemical treatment. Such small improvement in moisture regain value was possibly occurred due to removal of surface fatty layer as well as other impurities viz., dust, dirt and suint. The presence of many hydrophilic side polymer chains and peptide bonds of keratin polymer helps the sample to be hydrophilic in nature along with higher moisture regain value [12]. The optical and microscopic images of the different yak and jute fibres are represented in Fig. 1. An untreated yak hair fibre has a black colour; however, after chemical modification in alkaline condition in the presence of hydrogen peroxide (H_2O_2), its colour became blackish brown. With the increasing either alkali, hydrogen peroxide or time, the sample colour slowly turned into deep yellowish brown colour (golden) and the colour could be compared with ligno-cellulosic jute fibre. It can be seen from Table 2 that an untreated hair sample has linear density of 9.3 tex. The value slowly reduced with chemical treatment in the presence of alkali and bleaching agent. This results in 14% change in fibre linear density. The tenacity, breaking load, elongation percentage and initial modulus of yak hair fibres were measured in Instron Tensile Tester and the reported data are the average of sixty measurements. The stress–strain percentage curve of different yak hair fibres has been depicted in Fig. 2 using one of such measurement. From Fig. 2, it can also be seen that initial modulus of the modified fibre two is lower than the untreated sample. It was seen that after chemical treatment, elongation percentage gets reduced. The fibre strength was observed to decrease a little after chemical modification of yak hair fibre.

Fibre length was measured manually using a scale and the length was found to reduce to 11.2 cm in the chemically modified sample from 18.8 cm in the untreated. It may be noted that the fibre length of 11.2 cm was suitable for development of jute–yak fibres blended yarn. As protein fibre is soluble in alkaline solution; therefore, the presence of little amount of sodium hydroxide during chemical modification with temperature near about 80 °C might have removed the surface polymer chain molecules that also lead reduction in hair linear density along with breaking load (12% for sample 1). Indeed, the sample 1 with desired frictional value was suitable for production of jute and yak fibres blended textile yarn in jute fibre processing machines sequence as discussed in successive section.

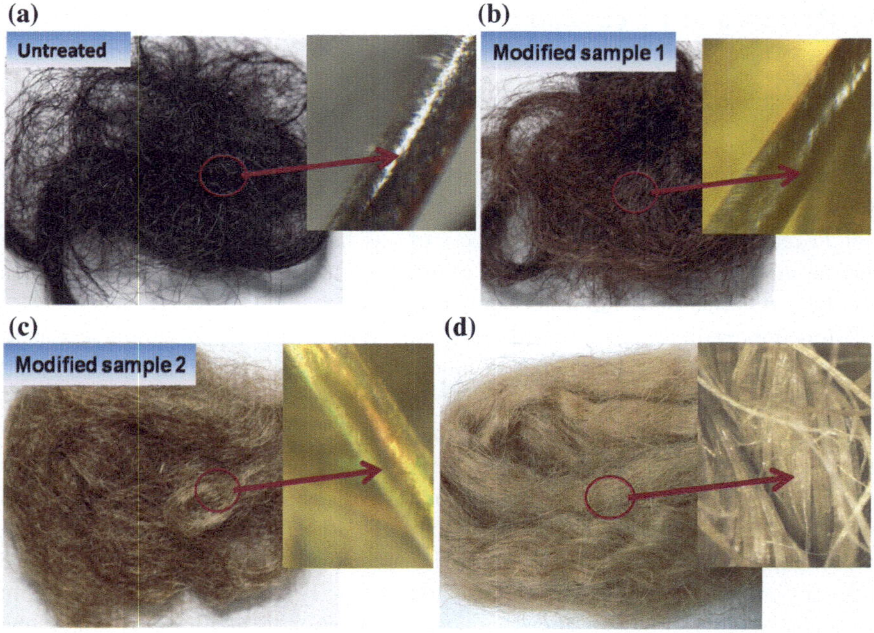

Fig. 1 Pictures of **a** untreated yak fibre, **b** treated yak fibre 1, **c** treated yak fibre 2 and **d** jute fibre

Table 2 Physical properties of jute and yak hair fibres [18]

Different samples	Linear density (tex)	Breaking load (N)	Tenacity (cN/tex)	Initial modulus (cN/tex)	Elongation (%)	Fibre length (cm)
Untreated yak fibre	9.3	0.97	10.4	428	25	18.8
Modified yak fibre: sample 1	7.8	0.85	10.8	455	19	17.6
Modified yak fibre: sample 2	8.0	0.65	8.0	408	6	11.2
Jute fibre (carded sliver)	0.95–4	–	40	500–600	1–2	9–15 (carded sliver)

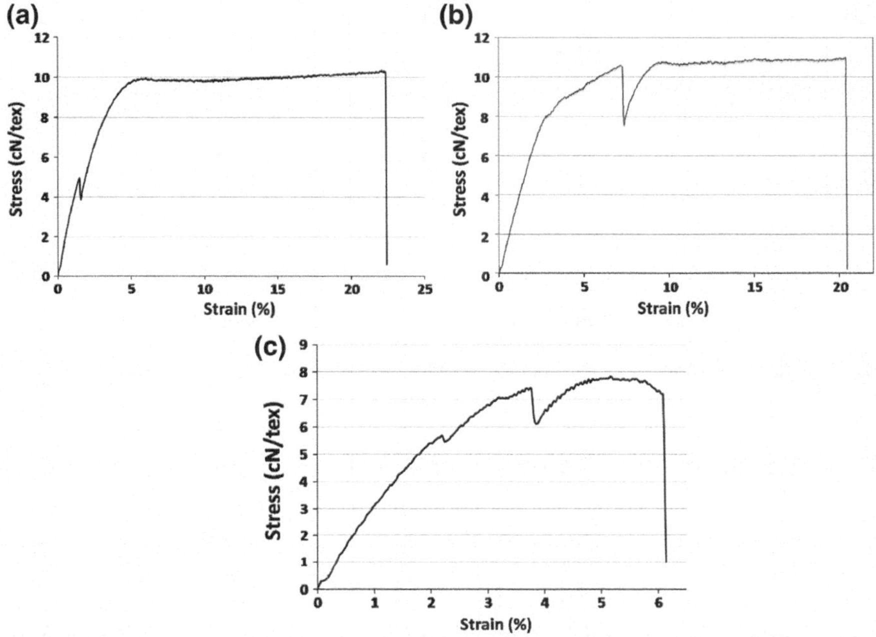

Fig. 2 Stress–strain percentage curves of different yak hair fibres **a** untreated, **b** modified sample 1 and **c** modified sample 2

It was observed that an untreated yak hair fibre possesses very low surface coefficient of friction value of 0.28 possibly due to presence of lipid film (polysaccharide) [10]. This indicates that the hair fibre surface is quite slippery and smoother. There was no noticeable increase in surface coefficient of friction after scouring in the presence of soap and mild alkali. In Sample 1, the coefficient of friction value increased to 0.368 possibly due to etching of fibre surface in the presence of alkali along with crimp development. This is considered as an important desirable requirement for spinning of proteinous yak hair with ligno-cellulosic jute fibre that has higher surface of friction of 0.45 and 0.39, respectively, in the parallel and perpendicular directions [7]. It was observed that with further enhancement in bleaching dosage, no further enhancement in coefficient of friction was noticed.

X-ray Diffraction Analysis

X-ray diffraction was performed in the untreated and modified (samples 2) yak hair fibres in order to investigate the crystalline and amorphous content in these hair fibres. Both the samples did not show any major diffraction peak due to the presence of crystallographic plane or structure as shown in Fig. 3. This implies that both these samples either treated or untreated are quite amorphous in nature. This is a quite different phenomenon as compared to other protein fibres like, silk and wool that have 25–66% crystalline phase. The X-ray diffraction of jute fibre shows clear peak at 2θ values of 14.96, 16.51 and 22.65 with crystallinity percentage of 55%.

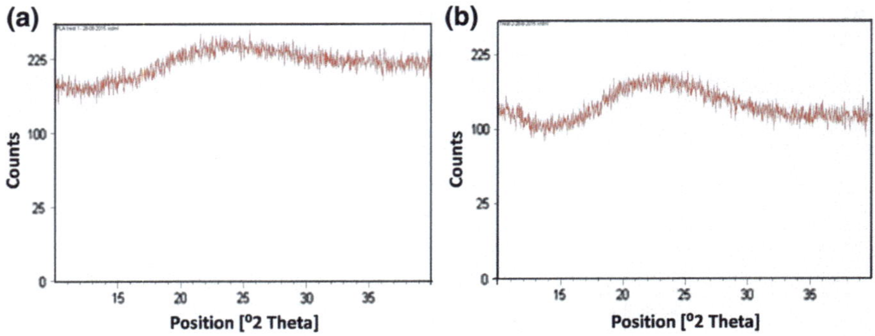

Fig. 3 X-ray diffraction curves of **a** untreated and **b** treated yak fibres

3.2 Measurement of Colour Parameters

As ascribed above, a large quantity of yak wool is obtained with inherent black colour, possibly due to the presence of melanin in the form of eumelanin and phaeomelanin pigments [2, 20]. It is quite fair to assume that black colour coarser yak hair would have a few characteristics alike to human hair. This is because both these hairs are protein fibre also black in colour, thus their properties were considered for comparison. In this study, the black colour coarser yak hair was subjected to chemical treatment to create requisite surface friction and also to change its black colour into a colour, similar to jute. Ultimately, this will help to produce jute–yak fibres blended woven fabrics. Figure 4 shows the spectroscopic reflectance percentage with wavelength for various yak hair fibres. As expected, the sample without any chemical treatment produces the lower reflectance percentage, and the corresponding spectroscopic image is shown in Fig. 5 (code 0), i.e. black in colour. Upon oxidative bleaching of yak fibre with hydrogen peroxide (H_2O_2), sample reflectance percentage was found to improve noticeably as shown in Fig. 4 (sample code 1) and the corresponding spectroscopic image showed in Fig. 5(1). With further addition of chemical agents as well as an increase in treatment time, sample colour slowly turned into blackish brown to strong reddish and deep yellowish brown as shown in Figs. 4(2) and 5(2). Similar results have also been observed in the camera-captured picture as depicted in Fig. 1. Table 3 shows the spectroscopic colour properties of various yak wool and ligno-cellulosic jute fibre samples. The lightness-darkness (L^*) was the lowest in the case of untreated yak fibre, resulting highest colour depth (K/S) of 24.3. As expected after bleaching with H_2O_2, the whiteness value notably enhanced to 34.8. The L^* value further improved to 41.4, with increasing chemical treatment parameters. Likewise in the chemical treated sample, the a^* (red–green) value increased to 8.9 from 1.4 in the untreated and the value could be compared with the a^* value of ligno-cellulosic jute fibre. Furthermore, the yellow-blue colour parameter (b^*) also notably increased as observed for the a^* value. As the a^* value (redness) was found to enhance, the red (R) value was also gradually increased from 41 to 83 and 122 in the untreated to

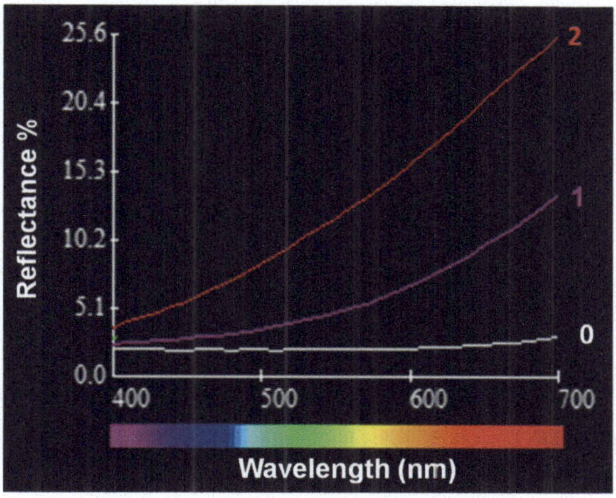

Fig. 4 Reflectance percentage in the untreated (0) and modified yak hair fibres (sample 1 and sample 2)

Fig. 5 Spectroscopic colours generated during colour measurement of various yak fibres: (0) untreated, (1) modified sample 1 and (2) modified sample 2

various treated samples. Similar trend was also noticed for green (*G*) and blue (*B*) colour parameters as shown in Table 3. Similarly, the *C* value (colour strength or saturation) increased from 1.6 in the untreated sample to 16.9 and 22.8 to modified hair samples, respectively. The *H* (a particular colour) values of the different fibre samples were measured using the spectroscopic software.

3.3 Scanning Electron Microscope Morphology

Almost all the hair fibres viz., human hair, angora, wool, pashmina, etc., are well known due to the presence of scale on their hair/fibre surface [21, 22]. Similar to the above hair fibre characteristic, an untreated yak fibre also shows the presence of scale on the surface under scanning electron microscope (Fig. 6a). After the chemical modification, the surface scale looks like smoother, but the features, i.e. the scales, were clearly visible at higher magnification (Fig. 6b) [9]. Therefore, it

Table 3 Colour parameters of the jute and various yak hair fibres [18]

S. No.	Colour parameters	Different yak fibres			Jute fibre
		Untreated sample	Modified sample 1	Modified sample 2	
		Code 0	Code 1	Code 2	
1	*K*/*S*	24.3	18.3	10.3	3.24
2	L^*	25.3	34.8	41.4	59.3
3	a^*	1.4	8.9	8.0	10.3
4	b^*	0.67	14.2	21.4	24.9
5	*C*	1.6	16.9	22.8	27.0
6	*H*	24.7	58.2	69.4	67.3
7	*R* (red colour)	41	83	122	175
8	*G* (green colour)	38	59	92	135
9	*B* (blue colour)	38	44	63	99

can be stated that the yak fibres were chemically treated to develop the requisite surface friction without much alteration in the surface properties. The change in scale morphology of yak hair has been generated possibly due negative interaction (surface etching) of alkali with hair polymer, as it was observed that in the 2.5% concentrated NaOH/KOH solution yak hair was soluble. It is interesting to note that both the untreated and the modified yak fibres cross-sections do not show presence of any medulla (internal open pore) (Fig. 6). The hair fibres are mostly elliptical in shape with notably variation in diameter due to inherent variation in hair diameter from bottom to top. The diameter of coarser yak fibre was measured to be about 50–100 μ, considering the top to bottom part of the fibre. It was also noticed that after chemical treatment with bleaching agent and alkali, fibre cross-sectional morphology remained unchanged. On the other hand, jute fibre showed pillar-like structure with polygonal shape as shown in Fig. 7. In each fibre cross-section, presence of several lumens (open pore) is also clearly visible.

3.4 EDX Elemental Analysis

Figure 8 depicts the energy dispersive X-ray images of the jute, and untreated and chemical treated yak hairs and the corresponding elemental atomic percentage of these fibre samples are reported in Table 4. Natural human hair consists of group of proteins, popularly known as keratins, which are interconnected to form stable fibrils. Such protein chains are very much complex in nature both histologically and chemically due to the presence of a multiplicity cross-linked protein molecules [18]. In hair or wool fibre, the protein part, i.e. the keratin, is composed of eighteen

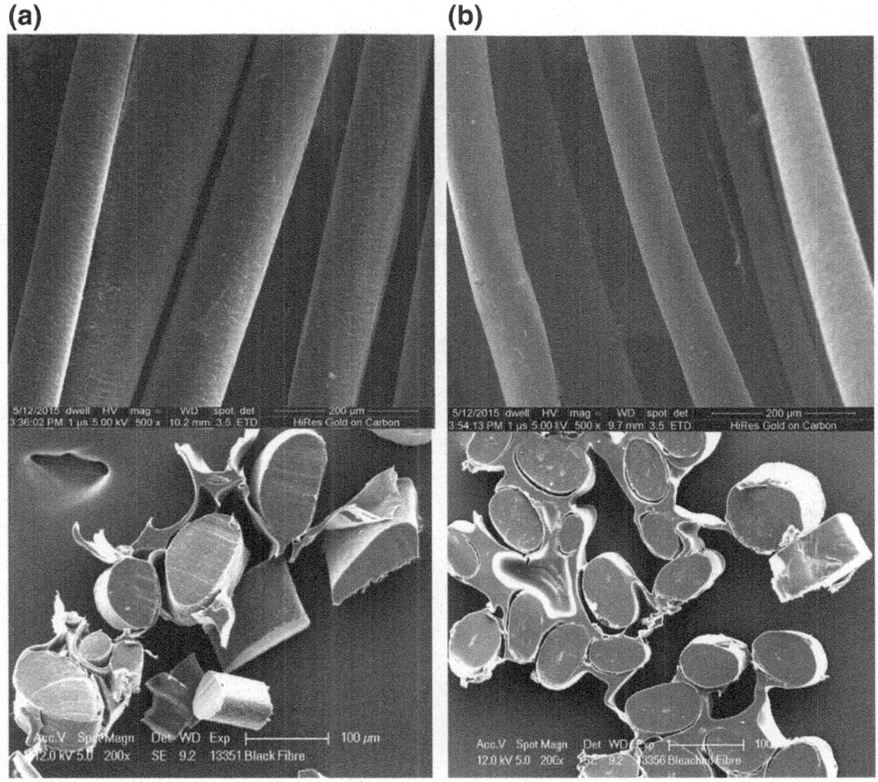

Fig. 6 Scanning electron surface and cross-sectional morphology of yak hair fibres: **a** untreated and **b** modified

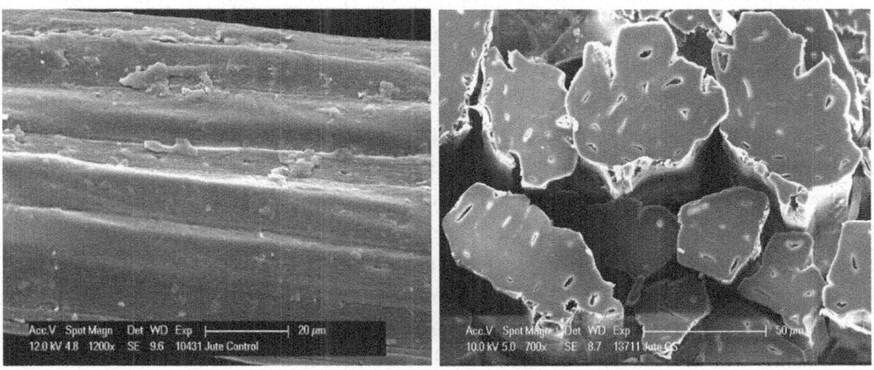

Fig. 7 SEM surface as well as cross-sectional pictures of ligno-cellulosic jute fibre [18]

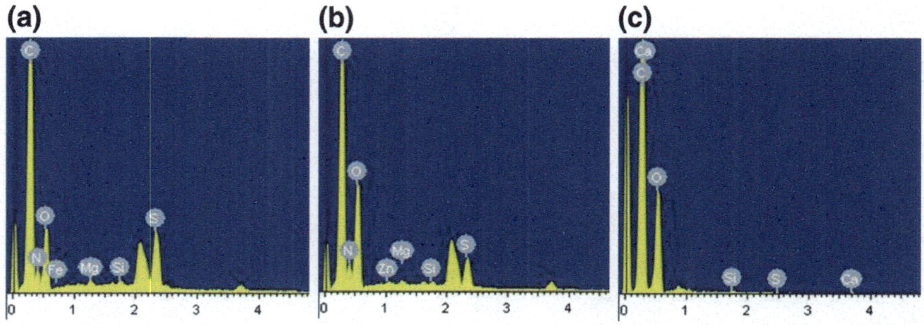

Fig. 8 EDX elemental spectra of **a** untreated yak hair, **b** modified yak hair and **c** jute fibre

Table 4 EDX atomic percentage in the various yak wools [18]

Elements	Yak fibres	
	Untreated	Treated sample 2
C	58.34	51.54
N	18.21	15.78
O	20.76	31.08
Mg	0.19	0.17
Si	0.16	0.12
S	2.26	1.26
Fe	0.10	0.00
Zn	0.00	0.05
Total	100.00	100.00

different amino acids, among those amino acids cysteine is rich in sulphur content [2]. Disulphide linkages (–S–S–) present between the adjacent keratin polymer chains improve the stability of keratin from chemical and biological degradation.

Table 4 depicts the presence of 58.34% carbon in an untreated hair sample and the value notably reduced to 51.54 in samples 2. In the yak hair polymer, elemental percentage could be arranged in the following order carbon > oxygen > nitrogen > sulphur. The EDX sample characterization technique could not detect the hydrogen element. Similarly, nitrogen percentage also decreased to 15.78 after chemical modification from 18.21 in the untreated hair sample. It may be noted that the oxygen atomic percentage profoundly increased to 31.08% after peroxide bleaching from 20.76% observed in an untreated sample, signifying an improvement in the hydrophilic characteristic of the sample. The wool and other similar hair fibres are known for presence of sulphur containing molecules as an important structural constituent of the polymer, thus the sulphur atomic percentage was analysed from the EDX data [19]. An yak hair possesses 2.26% sulphur that significantly decreased to 1.26 after chemical modification. This is happened, possibly due to chain scission of disulphide bond of cysteine and a similar phenomenon is reported for the bleached fine yak and hair fibres [13, 20]. A few more metal elements viz., zinc (Zn), magnesium (Mg), iron (Fe) and silicone (Si) were also observed, and similar elements were also observed for human hair owing to sweat deposition [12]. However, their total quantity is significantly lower as compared to other elements. On the other hand, in jute fibre only the carbon (62.7%) and oxygen (37.1%) elements were detected as major elements in EDX analysis as it is made of cellulose, hemi-cellulose and lignin.

3.5 *Development Jute–Yak Fibres Blended Textile*

Black colour coarser yak as well as ligno-cellulosic jute fibres were used to develop jute–yak fibres blended yarn (275 tex) with blend ratio of 50:50 in jute processing machineries and the tensile properties of the yarn were measured. Similar to the 50:50 jute–yak fibres blended yarn that was produced from the unmodified yak fibre, a blended yarn was also produced from the chemically modified yak, keeping yak fibre content at 75% and remaining is jute fibre. The aim was to develop jute–yak fibres blended yarn with more yak wool component. However, it was observed that during the spinning of yarn with more than 50% yak fibre in blend, there was a instability and slippage of sliver. After suitable chemical treatment of yak fibre with improved coefficient of friction, 75% yak wool fibre was possible to introduce with 25% ligno-cellulosic jute fibre in order to develop jute–yak (25:75) fibres blended

Fig. 9 Picture of a jacket produced from jute–yak fibre dyed yarns with blend ratio of 50:50

textile yarn. These yarns were thereafter dyed in different shades and utilized for production of 1 × 1 plain woven fabrics in handloom [18]. The jute-yak fibres blended woven fabrics were utilized for developing apparel product as shown in Fig. 9. In this fabric, black colour polyester yarn was used as a warp yarn and the jute–yak fibres blended yarn (50:50) was considered as a weft yarn. The woven fabric of 385 g/m^2 areal density prepared from the 50:50 jute–yak wool fibres blended yarn showed a thermal insulation value of 1.4 (Tog), whereas the twill design fabric made from the same yarn showed a little more thermal insulation value.

4 Conclusion

After the chemical intervention of black colour coarser yak wool fibre with alkali and bleaching agent, it was possible to convert it to strong brown yellow colour, a similar colour to ligno-cellulosic jute fibre. In the treated yak wool, fibre to metal coefficient of friction was observed to increase to 0.368 from 0.280 in an untreated yak hair. The enhancement in surface friction could meet one of the desirable requirements of fibre spinning in making blended yarn with jute fibre. In the treated fibre samples, tenacity at break remains almost unchanged as compared to the untreated sample. There was a little decrease in breaking load of the chemically modified yak hair fibre; a similar result was also observed for fibre linear density. Sulphur atomic percentage gets reduced from 2.26% in the untreated sample to 1.26% in the chemical treated samples, owing to chain scission of cystine linkage. Yak fibre was successfully blended with jute fibre in jute processing machineries to produce yarns with blend ratios of 50:50 and 75:25. Finally, these yarns were utilized for developing plain woven fabrics for apparel application. The jute–yak fibres blended textile products viz., overcoat, blazer and jackets could be used as warm garment.

References

1. Ramesha, K.P.: Yak-Moving Treasure of the Himalayas. 1st edn. Published by Director, National Research Centre on Yak, Arunachal Pradesh, India (2008)
2. http://www.hair-science.com/. Last accessed 20 May 2016
3. Danzan, B., Tsedev, K., Luvsandorj, N.: The shedding and growth dynamics of yak down wool and links to habitat ecological condition. Asian J. Agric. Rural Dev. **4**, 156–161 (2014)
4. Gongor, N., Ganbaatar, B., Tsogsaikhan, G., Khuanbai, Z.: Mongolian Yak—A Speciality Fiber: Developing Textile Made from Yak Fine Fibre, pp. 1–3 (2014). Last accessed 30 Jan 2017. http://180.211.172.109/ifost2014Pro/pdf/S6-P281.pdf
5. Liu, J., Weng, Y.: Effect of stretching slenderization treatment for microstructure of yak fibre. Adv. Mater. Res. **468–471**, 1231–1234 (2012)
6. Bergen, W.V.: Musk-ox wool and its possibilities as a new textile fibre. Melliand Text. Mon. **III**(8), 646–648 (1931)
7. Roy, A.N., Samanta, K.K.: Modification of coarser yak fibre for blending with jute fibre in jute spinning system. In: 4th Interface Meeting on Holistic Approaches to Sustain Livelihood of

Yak Rearers Through Scientific Intervention in India, pp. 61–76. National Research Centre on Yak, Arunachal Pradesh, India (2016)
8. Lu, W., Fei, J., Yang, J., Tang, M., Dong, Z., Zhou, Z., Ye, J., Wu, H.: A novel method to identify yak fiber in textile. Text. Res. J. **83**, 773–779 (2013)
9. Samanta K.K., Roy, A.N.: Modification of coarser yak hair fibre for jute blended textile production. In: Das, P.J., Bam, J., Paul, V., Medhi, D., Roy, A.N., Deb S.M. (eds.) The Yak Wool, pp. 37–75. National Research Centre on Yak, Arunachal Pradesh, India (2017). ISBN: 978-93-5291-419-7
10. Barton, P.M.J.: A forensic investigation of single human hair fibre using FTIR-ATR spectroscopy and chemometrics. Ph.D. thesis, pp. 1–337. Queensland University of Technology (2011)
11. Pudney, P.D.A., Bonnist, E.Y.M., Mutch, K.J., Nicholls, R., Rieley, H., Stanfield, S.: Confocal Raman spectroscopy of whole hairs. Appl. Spectrosc. **67**, 1408–1416 (2013)
12. Robbins, C.R.: Chemical and physical behavior of human hair, 5th edn., vol. XXIII, pp. 105–176. Springer-Verlag, Berlin, Heidelberg, USA (2012). ISBN: 978-3-642-25610-3
13. Akhtar, W., Edwards, H.G.M., Farwell, D.W., Nutbrown, M.: Fourier-transform Raman spectroscopic study of human hair. Spectrochim. Acta Part A Mol. Biomol. Spectrosc. **53**, 1021–1031 (2013)
14. Basu, G., Roy, A.N.: Blending of jute with different natural fibres. J. Nat. Fibers **4**(4), 13–29 (2008)
15. Roy, A.N., Basu, G.: Improvement of a traditional knowledge by development of jacquard shedding based handloom for weaving ornamental jute fabric. Indian J. Tradit. Knowl. **9**(3), 585–590 (2010)
16. Sengupta, S., Debnath, S.: A new approach for jute industry to produce fancy blended yarn for upholstery. J. Sci. Ind. Res. **69**, 961–965 (2010)
17. Ammayappan, L.: Eco-friendly surface modification of wool fibre for its improved functionality: an overview. Asian J. Text. **3**(1), 15–28 (2013)
18. Roy, A.N., Samanta, K.K., Patra, K.: Physico-chemical properties of black yak fibre and its modification for blending with jute fibre. J. Nat. Fibres **16**(2), 225–236 (2019)
19. Mohan, N.H., Debnath, S., Mahapatra, R.K., Nayak, L.K., Baruah, S., Das, A., Banik, S., Tamuli, M.K.: Tensile properties of hair fibres obtained from different breeds of pigs. Biosys. Eng. **119**, 35–43 (2014)
20. Yan, K., Hocker, H., Schafer, K.: Handle of bleached knitted fabric made from fine yak hair. Text. Res. J. **70**, 734–738 (2010)
21. Chattopadhyay, S.K., Bhaskar, P., Ahmed, M., Gupta, N.P., Plkharna, A.K.: Properties of indigenous angora rabbit hair and cotton blended yarns using short staple cotton spinning system. Indian J. Fibre Text. Res. **30**, 215–217 (2005)
22. Gupta, D., Basak, S.: Surface functionalisation of wool using 172 nm UV excimer lamp. J. Appl. Polym. Sci. **117**, 223–227 (2010)

Statistical Optimisation of Nano-Zinc Oxide-Based Fire-Protective Finish on Jute Fabric

Ashis Kumar Samanta, Reetuparna Bhattacharyay (Roy), Arindam Bagchi and Ranjana Chowdhuri

Abstract Emulsified mixture of nanoparticles of zinc oxide (ZnO) and poly-hydroxymethyl amino silicone (PHAMS) is padded on bleached jute fabric followed by drying and curing to develop fire-retardant finish on jute-based fabrics. After a preliminary study made and based on those results, the present study concentrates on statistical optimisation of two input variables such as concentrations of nano-ZnO and PHAMS applied on jute fabric by using response surface methodology (RSM) with user-defined quadratic model (UDQM) for determining optimum concentrations of said two input variables. Resultant variables considered for such optimisation are LOI, char Length, loss in fabric tenacity. Preliminary study shows that application of mixture of nano-ZnO particles (0.01% w/w) with PHAMS (10% w/w) on jute fabrics renders reasonable fire-retardant performance in terms of LOI value and char length. While after statistical optimisation, with UDQM experimental design method, optimum concentrations of nano-ZnO and PHAMS are determined to be 0.007 and 13.7%, respectively, which finally results LOI: 37.71, char length: 44.01 mm or 4.4 cm and loss in fabric Tenacity 17.35% with reasonable good wash fastness up to five cycles of washing. Nano-ZnO particle so produced is characterised in our earlier publication. In this part, presence of nano-ZnO and its particle size on the surface of treated jute fabrics are proven by IR spectra (FTIR), pXRD/EXD, SEM and analysis of particle size distribution. Thus, nano-ZnO finish with PHAMS renders acceptable level of fire-protective finish of jute fabrics at the said optimum concentrations.

Keywords Critical oxygen index (LOI) · Fire-protective finish · Jute · Length of char · Poly-hydroxymethyl amino silicone (PHAMS) · ZnO-nanoparticle (ZnO-NP)

A. K. Samanta (✉) · R. Bhattacharyay (Roy) · A. Bagchi
Department of Jute and Fibre Technology, Institute of Jute Technology, University of Calcutta, Kolkata 700019, West Bengal, India
e-mail: ijtaksamanta@hotmail.com

R. Chowdhuri
Chemical Engineering Department, Jadavpur University, Raja Subodh Mallik Road, Kolkata 700032, West Bengal, India

A. Majumdar et al. (eds.), *Functional Textiles and Clothing*,
https://doi.org/10.1007/978-981-13-7721-1_15

1 Introduction

Fire-protective finish by different fire-retardant agents for cotton has been reported during last few decades by many researchers in literature [1, 2]. However, reports on fire-protective or fire-retardant (FR) finish for jute-based fabrics are scanty and sporadic except few discrete recent reports [3–5] for fire-retardant finishing of jute. The main problems of fire-protective finish treatment of jute-based fabrics are higher chemical add-on, noticeable loss in tensile strength and yellowing, etc. Moreover, most of such fire-retardant formulations require high dosages of relevant chemicals and are non-durable or semi-durable [3]. Conventional fire-retardant jute fabrics (such as pandal fabrics for temporary structure and brattice cloth in mines) do not demand or does not require frequent washing and hence requirement of very high degree of wash fast fire-retardant finish is not felt essential. But now, after the development of finer variety of jute decorative and furnishing fabrics or jute–cotton union fabrics, it has become essential to find a semi-durable or durable fire-retardant formulation for jute for application as home textiles as furnishing materials or as kitchen apron, etc. Therefore, some degree of wash durability of a fire-retardant formulation applied on jute substrate should be specially looked into.

Recently, application of nanotechnology using nanoparticles of metallic oxides, i.e. zinc oxide (ZnO) has been started as a material used for solar energy conservation, semi-conductor and finishing including UV-protective coating, chemical sensor, etc. as per recent reports [6, 7] available in literature. All types of nanoparticles (1 nm = 10^{-9}m) are considered as special materials having unique and extensive applications in different fields [8]. The particles having size ranging from around 30 to 100 nm is considered to be nanoparticles [9] for textile application. Application of nano-sized metal oxides (like nano-ZnO) is an emerging area of research for fire protective finish of textiles. Such approach in jute fabrics is a new attempt.

The present work is a continuation of earlier preliminary report by same group of authors [10] on application of nanoparticles of ZnO with potassium salt of hydroxymethyl amino siliconate (PHAMS) binder emulsion for fire-retardant finish of Jute fabric. PHAMS, as a dispersive amino silicone binder, is able to react with jute through –OH or $–NH_2$ group of amino silicone binder, which is also capable of binding/incorporating with nano-ZnO particulate matter as an effective combination fort imparting fire-retardant finish of jute fabric. So far, there is no report on application of nanoparticles of ZnO applied with potassium salt of hydroxymethyl amino siliconate (PHAMS) emulsion (as dispersive binder media) for imparting fire-retardant finish on jute fabric except the above said recent report [10] by same group of authors on it. Based on the above said preliminary work by same group of authors, it is felt essential to optimise the effects of different chemical dosages (concentration of nano-ZnO and concentration of PHAMS binder) on fire-retardant performance on jute fabric. Hence, the present study mainly concentrates on statistical optimisation of concentration of nanoparticles of ZnO and concentration of PHAMS binder-based fire-retardant formulation for jute fabric. User-defined quadratic model (UDQM) of statistical experimental design is used in the present work to optimise the two input variables (concentration of nano-ZnO and concentration of PHAMS) to obtain maximum LOI value, minimum char length and loss of fabric tenacity as three output variables (char length, loss of fabric tenacity and LOI values), by using response surface methodology (RSM).

2 Materials and Methods

2.1 *Fabrics*

Conventional 3% H_2O_2 bleached fine Hessian type plain weave 100% jute fabrics having 63 ends/dm, 59 picks/dm, with 220 g/m^2 (areal density) with 195 tex warp count, 214 tex weft count, having 0.80 mm thickness was used in this experiment.

2.2 *Chemicals*

Dehydrated zinc acetate and sodium hydroxide were obtained from Loba chemicals and potassium salt of hydroxymethyl amino siliconate (PHAMS) was obtained from Wacker silicones and Saraflam CWF (Tetra Kis-hydroxymethyl (methylol) phosphonium chloride, i.e. THPC-based) was obtained from Sarex Chemicals, India.

2.3 *Preparation of Nanoparticles of ZnO*

50 g of zinc acetate dihydrate was dissolved into 1000 ml of demineralised water with stirring. After few minutes of stirring, aqueous solution of 2 M NaOH was put into the above dissolved Zn-acetate solution, obtaining a slurry like white milky emulsion, which was further put on plate of magnetic stirrer for continuous stirring for 2.5–3 h. ZnO formed was then started precipitating. ZnO precipitate thus obtained was then washed with demineralised water followed by washing with ethanol to remove all impurities from ZnO precipitate. Then, this freshly prepared ZnO was dried at 100 °C under vacuum for 6 h. Finally, ZnO precipitate thus obtained was then calcinated by heating at 600 °C for 4–5 h using a muffle furnace to obtain purified ZnO nano-sized particles [11].

The preparation of nano-sized particles of ZnO was made here by co-precipitation technique followed by calcination at 600 °C, during preparing fresh ZnO from Zn-acetate dihydrate by the following reactions [12] as given below.

$$Zn(CH_3COO)_2 + NaOH \rightarrow Zn(CH_3COO)(OH) + Na(CH_3COO) \quad (1)$$

$$Zn(CH_3COO)OH + NaOH \rightarrow Zn(OH)_2 + Na(CH_3COO) \quad (2)$$

$$Zn(CH_3COO)OH + NaOH \rightarrow ZnO + Na(CH_3COO) + H_2O \quad (3)$$

$$Zn(OH)_2 \rightarrow Zn^{++} + 2OH^- \rightarrow ZnO \quad (4)$$

However, this may be noted that ZnO particles (nano-form or normal form) cannot make any chemical binding reaction with cellulose or jute and hence there is a need of a suitable reactive binder to apply it in wash stable form. If the selected binder is reactive to the fibre substrate, it is an additional advantage and may form non-washable film of the reactive binder anchoring to the reactive sites of the fibre embedding nano-ZnO powder in the anchored binder film. So, PHAMS as a reactive binder (i.e. hydroxy amino silicone emulsion binder cum dispersive medium) was used in this case as amino group of binder and aldehyde group of jute hemicellulose may react and form aldemine adduct. Also hydroxy group of PHAMS can react with –OH group of cellulose in jute for better anchoring of the said reactive binder on the fibre substrate. Moreover, any nanoparticles of metal oxide (like ZnO) always have a tendency of agglomeration during storing before application and hence use of a suitable dispersing medium and vigorous stirring before application is essential to avoid agglomeration as much as possible. However, to some extent, agglomeration cannot be avoided. So, ZnO can be agglomerated by forming complex with another ZnO and can form –Zn–O–Zn– linkages besides some degree of crosslinking between fibres and PHAMS. That is why poly-hydroxymethyl amino poly-silicone (PHAMS) was chosen in the present work for its dual nature as reactive binder (due to its amino group and –OH group content being able to react with different functionality of jute). PHAMS was selected as a reactive binder cum dispersion medium (to avoid agglomeration of nano-ZnO dispersed in PHAMS) acting as an effective reactive binder as well as dispersive medium for application of nano-ZnO on selective jute-based fabrics for effective fire-protective finish.

2.4 Application of ZnO Nanoparticles

As nanoparticles of ZnO has tendency to agglomerate, it is wise to use freshly prepared nano-ZnO particle in potassium salt of hydroxymethyl amino siliconate (PHAMS) binder emulsion. So, requisite amount of ZnO nano-powder was weighed and made it dispersed with requisite amount of PHAMS emulsion with vigorous stirring to avoid agglomeration. Requisite amount of distilled water was added in this and the volume of the binder + ZnO nano-powder mixed emulsion was made up to 100 ml. The pre-wetted bleached jute fabric was then dipped/soaked in the said mixture of nano-ZnO and PHAMS emulsion in different concentrations and ratio (as per experimental design and need of the experiments) by dipping the jute fabric in that PHAMS binder + ZnO nano-powder mixed emulsion by batching for 30 min for soaking. Then the selective fabric was finally padded with the laboratory two bowl padder by 2 dip-2 nip for maintaining requisite weight pick up %. The treated jute fabric was then taken out and dried at 100 °C and cured at 120 °C for 5 min. Finally, the fabrics were washed and dried in air.

2.5 Method of Optimisation of Fire-Retardant Formulation Using Response Surface Methodology

Response surface methodology (RSM) is a useful statistical tool for understanding and optimising interactive effects amongst different process variables for optimising the process conditions. Response surface methodology (RSM) thus determines interactive effects between selective experimental input and selective output response careful selection of suitable statistical design of experimental model like user-defined quadratic model (UDQM).

In statistics, design of experiment explores series of systematic experimental test run, designed in a planned and scientific manner with equal differences of variables as varied input variables in order to identify assessment of corresponding changes in the output response. Based on preliminary experiments reported by the same group of authors elsewhere [10], the average dosages of fire-retardant chemicals and PHAMS binder were chosen and 15 nos. of experimental sets under fixed conditions of treatment were identified for determining the optimised input process variables (concentration of nano-ZnO and concentration of PHAMS binder) for maximising limiting oxygen index (LOI) value, minimising char length and loss of tenacity of corresponding jute fabric, for application of nano-ZnO powder and PHAMS binder combination as effective fire retardant formulation for jute fabric.

As per UDQM statistical experimental design for the said 2 input variables i.e. x_i (concentration of nano ZnO i.e. x_1 or x_i or A and concentration of PHAMS binder i.e. x_2 or x_j or B) at 3 levels needed total 15 experimental run for determining three resultant response variables (Y_i) considered i.e. Average LOI value (Y_1 or R_1), char length (Y_2 or R_2) and Fabric Tenacity retention % (Y_3 or R_3) as shown in results and discussion part in relevant Table 3 which is guided by the following quadratic equation as per UDQM statistical optimization RSM equation for each response variables (Y_i or R_i) against specific input variable (x_i or x_j) to be determined/predicted by the following RSM equations:

$$Y_i \text{ or } R_i = b_o + \sum_{i=1}^{i=k} b_i x_i + \sum_{i=1}^{i=k} b_{ij} x_i x_j + \sum_{i=1}^{i=k} b_{ii} x_i^2 \quad (5)$$

where b_o, b_i, b_{ij} and b_{ii} are corresponding regression coefficients.

2.6 Testing Methods

All the selected untreated and treated jute fabric specimens were conditioned by exposure to an atmosphere of 65 ± 2% relative humidity for at least 8 h before following tests.

2.6.1 Flammability Test

The selected jute-based fabric specimens, 2.75 × 10 in., were mounted in a suitable clamp in vertical flammability tester. The bottom edge of the fabric was then put in front of a standard flame for exactly 10 s time under controlled standard conditions. After removal of the flame, time of flaming and afterglow were observed. This test was done strictly following ASTM D6413/D6413M-13b test method [13]. This test method determines afterglow time (sec) and length of char (cm). Also afterglow time was evaluated using a stop watch. Length of char was also evaluated by measuring length of the part of the charred portion in a centimetre scale.

2.6.2 Test of Tensile Properties

Warp-way breaking tenacity (cN/tex) of jute fabrics was determined by ravelled strip technique as per IS-1969-1968 [14] using an Instron (Model-1445) CRT-Universal tensile strength tester, using a traverse speed of 100 mm/min and a pretension of 0.5 N. The final gauge length (sample size) was 50 mm × 20 mm size. Test results were obtained by an average of 10 results for each case. Fabric tenacity was calculated as

$$\frac{\text{Breaking load (cN)}}{\text{Fabric wt}\left(\text{g/mt}^2\right) \times \text{fabric width (mm)}} \tag{6}$$

The retention % of the tensile strength or tenacity for selected treated and untreated jute fabric specimen was calculated using following relationship:

$$\frac{\text{Treated fabric tenacity}}{\text{Untreated fabric tenacity}} \times 100 \tag{7}$$

2.6.3 Evaluation of Bending Length

The bending length of selective jute-based fabric samples was measured following IS-6490-1971 method [14] with standard a specimen size (200 mm × 25 mm). Higher is the bending length, stiffer is the fabric and vice versa. The test results were obtained by an average of using of five tests for each sample.

2.6.4 Determination of Whiteness and Yellowness Index

Whiteness index of the selected jute fabric samples was measured as per Hunter Lab-Scale formula [15] directly evaluated by a computer aided reflectance spectrophotometer (Macbeth 2020 having D_{65} standard illuminant and 10° standard observer set) and associated colour lab plus colour measurement software. Yellowness index (15) was similarly evaluated by using ASTM formula for it.

2.6.5 Determination of Critical Oxygen Index (LOI) Values

Critical oxygen Index or limiting oxygen index (LOI) value indicates the relative measure of fire retardancy. It is known to be fire-retardant textile material if LOI value of any material is above 27. LOI value is determined as the minimum value of volume concentration of O_2 (in a mixture of O_2 and N_2), to sustain the flame for combustion of the textile material and below which level of volume concentration of oxygen, the fire will extinguish. Critical/limiting oxygen index of untreated and treated jute fabric samples was determined following ASTM-D-2863-77 method [13] as per the equation given below:

$$\text{Critical Oxygen Index } (n) = \frac{\text{Volume concentration of } O_2 \text{ only}}{\text{Volume concentration of } (N_2 + O_2)} \times 100 \quad (8)$$

2.6.6 (FTIR) Analysis

Selected jute fibre samples (3 mg) taken out from corresponding fabrics were crushed to powder form and were exposed to a double beam FTIR spectrophotometer with attenuated total reflectance (ATR) to obtain surface FTIR scan of the samples having wave function scan of different functional group of the said samples.

2.6.7 PXRD Analysis

X-ray diffraction (XRD) pattern of the selective specimen of fibres taken out from fabrics were analysed using X'PERT PRO XRD instrument using Cu Kα1 at 40 kV and 30 mA at 1.54 Å. Finally, XRD densitometer scan was obtained and used for analysis.

2.6.8 Analysis of Scanning Electron Microscopy (SEM) Picture and Determination of Particle Size

The jute fibre samples for both untreated and treated jute fabrics were repeatedly washed in water followed by preparing samples for washing by ethyl alcohol and acetone before for the SEM scan after gold–palladium alloy coating of the samples. Selective fibres taken out for same fabric samples were scanned under scanning electron microscope (SEM) at $\times$100 and $\times$1000 magnifications also (to determine size of deposited nano-ZnO particles microscopically). Also the distribution of size of nanoparticles of ZnO deposited on the surface of the said fabrics was determined using Malvern Master sizer 2000 model particle size analyser.

3 Results and Discussions

3.1 Preliminary Study on Application of Mixtures of ZnO Nanoparticles and PHAMS Binder

Changes in selective textile-based properties for application of mixtures of ZnO and PHAMS binder (as a dispersing medium and reactive binder) combination were evaluated initially in terms of tensile strength, stiffness, whiteness, yellowness index, etc. and fire protection performances in terms of critical oxygen index, length of char and loss in fabric tenacity, etc. Relevant results are shown in Table 1, which shows the consequent result of application of said mixture of nano-ZnO particle and PHAMS binder in different concentration (Table 1).

Table 1 Physical properties of jute fabric after application of different proportion of mixture of PHAMS and nano-ZnO (with minimum and maximum range)

S. No.	Treatments	Physical properties		
		% loss in fabric tenacity	Bending length (cm)	Whiteness index (Hunter scale)
(i)	Untreated (control)	–	3.8 ± 0.2	75.20 ± 2.0
(ii)	1% PHAMS + 0.01% nano-ZnO	5.5 ± 1.0	4.0 ± 0.2	70.68 ± 2.0
(iii)	5% PHAMS + 0.01% nano-ZnO	8 ± 1.5	3.9 ± 0.3	63.52 ± 3.0
(iv)	**10% PHAMS + 0.01% nano-ZnO**	**10 ± 2**	**4.5** ± 0.3	**60.25** ± 3.0
(v)	15% PHAMS + 0.01% nano-ZnO	21 ± 3	4.8 ± 0.2	58.96 ± 2.5
(vi)	10% PHAMS + 0.005% nano-ZnO	10 ± 0.5	4.4 ± 0.3	60.32 ± 2.5
(vii)	10% PHAMS + 0.02% nano-ZnO	10 ± 0.75	4.7 ± 0.2	61.64 ± 3.0
(viii)	10% PHAMS + 0.01% nano-ZnO (after 5 cycle of washing)	10 ± 2.5	4.0 ± 0.2	67.35 ± 3.0
(ix)	20% SARA fire CFW (THPC-based commercial FR agent)	28 ± 3.5	4.7 ± 0.3	67.52 ± 3.0

All the above reactions are carried out with $MgCl_2$ in 1/5th of concentration of PHAMS binder (though PHAMS binder itself is alkaline in nature), ± data shown above indicate range of maximum and minimum values

As per results of preliminary experiments shown in Table 1, there is reduction of fabric tenacity around 10–20% after application of different proportions of nano-ZnO in presence of varying or fixed concentration of PHAMS binder. This may be viewed as an effect of alkaline degradation of major jute constituents for application of PHAMS which is alkaline amino silicone (PHAMS) dispersion cum binder, particularly during drying (heating) process, corroborating such tenacity loss in jute by amino silicone treatment after curing reported earlier [16]. Due to alkaline pH of PHAMS, removal of part of hemicellulose and a small part of removal of the lignin from jute is obvious, which also additionally results some loss of tensile strength of such treated jute fabrics. Moreover, acidic degradation of cellulosic chains of jute due to $MgCl_2$ catalyst cannot be excluded.

Normal bending length value of control untreated jute fabric is 3.8 ± 0.2 cm. Bending length values are found to increase up to 4.8 ± 0.2 cm, showing increased stiffness after different dosages of treatment with nanoparticles of ZnO. This may be due to possibility of some degree of crosslinking between reactive amino silicone binder (PHAMS) and cellulose of jute fabric, which is corroborated by earlier report [16]. Bending length value is found to be reduced to a small extent after five cycles of washing of nano-ZnO finished jute fabrics, which may be due to partial removal of unreacted PHAMS remained adhered on the surface of the said treated jute fabrics. Data in Table 1, therefore, indicate that the application of nano-sized particle of ZnO only does not contribute towards the loss in tenacity and an increase in bending length, but when nano-sized zinc oxide is treated with PHAMS, there is reduction of fabric tenacity and to some extent increase in bending length.

Hunter scale whiteness index (WI) of control bleached jute fabric is 75.20 ± 2.0, and the same after application of 10% PHAMS + 0.01% nano-sized ZnO, it is found to be decreased to 60.25 ± 3.0, which again shows a small increase up to 67.35 ± 3.0 after five cycles of washing. Resultant reduction of whiteness index (WI) may be viewed as effects of both acidic (due to $MgCl_2$ catalyst) degradation and also due to some alkaline degradation and browning of jute in alkaline pH (due to alkaline PHAMS). Thus, from this preliminary study, there is an indication of better balanced results in terms of data on physical properties for treated fabrics as shown in expt no. IV, i.e. application of mixture of 10% PHAMS + 0.01% nano-sized ZnO.

Results of fire-protective performance from the said preliminary study are shown in Table 2. Corresponding results shown in Table 2 for expt no. IV formulation (10% PHAMS + 0.01% nano-ZnO) show that in this case, the length of char is around 1.0 ± 1.0 cm and fire almost does not spread; however, afterglow remained for 37 ± 3.0 s. The presence of metallic impurities in jute including silica contamination and silica metal from PHAMS, as a metallic residue, perhaps caused higher afterglow time for the said treated jute fabric, showing much above 10 s afterglow time, as acceptable value (33). Amongst all the formulations used, the maximum critical oxygen index (LOI value) is found to be 35 ± 1.0, for expt no. IV (10% PHAMS + 0.01% nano-sized ZnO). There is a further increase in the concentration of nano-sized ZnO from 0.01 to 0.02%, but there is no increase LOI value any more. After five cycles of washing, there is a marginal decrease in fire-protective performances, showing a small decrease in LOI value (30 ± 1.0 from 35 ± 1.0), with small decrease

Table 2 Fire-protective performance properties of nano-ZnO-treated jute fabric along with PHAMS binder

S. No.	Treatments	Vertical flammability test			Limiting oxygen index (%)
		Fire spread time (s)	After glow time (s)	Char length (cm)	
(i)	Untreated	49.0 ± 2 (BEL*)	55.0 ± 3	12.5 ± 2.5 (BEL)	20.5 ± 0.5
(ii)	1% PHAMS + 0.01% ZnO	5 ± 1,SE*	35 ± 2	2.2 ± 1.5	30 ± 2
(iii)	5% PHAMS + 0.01% ZnO	7 ± 1,SE	35 ± 2	1.3 ± 1.5	32 ± 2
(iv)	**10% PHAMS + 0.01% ZnO**	**Nil, SE, NFS**	**37** ± 3	**1** ± 1	**35** ± 1
(v)	15% PHAMS + 0.01% ZnO	Nil, NFS	37 ± 3	1 ± 1	35 ± 1
(vi)	10% PHAMS + 0.005% ZnO	10 ± 1, SE	40 ± 3	2 ± 1	29 ± 1
(vii)	10% PHAMS + 0.02% ZnO	Nil, NFS	45 ± 2.5	1 ± 1	35 ± 1
(viii)	10% PHAMS + 0.01% ZnO (after 5 cycle of washing)	5 ± 1, SE	30 ± 2	2 ± 1	30 ± 1
(ix)	20% SARA fire CSW (THPC-based commercial FR agent)	11 ± 1.5, SE	8 ± 2	4.5 ± 1.5	30 ± 2

BEL—burnt entire length, SE—self extinguished, NFS—no fire spread, ± data are indicating maximum and minimum values

in afterglow time (30 ± 2.0 s from 37 ± 3.0 s), with increase in char length (2 ± 1.0 cm from 1 ± 1.0 cm).

These results are compared with results on application of same dosages of commercial fire-retarding chemicals application of 20% SARA Fire—CFW, a THPC-based commercial fire-retardant agent, experiment no. IX. For the said widely used commercial fire-retardant formulation, for getting LOI value of 30 to 35 ± 1 to 2, application, around 200–300 gpl (i.e. 20–30%) concentration is required. While by application of nano-zinc oxide, much lower concentration of nano-ZnO is sufficient. Characterisation of freshly prepared nanoparticles of ZnO and untreated and treated jute fabrics after treatment with 0.01% nano-ZnO + 10% PHAMS (experiment Sl no iv) is studied in detail in the earlier preliminary report by the same group of authors [10].

LOI value of control jute is found to be 20.5. While for expt no-IV after washing for five cycles (i.e. expt no.-VIII), jute fabric having been treated with mixture of 0.01% nano-ZnO and 10% PHAMS shows LOI value remains 30 even after five cycles of washing. This Value of LOI (30) is considered as a moderate to good fire-retardant finish in terms of LOI value for producing fire-retardant jute fabric, and is considered suitable as fire-retardant-treated jute brattice cloth used in mines are seldom washed. This show the ZnO nanoparticles admixtured with the said PHAMS binder are fairly fixed on jute fabric surface, providing a moderately durable fire-protective action. After five cycles of washing, some loosely held and unreacted PHAMS macromolecules are washed out along with loss of part of nano-ZnO particles from the surface of the said treated jute fabric, for which critical oxygen index (LOI) values are reduced by a little extent. A treatment SARA Fire—CFW, a commercially available THPC-based FR agent need its application of concentration on jute fabric for obtaining LOI value of 30. So, as compared to this commercial fire-retardant chemicals that too after application at very high concentration, the fire-retardant performance of this newer formulation expt. no. iv (0.01% nano-ZnO + 10% PHAMS) is found to be much better than the said commercial FR chemical. Thus, from this preliminary study, the formulation in expt. no. iv (0.01% nano-ZnO + 10% PHAMS) as applied on jute fabric appears to be a better option for imparting moderately permanent and reasonably fair wash stable (up to repeat five cycles of washing) fire-protective finish of jute fabric. This use of 0.01% nanoparticles of zinc oxide dispersed in 10% PHAMS with $MgCl_2$ catalyst shows an overall and better balanced results of both physical properties and fire-protective performance as obtained in Expt no-IV, vide Tables 1 and 2.

However, to analyse the interactive effect of each of nano-ZnO and PHAMS in different combinations and concentrations (dosages), it need to be optimised by statistical experiment of design using a suitable model fitting to the problem defined. So, in the next part, statistical optimisation of the fire-retardant finish formulation with nanoparticles of zinc oxide and PHAMS combination is carried out by using user-defined quadratic model (UDQM) of experimental design software following response surface methodology, as discussed below.

3.2 Optimisation of the Fire-Retardant Formulation Using Nano-ZnO with Potassium Salt of Hidroxy-Methyl Siliconate (PHAMS) with User-Defined Quadratic Model (UDQM) by Response Surface Methodology

Statistical optimisation of newer formulations with nanoparticles of ZnO and PHAMS combination is optimised by using response surface methodology (RSM), as a useful statistical tool for optimisation by determining the interactive effects between and amongst process variables for assessing the optimal formulation at prefixed process conditions. It is customary to get the optimal formulations using statistical experiment of design with the RSM design on the basis of choosing process variables from the results of the preliminary study conducted and reported above.

The user-defined quadratic model (UDQM) of experimental design used is used here to determine the statistically optimised fire-retardant formulations of nanoparticles of ZnO and PHAMS combination with prefixed dosages of $MgCl_2$ catalyst (1/5th of the weight of PHAMS binder used) to predict the optimum level of LOI, char length and loss in fabric tenacity for the said fire-retardant treatment of jute fabrics against different concentrations of nano-ZnO and PHAMS as input variables under same specific conditions of treatment.

From the preliminary study, selection of the justified application dosages of nano-ZnO particles and PHAMS emulsion used (as input variables) are finalised, keeping other process parameters constant/fixed, to get optimum output variables such as LOI (R_1), char length (R_2) and loss in fabric tenacity (R_3) for application nano-ZnO and PHAMS in varying percentage as input variables (dependant process variables) by using suitable statistical experimental design (using UDQM model under statistical design expert software) to obtain RSM responses for resultant output variables.

Data in Table 3 represent the values of response variables R_1 (LOI), R_2 (char length) and R_3 (loss in fabric tenacity) against the respective independent process variables A (for nanoparticles of ZnO) and B (for PHAMS) as per the UDQM model generated by the design expert software.

Table 3 given below shows the value of resultant response variables (R_1, R_2, R_3) against the respective dependent process variables like A (for nano-ZnO) and B for potassium salt of methyl siliconate (PHAMS) emulsion is selected at five level (i.e. $-2, -1, 0, +1, +2$ level) for statistical experimental design technique. Table 3 shows the results of response surface variables (R_1, R_2, R_3, i.e. LOI, char length and loss of fabric tenacity, respectively) against the independent process variables (A and B, i.e. concentration of nano-ZnO and PHAMS) as given by the response surface method used.

The corresponding annova data generated for LOI is shown in Table 4, annova data for char length is shown in Table 5 and annova data for loss in fabric tenacity is shown in Table 6.

Annova Tables 4, 5 and 6 for all the three response variables R_1, R_2, R_3, respectively, elucidate the lack of fit in terms of p ($p > 0.05$), which is, therefore, neglected and used to obtain corresponding resultant equation in coded form as given below in following Eqs. 9–11 for R_1 (LOI), R_2 (char length) and R_3 (loss of fabric tenacity), respectively, showing the corresponding values of response coefficients for predicting the resultant response variables:

$$R_1 = 19.40 + 153.33 * A + 0.04 * B + 0.60/100000 * A * B - 516.54 * A * A - 6.26 * B * B \quad (9)$$

$$R_2 = 131.20 - 1579.90 * A_1 - 0.14 * B_1 + 0.98 * A_1 * B_1 + 6192.76 * A_1 * A_1 + 1.53/10000 * B_1 * B_1 \quad (10)$$

$$R_3 = 9.74 - 37.52 * A_1 - 0.10 * B_1 + 0.16 * A_1 * B_1 + 130.43 * A_1 * A_1 + 1.15/1000 * B_1 * B_1 \quad (11)$$

Table 3 Values of response variables against the respective independent process variable as per the UDQM model of experimental design run in RSM technique

Concentration of nano-ZnO (g/l)	Concentration of PHAMS (g/l)	LOI (%)	Char length (mm)	Loss in tenacity (%)
A	*B*	R_1	R_2	R_3
0	80	22	125	6
0	10	21	125	10
0	150	24	125	22
0.05	115	29	20	11
0.05	45	25	100	5
0.05	80	28.8	25	11
0.1	45	32	13	8
0.1	115	35.4	10	10
0.1	80	35	12	9
0.1	10	30	22	5
0.1	150	35.2	11	21
0.15	115	35.2	10	11
0.15	80	34.9	10	10
0.15	45	30.5	55	6
0.2	80	33.8	35	10
0.2	10	30.4	49	6
0.2	150	35	11	23

Table 4 ANOVA table for response variable (LOI) R_1

Table for analysis of variance (partial sum of squares—type III)						Remarks
Source	Sum of squares	d*f*	Mean square	*F*-value	*p*-value Prob > *F*	
Model	358.42	5	71.68	34.18	<0.0001	Significant
A-A	225.23	1	225.23	107.4	<0.0001	
B-B	47.38	1	47.38	22.59	0.0006	
AB	0.74	1	0.74	0.35	0.5642	
A^2	74.1	1	74.1	35.34	<0.0001	
B^2	0.26	1	0.26	0.12	0.7309	
Residual	23.07	11	2.1			
Cor total	381.49	16				

Table 5 ANOVA table for response variable (char length) R_2

Table for analysis of variance (partial sum of squares—type III)						Remarks
Source	Sum of squares	D*f*	Mean square	F-value	*p*-value Prob > *F*	
Model	26,982.05	5	5396.41	11.62	0.0004	Significant
A-A	13,230	1	13,230	28.48	0.0002	
B-B	1702.53	1	1702.53	3.66	0.0819	
AB	201.31	1	201.31	0.43	0.5239	
A^2	10,650.97	1	10,650.97	22.93	0.0006	
B^2	1.57	1	1.57	3.37E−03	0.9547	
Residual	5110.07	11	464.55			
Cor total	32,092.12	16				

Table 6 ANOVA for response variable (loss in tenacity) R_3

Table for analysis of variance (partial sum of squares—type III)						Remarks
Source	Sum of squares	D*f*	Mean square	F-value	*p*-value Prob > *F*	
Model	476.31	5	95.26	20.09	<0.0001	Significant
A-A	0.13	1	0.13	0.028	0.8699	
B-B	353.63	1	353.63	74.58	<0.0001	
AB	5.31	1	5.31	1.12	0.3127	
A^2	4.72	1	4.72	1	0.3397	
B^2	88.88	1	88.88	18.74	0.0012	
Residual	52.16	11	4.74			
Cor total	528.47	16				

It is apparent from the first two annova Tables 4 and 5 that *A* (nano-ZnO) is the most prominent factor followed by *B* (PHAMS) affecting the first two response variables (R_1 and R_2). But Annova Table 6 indicates that *B* is the most prominent factor followed by *A* affecting R_3 (loss of fabric tenacity) as response variables. Both the independent process variables *A* and *B* have thus synergetic effects on these response variables. Corresponding three contour plots for three different response variables (R_1, R_2, R_3) are shown in Figs. 1, 2 and 3.

Data in Table 4 show the values of process variables (*A* and *B*) used in statistical experimental design with UDQM model run in design expert software and the corresponding response variables (R_1, R_2, R_3) obtained from the test results of these treated fabrics under the corresponding experimental set for each case.

These results are then processed for analysis of variance (Annova Tables) to obtain the *F*-values and *P*-values (as shown in Annova Tables 4, 5 and 6). To establish relationship between the said variables, analysis of variance and regression analysis are also done. The coefficient generated by annova data is to determine the RSM equation as shown in Eq. 9 for R_1 (LOI), Eq. 10 for R_2 (char length), Eq. 11 for R_3 (loss in fabric tenacity), for determination of predicted values of that particular

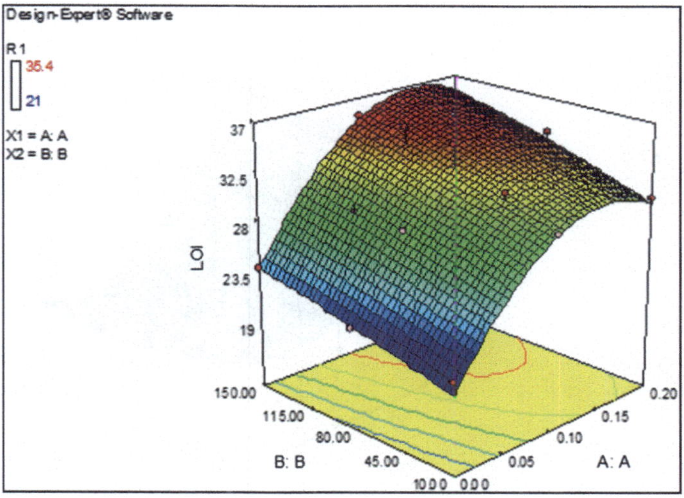

Fig. 1 RSM plot in respect to LOI

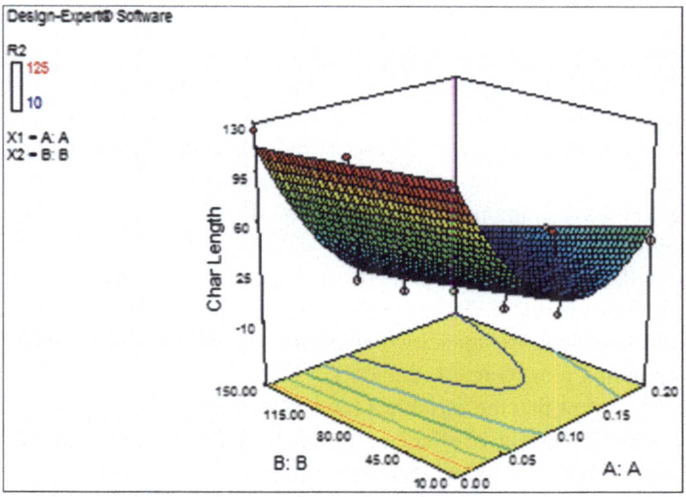

Fig. 2 RSM plot in respect to char length

property data (response variables). The significance of the effect of the variables is treated by *F*-values. These quadratic second-order equations generated by response surface methodology analysis using design expert software are used to get the best possible treatment conditions for achieving each response variables to a minimum or maximum desired value suitable for its textile application.

The plots show the effects of independent process variables individually or in combination on the resultant value of LOI (Fig. 1), char length (Fig. 2) and loss in fabric tenacity (Fig. 3) are given here for understanding both the effects of single

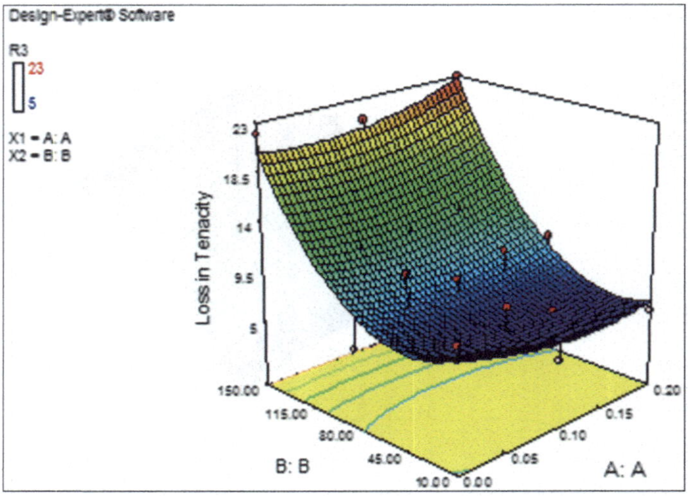

Fig. 3 RSM plot in respect to loss in tenacity

process variables (*A* and *B* separately) and also for effect of combination (*AA*, *BB*, *AB*), i.e. interdependent factor also. The coefficient obtained from optimisation by the software either have positive or negative values and so have a positive or negative effects on experimental results. For a process variable to have a significant effect, the coefficient must be greater than twice of the standard error. However, non-significant coefficients also have some inputs and not to be dismissed/neglected, as there is a very small effect but is important too.

Thus, from these three contour plots (Figs. 1, 2 and 3), it can be said that the effect of the independent process variables or their interactive effects on R_1 (LOI), R_2 (char length) and R_3 (loss in fabric tenacity) can be interpreted easily and can be explained from corresponding plots (Figs. 1, 2 and 3).

Plots in Fig. 1 show that the variable *A* (concentration of nanoparticles of ZnO) has much more and higher positive effect on increase of LOI than the effect of variable *B* (concentration of PHAMS) showing sharper steep increase in LOI value for increase of dosages of nano-ZnO from 0.05 to 0.15 gpl application. This effect is flattened after 0.15 gpl concentration value of variable *A* (nano-ZnO) and starts declining the LOI value at 0.20 gpl of nano-ZnO application, while variable *B* (concentration of PHAMS) also show gradual marginal increase in LOI with increase in application of PHAMS percentage steadily from 10 to 150 gpl.

Thus, corresponding Eq. 9 generated for predicting the values of *R* (LOI) for specific dosages/concentration of *A* (nano-ZnO) and *B* (PHAMS) variables indicates that positive value of both *A* and *B* individually and in combinations for increase in dosages of *A* (nano-ZnO) and *B* (PHAMS) up to a certain limit and showing positive impact for interactive effect of *A* and *B* applied in combination. While negative values of both the coefficients of *AA* and *BB* indicates in this case shows that further very high increase in concentration of both *A* and *B* either reduce or have less or no impact as can be seen in the plot (Fig. 1). So, it may be predicted that concentration of nano-ZnO is more important variable in this case and concentration of PHAPMS has less but positive effect on increase in LOI values and hence both the chemicals are interactively required for this fire-retardant finishing of jute.

Plot in Fig. 2 indicates that the char length is shortly reduced by increase in dosages of input variable *A* (nano-ZnO particle) from 0.05 to 0.15 gpl and therefore it further starts increasing up to 0.20 gpl of nano-ZnO, while char length remains almost same or marginally reduced only showing a very less and small, i.e. insignificant effect though it is also important too considering its ability to increase in LOI value to some extent by creating a barrier effect by the binder coating of PHAMS restricting the textile surface for catching fire.

Corresponding RSM Eq. 10 generated for predicting the resultant values of R_2 (char length) response variable for use of specific dosages/concentration of *A* (nano-ZnO) and dosages/concentration of *B* (PHAMS) indicate that both the coefficients of individual *A* and *B* are negative, and amongst *A* and *B*, the input variable *A* shows very high negative value coefficient and the input variable *B* shows very low negative value, i.e. there is much higher reduction in char length for increase in dosages of *A* (nano-ZnO) and there is very less reduction on char length for increase in dosages of *B* (PHAMS concentration). While positive values of coefficients for interactive efforts of *A* and *B* (*AB*) and also for higher increase in *A* (A^2 term), i.e. higher increase in *A* (nano-ZnO) may lead to further reversing the trend, i.e. it may increase the char length above a certain level and so is the indication in Fig. 2. Thus, nano-ZnO should not be used above 0.15 gpl with approximate dosages of PHAMS (i.e. 115–150 gpl of PHAMS).

Plot in Fig. 3 indicates that for loss in fabric tenacity (one of the side effects of fire-retardant chemical treatment) of treated jute fabric dosage/concentration of *B* (PHAMS) is the most prominent factor and with increase in dosages/concentration of B (PHAMS), loss in fabric tenacity is increased significantly, which is considered to be alkaline degradation of hemicellulose and lignin partially from jute, due to the alkalinity of amino silicone PHAMS binder emulsion, while the increase in dosages/concentration of nano-ZnO (*A*) individually has little or small effect on loss of fabric tenacity. But the interactive effect of *A* and *B* (*AB*) is positive, as understood from the contour curve shown in Fig. 3, showing positive interactive effects for this, while further and much higher increase of *A* (in A^2 term) and *B* (in B^2 term) are found to have small positive effect, i.e. higher increase in concentration of PHAMS or nano-ZnO had not further much degradation effect on treated jute fabric.

Corresponding RSM Eq. 11 generated for predicting the resultant values of R_3 (loss in fabric tenacity) as another important response variable for use of specific dosages/concentration of *A* (nano-ZnO) and dosages/concentration of *B* (PHAMS) indicates that both the coefficients of individual *A* and *B* are negative, and amongst *A* and *B*, *B* variable shows higher effect than *A* variable including the coefficient of *AB* term, i.e. interactive effects are found to be positively small. Values of coefficients A^2 and B^2 terms are also found to be small but positive. Thus, the loss in fabric tenacity is increased by increase in concentration of PHAMS up to a certain level of nano-ZnO as per the coefficient found for Eq. 3.

The numerical optimisation program is run to evaluate the optimal parametric values which are correspondingly represented in the three respective contour curves in Figs. 1, 2 and 3 corresponding to the results discussed above.

At the optimal value of *A* and *B* as generated by the RSM–UDQM experimental design software, i.e. *A* (concentration of nano-ZnO) = 0.07 gpl and *B* (concentration of PMS) = 137.80 gpl, the optimal value of R_1 (LOI), R_2 (char length) and R_3 (loss in tenacity), respectively, showing 37.71 (%) as LOI value, 44.01 mm as char length and 17.35% loss in fabric tenacity, as predicted values of three response variables, against actual values for varying parameters, which are given in Table 3. Thus, this optimum values of process variables for dosages/concentration of nano-ZnO and PHAMS obtained by using experimental design are found to be nearer to these earlier found in preliminary experiment done here, i.e. nano-ZnO dosages is 0.07% (as per statistical optimisation results from RSM of experimental design following UDQM design software) instead of 0.1% (as is found in preliminary experiment) and optimum dosages/concentration of PHAMS is 13.7% (as per statistical optimisation results from RSM of experimental design following UDQM design software) instead of 15% (as is found in preliminary experiment). Thus, this optimal results obtained by the statistical optimised technique indicate very close values of optimised dosages/concentration of these two independent process variables found in the preliminary experiment, confirming the efficacy of this UDQM model of RSM software for statistical optimisation work.

3.3 Comparison of Effect of Application of Normal ZnO Particles Versus Nanoparticles of ZnO for Obtaining Fire-Retardant Finish on Jute Fabric

Additional experiments are carried out with similar dosages of normal bulk ZnO powder application (instead of nanoparticles of ZnO) to understand the differences between the effect of application of normal ZnO and nanoparticle of ZnO on said jute-based fabric for comparing their fire-protective finish performance on jute fabrics.

Table 7 represents relevant data for application of different dosages of normal bulk ZnO powder (instead of nanoparticles of ZnO) with 100 g/l, i.e. 10% PHAMS binder application. Results show that 0.01–0.07% normal ZnO powder with 10% PHAMS respective LOI value lies 22–25, respective char length varies from 100 to 125 mm, i.e. 10–12.5 cm with loss in fabric tenacity ranging 27–30%. These results are found to be much inferior to those obtained by application of optimised concentrations of nano-ZnO, which thus proves that application of nano-ZnO with PHAMS at prescribed optimised concentrations with prefixed dosages of $MgCl_2$ catalyst is much superior technique for developing semi-permanent fire retardancy on jute fabric than applying normal nano-ZnO.

Table 7 Application of normal ZnO with PHAMS for obtaining fire retardancy on jute fabrics

Concentration of normal ZnO (%)	Concentration of PHAMS (%)	LOI (%)	Char length (mm)	% loss in fabric tenacity
0.01	10	22.0	125	30.00
0.03	10	23.25	110	29.96
0.05	10	24.20	105	28.20
0.07	10	25.10	100	27.04
8.0	10	34.50	20	22.05
10.0	10	35.00	10	20.00

It is very much evident from the optimum values of normal ZnO-treated fabric that for application of the same range of concentration as normal ZnO particle did not show good fire-retardant property as char length is much higher and LOI is too low and hence truly not appreciable fire-retardant performance is achieved by same dosages of normal ZnO application with same amount of PHAMS. However, on further trial, it is observed that 80 g/l normal ZnO, i.e. 8% normal ZnO with 100 g/l of PHAMS, i.e. 10% of PHAMS is at least required for achieving the same level of fire-protective performance, as that is obtained in case of expt No. IV, Tables 1 and 2 (0.01% nano-ZnO particle with 10% PHAMS). The good fire-protective performance for application of nano-ZnO as compared to the same dosages of normal ZnO may be attributed by the better coated film of nano-ZnO with PHAMS rendering more surface area and better coverage of the nano-ZnO particles than the normal ZnO particles. Thus, the superiority of nano-ZnO along with PHAMS is found to be encouraging and useful for imparting fire-retardant finishing of jute fabric.

3.4 *Surface Appearance by Analysis of SEM Pictures for Control Jute and Treated Jute (Nano-ZnO and PHAMS Treated)*

Scanning electron micrographs (SEMs) (a) and (b) in Fig. 4 show the surface characteristics of bleached jute and 0.007% nano-ZnO with 13.7% PHAMS-treated jute fabrics. SEM micrograph (a) of bleached jute show the small serrations with multicellular appearance of surface structure, while SEM micrograph (b) for jute fabrics treated with nano-ZnO clearly reveals deposition of some distinct minute particles (nano-ZnO particles) embedded in a deposited film (film of PHAMS anchored to surface of jute) having varying degree of extraneous deposition of minute particles (nano-ZnO) embedded in the binder film of PHAMS on the surface of jute fabrics. Enlarged view of SEM micrograph (b) shown in SEM (c) clearly indicate nano-rod like appearance on jute fibre surface with or without some agglomeration. Also, there is indication some partial agglomeration of nano-ZnO particles over the fibre surface

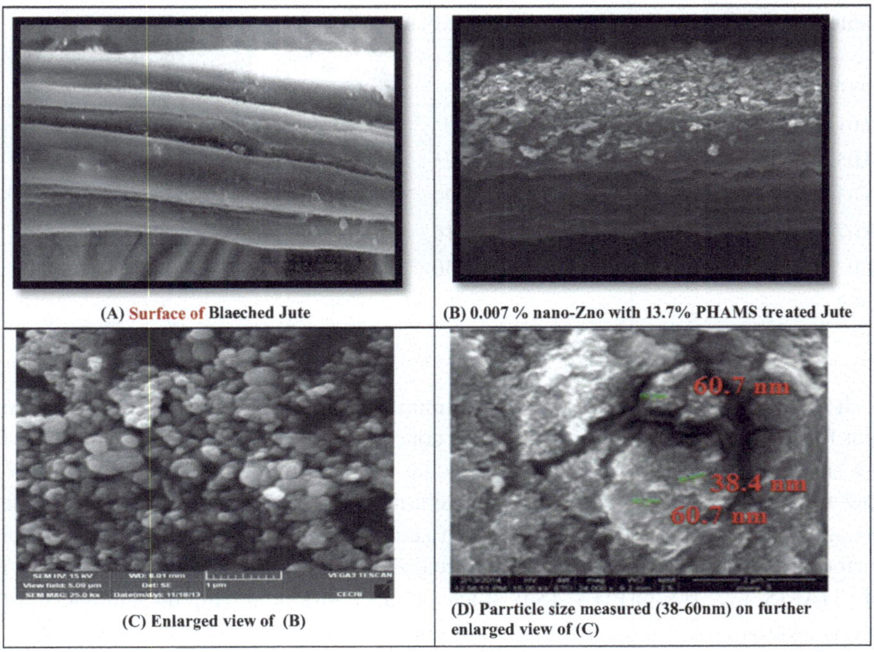

Fig. 4 SEM micrograph of bleached jute and 0.007% nano-ZnO with 13.7% PHAMS-treated jute and enlarged view of nano-ZnO coating/finishing on jute with particle size analysis

which is revealed from enlarged view of SEM micrograph (b) as SEM micrograph (c). As this SEM pictures shown in Fig. 4. SEM micrograph (d), i.e. further enlarged view of SEM micrograph (c). As shown in SEM micrograph (d), it indicates the sizes of some of the nano-ZnO particles deposited being 38–60 nm or higher (visually showing few higher sizes nano-ZnO particles due to agglomeration), as understood after microscopical measurement of particle size from SEM picture after 1000 or higher magnification.

3.5 Reaction Mechanism

The results/observations on property data and analysis of subsequent analysed FTIR peaks will be clearer and better understood, if following reaction mechanism from 1 to 6 given below are looked into.

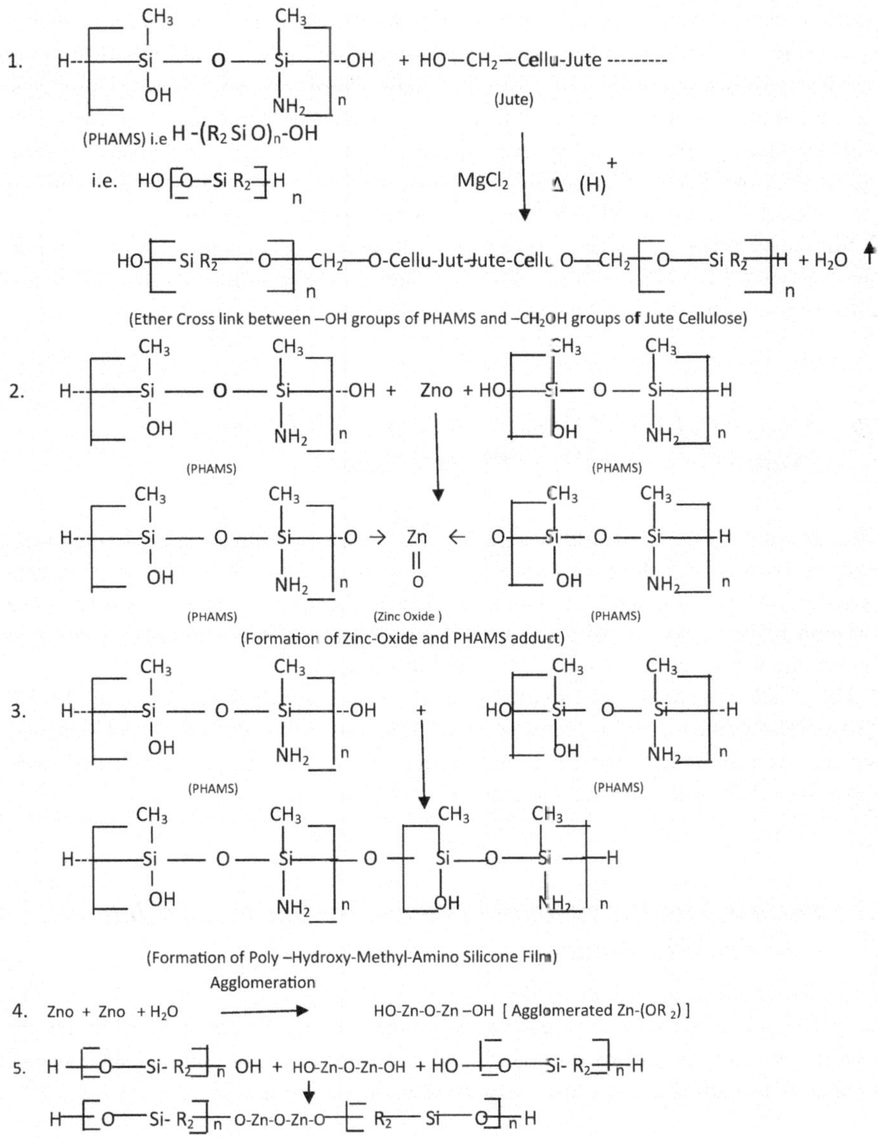

6. $H\text{-}(H_2N\text{-}R_1\,SiO)\text{-}OH$ (PHAMS) + ZnO ⟶ $Zn\,(O\,Si\,R_1\text{-}NH_2)_2$

7. .Jute Cellu-CHO + $H\text{-}(\text{-}H_2N\text{-}R_1\,SiO)\text{-}OH$ ⟶
---Jute- Cellu – CH= NH - R_1 Si O - Zn - O Si R_1-NH=CH-Cellu-Jute---

The above reaction mechanism shows that –OH end groups in PHAMS can form ether linkages with –OH group of jute cellulose chain to form crosslinks as shown in reaction 1. Adduct of ZnO-Phams may also be formed as per reaction 2 by utilising multivalency of Zn in ZnO with electron loan pair donated from end –OH group of PHAMS in presence of acidic catalyst ($MgCl_2$) by application of ZnO and PHAMS together on jute fabric by pad-dry-cure technique. Formation of poly-hydroxymethyl amino silicone film by self-polymerisation between multiple nos. of PHAMS joining together is shown in reaction 3 by similar reaction between two PHAMS molecules. Reaction 4 also cannot be excluded forming Zn–$(OR)_2$ agglomerated form, which is evidenced partly in SEM picture given above. This agglomerated ZnO can also form adduct with two PHAMS molecules as shown reaction 5. Formation of $Zn(OSiR_2)_2$ from adduct of ZnO and PHAMS are common, as shown in reaction 6.

Similarly, utilising –CHO groups present in jute hemicelluloses and amino silicone-based PHAMS can form aldemine adduct with two molecules of jute hemicelluloses in a crosslinking fashion as shown in reaction 7 given above.

3.6 Analysis of FTIR Peak of Untreated and Treated (Nano-ZnO with PHAMS) Jute Substrate

The common FTIR transmittance peaks with corresponding assigned functional group responsible for the corresponding FTIR peak(s) for controlled jute fabric and treated jute fabrics are mentioned in our earlier publication (9). Hence, besides these common FTIR peaks, the difference FTIR spectra peaks between control jute and treated jute fabric are shown in Table 8 and discussed.

The FTIR scan of the jute treated with 0.007% nano-ZnO particle with 13.7% PHAMS shows the following differences in FTIR pattern than that of control bleached jute, as given in Table 8, from which presence of both Zn and Si is confirmed from respective FTIR peaks in the nano-ZnO + PHAMS-treated jute.

3.7 Particle Size Distribution in Nano-ZnO on Treated Jute with PHAMS Binder

The particle size distribution is also analysed and found to be about 49–50% having sizes within 100 nm and rest are agglomerated to higher sizes up to 1000 nm but having some fraction also as low as 10 nm size, as shown in Fig. 5.

3.8 Elemental Analysis by EDX and AAS of Nano-ZnO with PHAMS-Treated Jute Substrate

EDX data gives Zn content on the surface of the treated jute fabric, whereas AAS data show Zn content in the bulk of treated jute fabric on the weight % of dry jute substrate. The analysis of EDX data also shows distinct presence of O, C, Si, K and Zn in quantitative term. Quantitative estimation of presence Zn on the surface of

Table 8 Specific/additional transmittance FTIR peaks in nano-ZnO plus PHAMS-treated jute

S. No.	FTIR transmittance peaks	Assigned functional group responsible for peaks
1	FTIR peak at 500–550 cm^{-1}	The peak at 500–550 cm^{-1} or below show the distinct lattice vibrations of ZnO present in the nano-ZnO + PHAMS-treated jute
2	FTIR small peak at 760 cm^{-1}	Stretching vibration of Zn–O–Zn bridging evidencing some agglomeration of nano-ZnO
3	Sharper FTIR sharp peak at 1050 cm^{-1}	Merger of FTIR peak of jute at 1030 cm^{-1} and original peak of Si–O stretching vibration at 960 cm^{-1} showing a sharp FTIR peak at 1050 cm^{-1} indicating Si–O stretching, which is shown in all reactions shown from reaction 1 to 6
4	FTIR peak at 1290–1300 cm^{-1}	Si–C stretching vibration of the silicone compound in the nano-ZnO + PHAMS-treated jute is obvious for all reactions 1–6
5	FTIR peak at 1735 cm^{-1} is reduced in the nano-ZnO + PHAMS-treated jute	Due to utilisation/reduction in jute –CHO in jute hemicellulose by formation of aldemine adduct for reaction between –CHO groups of jute hemicellulose and $–NH_2$ groups of amino silicone-based PHAMS in treated jute substrate as shown in reaction 7

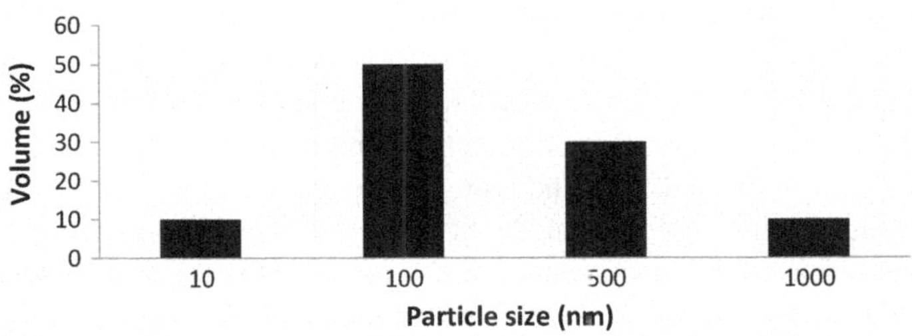

Fig. 5 Particle size distribution of ZnO nanoparticles deposited on the surface jute on treatment with 0.007% nano-ZnO with 13.7% PHAMS binder by pad-dry-cure technique

0.007% nano-ZnO plus 13.7% PHAMS silicone binder-treated jute substrate by EDX analysis is shown to be 1.05% on the basis of atomic weight % of Zn, i.e. 0.0093% by molecular weight percentage. Total quantity of Zn present in the said nano-ZnO-treated jute assumed in a AAS spectrophotometer and it is found to be 588.4 mg per kg, i.e. 0.0588% in weight % of dry jute substrate (against the application of 0.007% nano-ZnO particle along with 13.7% PHAMS binder). Si and potassium contents (from 13.7% PHAMS application along with 0.007% nano-ZnO) found in are 13.70%. There is 8.47% Si content as atomic weight %, i.e. around to 6.66% by molecular weight % of Si is evidenced to be present on the surface of treated jute and rest may be in the bulk of the treated jute substrate.

4 Conclusions

1. Preliminary experiments show that application of 0.01% nano-ZnO and 10% PHAMS binder in presence of $MgCl_2$ catalyst (1/5th of the weight of PHAMS binder, which itself is alkaline in nature) applied on bleached jute fabric (by pad-dry-cure method) shows a good level of fire-protective performance achieving LOI value 35 and char length 1 cm having nearly 10% loss in fabric tenacity and reduction of 15 unit loss in hunter whiteness index along with some minor stiffening effect.
2. By using the statistical design of experiment software following user-defined quadratic model (UDQM) model and following response surface methodology (RSM), the statistically optimal values of application of nano-ZnO and PHAMS binder (with prefixed dosages of $MgCl_2$ catalyst) are found to be 0.007% nano-ZnO (*A*) and 13.78% PHAMS (*B*) predicting values of resultant/output variables, i.e. R_1 (LOI), R_2 (char length) and R_3 (% loss in fabric tenacity), respectively, are 37.71% for LOI, 44.01 mm (4.4 cm) for char length and 17.35% for loss in fabric tenacity. Thus, statistical design expert software using UDQM model following RSM technique, the optimum values of concentrations of nano-ZnO and PHAMS render better and optimum results of response variables for maximum LOI, moderate char length and minimum loss in fabric tenacity.
3. Additional test of wash stability of 0.007% nano-ZnO and 13.78% PHAMS binder-treated jute fabric indicate that the LOI value become 30 from 35 after five normal wash cycles indicating reasonably good wash stability.
4. Further, from the additional experiment to compare the effects of application of same dosages of normal ZnO and nano-ZnO particle on jute fabrics, it is found that the same dosages of normal ZnO (with 10% PHAMS) treated at otherwise same conditions of treatment cannot render required fire retardancy performance showing a much lower LOI (22), higher char length nearly about 10 cm (100 mm) with reasonable higher loss (more than 20%) in fabric tenacity. For obtaining same level of fire retardancy as given by treatment with 0.007% nano-ZnO particle with 13.7% PHAMS (with magnesium chloride catalyst: 1/5th of the weight of PHAMS binder), it is required to use 8% or more normal size bulk ZnO powder along with 10% PHAMS binder.
5. The overall results of the present work, therefore, clearly shows the high efficacy of nano-ZnO plus PHAMS combination at their optimum concentrations for fire-retardant finish of jute fabrics for its variety different uses under fire-protective technical textiles applications.

Acknowledgements The authors are grateful to Govt of India Ministry of Textiles, and Office of the Jute Commissioner, Govt of India for required financial support in the form of R & D Project grant-in-aid for the project entitled 'Standardization of Fire Retardant Finish for Jute Fabrics' provided to this Department of University of Calcutta (DJFT, IJT, CU) vide Ref Sanction letter no-nil dt 08-03-2016 from MOT, GOI.

References

1. Horrocks, A.R.: Flame-retardant finishing of textiles. Rev. Prog. Color. **16**, 62–70 (1986)
2. Zhang, Hongkun: Textile flame retardant review. Mod. Appl. Sci. **3**(2), P129–P133 (2009)
3. Samanta, A.K.: Chemical finishing of jute and jute blended textiles: a review. Colourage **42**, 37–42 (1995)
4. Samanta, A.K., Biswas, S.K., Bagchi, A., Bhattacharya (Roy), R.: Semi durable fire-retardant finishing of jute fabric and its thermal behaviour. IE(I) J. Text. Eng. **91**, 18–28 (2011)
5. Samanta, A.K., Bagchi, A., Biswas, S.K.: Fire-retardant finishing of jute fabrics and its thermal behaviour using phosphorous and nitrogen based compound. J. Polym. Mat. **28**(2), 149–158 (2011)
6. Chen, D.Q., Wang, Y.Z., Hu, X.P., Wang, D.Y., Qu, M.H., Yang, B.: Flame-retardant and anti-dripping effects of a novel char-forming flame retardant for the treatment of poly(ethylene terephthalate) fabrics. Polym. Degrad. Stabil. **88**, 349–356 (2005)
7. Chen, W., Li, F., Yu, J., Li, Y.: Novel salt-assisted combustion synthesis of high surface area ceria nano powders by an ethylene glycol-nitrate combustion process. J. Rare Earth **24**, 434–439 (2006)
8. Matei, A., Cernica, I., Cadar, O., Roman, C., Schiopu, V.: Synthesis and characterization of ZnO—polymer nanocomposites. Int. J. Mater. Form. **1**, 757–770 (2008)
9. Reddy, K.M., Kevin, F., Jason, B., Denise, G.W., Cory, H., Alex, P.: Selective toxicity of zinc oxide nanoparticles to prokaryotic and eukaryotic systems. J. Appl. Phys. Lett. **90**(21), 1–3 (2007)
10. Samanta, A.K., Bhattacharyay, Reetuparna, Jose, Seiko, Basu, G., Chowdhury, R.: Fire retardant finish of jute fabric with nano-zinc oxide. Cellulose **24**, 1143–1157 (2017)
11. Thilagavathi, T., Geetha, D.: Low-temperature hydrothermal synthesis and characterization of ZnO nanoparticles. Indian J. Phys. **87**, 747–750 (2013)
12. Ghule, K., Ghule, A.V., Chen, B.J., Ling, Y.C.: Preparation and characterization of ZnO nanoparticles coated paper and its antibacterial activity study. J. Green Chem. **8**, 1034–1041 (2006)
13. Annual Book of ASTM Standards (American Society for Testing and Materials, Philadelphia Pennsylvania, USA) (1994)
14. ISI (BIS) Handbook of Textile Testing, SP: 15-1981 (Bureau of Indian Standard, New Delhi) (1982)
15. Hunter, R.S.: Measurement of Appearance, p. 174. Inter Science Publishers, New York (1975)
16. Samanta, A.K., Basu, G., Ghosh, P.: Effects of glycol and acrylamide treatments on textile-related properties of jute fibre. J. Nat. Fibres **4**, 59–72 (2007)

A Study on the Efficiency of Lavender Microcapsules on Silk/Lyocell Blended Fabrics

Mariyam Adnan and J. Jeyakodi Moses

Abstract In this study, silk/lyocell 50:50 blended fabrics were treated with lavender microcapsules and its efficacy was evaluated by two methods, namely psychophysical assessment to test the fragrance properties and organoleptic evaluation of odour control for anti-odour assessment. Further, a panel of 25 judges were made to assess the wash durability of the finish by rating the intensity of the aroma after 5, 10 and 15 washes. Lavender treated silk/lyocell 50:50 blended fabrics performed well in terms of aroma finish. In the evaluation of aroma finish by psychophysical assessment, very good ratings were given by the participants in the survey with respect to aroma's pleasantness and intensity. The participants also showed interest in buying perfumed products by giving positive ratings in the survey. Organoleptic evaluation of odour control also gave good anti-odour results. When wash durability of the finish was evaluated subjectively after 5, 15 and 25 washes, the participants rated the intensity of the finish as good up to 25 washes. Characterization of lavender treated fabrics with the untreated ones by SEM, EDX and FTIR analyses confirm the presence of lavender finish on the silk/lyocell 50:50 blended fabrics.

Keywords Silk · Lyocell · Aroma · Lavender · Organoleptic evaluation

1 Introduction

Silk fabrics are priced for their vanity, versatility, wearability and comfort. It absorbs moisture, which makes it cool in summer and warm in winter. It retains the shape, drapes well and sparkles with a dazzling lustre. In spite of all the wonderful properties silk possesses, it is extremely costly. Due to constant increase in price of silk fabrics,

M. Adnan (✉)
Department of Apparel and Fashion Design, PSG College of Technology, Coimbatore 641004, Tamil Nadu, India
e-mail: mariyam.ag@gmail.com

J. Jeyakodi Moses
Department of Applied Science, PSG College of Technology, Coimbatore 641004, Tamil Nadu, India

A. Majumdar et al. (eds.), *Functional Textiles and Clothing*,
https://doi.org/10.1007/978-981-13-7721-1_16

demand for blended silk fabrics is increasing thereby making silk and synthetic blended fabrics more popular in the market [1]. Lyocell fibre is the latest cohort of man-made cellulosic fibre shaped by extremely advanced environment friendly and engineered technical process [2]. Blending of silk with lyocell has many advantages over blending of silk with cotton or any other synthetic fibres. Lyocell is more versatile and uniform than cotton as it is a man-made fibre. It scores over synthetics because it is a biodegradable fibre and thus causes fewer burdens to the ecosystem [3]. Thus, considering the properties of silk and lyocell an attempt is made in the present study to blend silk and lyocell in different proportions and explore its properties so that even a common person can enjoy the unique richness of silk with excellent softness of lyocell.

In recent times, fragrances which can be readily added on textiles are available. The finished fabrics emit a pleasant aroma which is used as a medicine renowned as aromachology [4]. Because the sense of smell cannot be turned off and it gives an immediate emotional response, marketers are getting aware of its usefulness in communicating with consumers. Therefore, since the last few years consumers have started to become increasingly influenced by scents [5].

Lavender is the most used of all the essential oils. It is distilled from Lavandua Angustifolia, native of the Alpine slopes of the Mediterranean. It is very useful oil, especially when used against symptoms of a nervous problem. The major compositions of lavender essential oil are linalyl acetate, linalool, lavandulol and leaf alcohol isobutyrate [6]. The present work addresses the application of lavender microcapsules on silk/lyocell blended fabrics to develop an aroma-finished fabric.

2 Materials and Methods

2.1 Materials

Commercially available mulberry silk cut filaments were brought from M. Jiju Silk Mills, Bengaluru, Karnataka, India and lyocell fibres were purchased from KG Mills, Coimbatore, Tamil Nadu, India. Lavender fragrance microcapsules and binder supplied by Resil Chemicals, Bengaluru, India were used for aroma finish.

2.2 Methods

Fabric Development A blended yarn containing silk/lyocell in the ratio of 50:50 (S/L 50:50) was manufactured in a spinning unit and processed to produce 60 s Ne (9.84 tex) yarn. The blended yarn was woven into fabric having the specifications of plain weave, with warp 30 ends/cm, weft 25 picks/cm and 71.8 GSM fabric weight.

Table 1 Sample questionnaire of psychophysical assessment of lavender finish on S/L 50:50 blended fabrics

Question No.	Question
1	Please rate the intensity of the fragrance in the fabric
	1. Excellent 2. Good 3. Average 4. Poor 5. Nil
2	Please rate the pleasantness of the fragrance in the fabric
	1. Excellent 2. Good 3. Average 4. Poor 5. Nil
3	Please rate to what extent does the fragrance on this fabric influence your mood
	1. Excellent 2. Good 3. Average 4. Poor 5. Nil
4	Please rate the intensity of your interest/preference in purchasing innovative products such as a shoe insole, a baby diaper or a handkerchief from scented fabrics
	1. Excellent 2. Good 3. Average 4. Poor 5. Nil
5	Please rate the intensity of the fragrance of this sample if a company were to create clothing or textile goods (shoe insole, a baby diaper or a handkerchief) from this fabric
	1. Excellent 2. Good 3. Average 4. Poor 5. Nil

Aroma Finishing Using Lavender Microcapsules To acquire aroma finish on S/L 50:50 blended fabric, commercially available lavender fragrance microcapsules (3% of) were mixed with binder in a dry beaker with a ratio of 4:1. The homogenous mixture was added to distilled water at a material to liquor (M:L) ratio of 1:10. Samples were then treated in the solution for thirty minutes at pH 5.5 set by acetic acid and passed in a padding mangle, dried at 80 °C for 15 min and cured at 130 °C for 3 min.

Aroma Finish Evaluation S/L 50:50 blended fabrics were treated with lavender microcapsules and the efficiency of aroma finish was tested subjectively by psychophysical assessment of aroma finish and organoleptic evaluation of odour control.

Psychophysical Assessment A psychophysical assessment was carried out by 50 participants for evaluating the lavender treated S/L 50:50 blended fabrics. Each participant rated the treated fabric in a single evaluation session. Table 1 gives the questions asked to all participants for evaluation. Participants rated each sample for intensity and pleasantness. Participants were asked to rate their interest in aroma-finished products also. All ratings were made with a 5-point Likert scale; 5 being the maximum fragrance and 0 being nil fragrance [7].

Organoleptic Evaluation of Odour Control The effectiveness of the treatment was assessed using the organoleptic evaluation of odour control of untreated and lavender treated S/L 50:50 blended fabrics. Five-male panellists were each given an untreated and lavender treated shoe insole daily over the five-day test period. Each shoe insole was supposed to be worn on a specific foot. During the end of a workday, panellists reported to the lab to remove the shoe insoles, the shoe insoles were sealed in plastic bags, and the panellists received shoe insoles for the next day. Four judges made odour evaluations after 14 h from the removal of the insoles on each test day. The judges were given individual scoring sheets and new sheets were given every day for the evaluation. The odour rating scale was 10–0 (with 10 being "no odour" to 0 being "very intense and disagreeable odour").

Wash Durability The durability of the aroma finish for lavender treated fabrics was evaluated by subjective method after 5, 15 and 25 wash cycles as per the AATCC 135-2004 method. Twenty-five judges evaluated the intensity of the fragrance by ratings made along a 5-point Likert scale with 0 being very poor fragrance and 5 being excellent fragrance.

Characterization by SEM, EDX and FTIR Analysis A Scanning Electron Microscope (SEM), model JEOL SEM JSM-6360, was used to observe the morphology. Energy Dispersive X-ray (EDX) was carried out using a system fitted on the SEM to allow identification of the elements present in the surface of the fabrics. Infrared transmission spectra of the fabrics were recorded using a Shimadzu FTIR spectrophotometer.

3 Results and Discussion

Scents can improve mood, promote optimism, reduce anxiety and facilitate creative thinking [7]. Silk/lyocell 50:50 blended fabrics were treated with lavender microcapsules and its efficacy was evaluated by two methods, namely psychophysical assessment to test the fragrance properties and organoleptic evaluation of odour control for anti-odour assessment. Further, a panel of 25 judges was made to assess the wash durability of the finish by rating the intensity of the aroma after 5, 10 and 15 washes. The results obtained are discussed below.

3.1 Psychophysical Assessment

Fifty participants in total took part in the assessment of aroma of lavender treated S/L 50:50 blended fabrics. The questions were evaluated for their aroma along a 5-point Likert scale, with 5 being excellent fragrance and 0 being very poor fragrance. Table 2 gives the mean ratings of the lavender treated fabrics.

From Table 2, it is seen that the participants gave uniformly positive ratings to the lavender finished silk/lyocell blended fabric in the proportion of 50:50. The mean rating for intensity was found to be 4.4 which signify that participants rated the intensity of the fragrance as very good, though not excellent. The mean rating for pleasantness of fragrance received similarly positive ratings (4.3). Participants also gave a mean rating of 3.9 for fragrance influencing mood which signifies a very good

Table 2 Psychophysical assessment of lavender treated S/L 50:50 blended fabrics

Questions to the participants	Mean rating[a]	Standard deviation
Intensity of the fragrance	4.4	0.59796
Pleasantness of the fragrance	4.3	0.66425
Fragrance influencing mood	3.9	0.80407
Interest in buying a scented product	3.9	0.66517
Intensity of lavender fragrance in buying	4	0.75593

[a]Interpretation: 1. Very poor, 2. Poor, 3. Good, 4. Very good, 5. Excellent

rating. In the survey, participants rated highly (3.9) their interest in purchasing scented products. The results also showed consistent positive rating of 4 by participants for buying lavender-infused fabrics and other products such as a diaper or a handkerchief.

3.2 *Organoleptic Evaluation of Odour Control*

The average rating for anti-odour assessment performed on five-male subjects and evaluated by four judges for lavender treated S/L 50:50 blended fabrics are given in Table 3.

Four judges evaluated the shoe insole for organoleptic evaluation of odour control. From Table 3, it is observed that among the five subjects, the maximum average anti-odour rating given was 8, for Subject 5, which signifies very good anti-odour behaviour of lavender treated fabrics, followed by a rating of 7, which was given for Subject 1 and signifies good anti-odour behaviour of lavender treated S/L 50:50

Table 3 Organoleptic evaluation of odour control on lavender treated S/L 50:50 blended fabrics

Subjects	Ht (cm)/Wt(Kg)	Average rating[a]
Subject 1 (Male/39 yr)	175/62	7
Subject 2 (Male/40 yr)	170/78	6
Subject 3 (Male/28 yr)	172/81	6
Subject 4 (Male/29 yr)	165/75	6
Subject 5 (Male/24 yr)	163/72	8

[a]Interpretation: 10. Ideal, 9. Excellent, 8. Very good, 7. Good, 6. Fairly good, 5. Acceptable, 4. Fair, 3. Poorly fair, 2. Poor, 1. Very poor

Table 4 Wash durability of lavender treated S/L 50:50 blended fabrics by subjective evaluation fabrics

Number of washes	Mean rating[a]	Standard deviation
5	4	0.53852
15	3.3	0.45826
25	2.8	0.55377

[a]Interpretation: 1. Very poor, 2. Poor, 3. Good, 4. Very good, 5. Excellent

blended fabrics. The least odour rating given by the judges is 6, for Subject 2, 3 and 4, which signify fairly good odour control of the lavender treated S/L 50:50 shoe insole given to the subjects. The overall average rating is 7, which signify good anti-odour behaviour of lavender treated shoe insoles, thereby giving a positive result on further diversification of scented products such as a shoe insole, handkerchief or baby diaper.

3.3 *Wash Durability*

The wash durability of lavender microcapsules on S/L 50:50 blended fabrics were evaluated subjectively. A total of 25 participants were asked to evaluate the aroma of the lavender treated test samples after 5, 15 and 25 washes along a 5-point Likert scale with 0 being very poor fragrance and 5 being excellent fragrance. Table 4 gives the mean ratings of the lavender treated fabrics after 5, 15 and 25 washes.

From Table 4, it is observed that the mean rating of lavender treated S/L 50:50 blended fabrics were 4 after 5 washes, which imply very good intensity of aroma. As the number of washes increased to 15, the mean rating came down to 3.3 which still imply a good aroma. For lavender treated S/L 50:50 blended fabrics after 25 washes, the mean rating went down to 2.8 which imply the average intensity of aroma. Hence, it can be concluded that S/L 50:50 blended fabrics gave good aroma up to 25 washes. This excellent wash durability of the aroma can be attributed to the microencapsulation technology used for the application of lavender finish in the fabrics.

3.4 *Characterization of Untreated and Lavender Treated S/L 50:50 Blended Fabrics*

Characterization of untreated and lavender treated S/L 50:50 blended fabrics by SEM, EDX and FTIR analysis are presented below.

SEM Analysis The SEM images of untreated and lavender treated S/L 50:50 blended fabrics are given in Fig. 1a, b, respectively. For the untreated fabric, the fabric surface is smooth without any coating. In comparison, the lavender treated

(a)

(b)

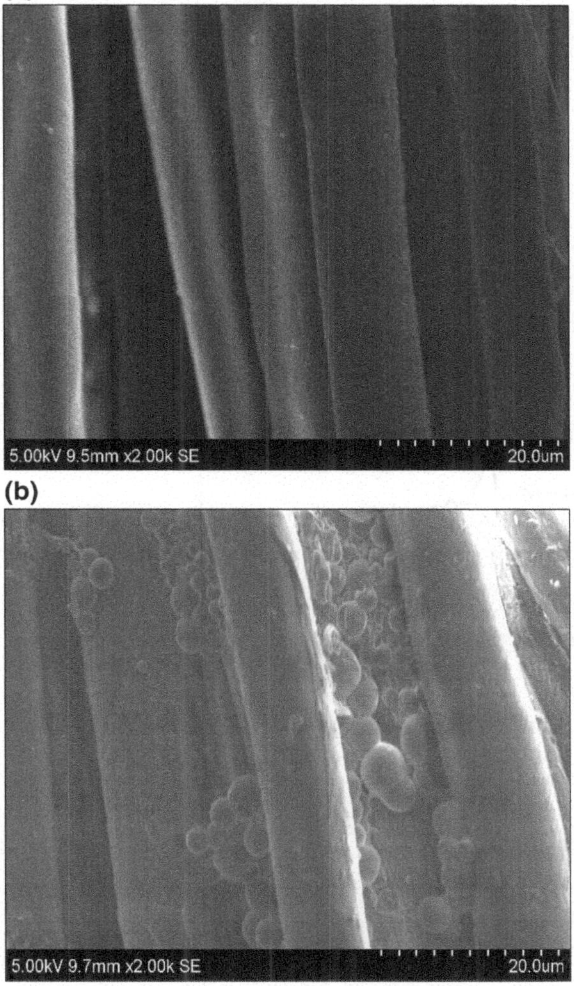

Fig. 1 SEM images of **a** untreated, **b** lavender treated S/L 50:50 blended fabrics

fabrics show clear microcapsules. The microcapsules are clear spheres and spread uniformly over the fabric surface.

EDX Analysis The EDX spectrum of untreated S/L 50:50 blended fabric (Fig. 2a) exhibits significant strong peaks for carbon and oxygen, the main constituents of the cellulose substrate. In treated fabrics (Fig. 2b), distinct peaks are observed for silica (Si), sulphur (S) and chlorine (Cl) along with those for carbon (C) and oxygen (O). This confirms the presence of lavender finish on S/L 50:50 blended fabrics.

FTIR Analysis Figure 3 shows the FTIR spectra of both the untreated and the lavender treated S/L 50:50 blended fabric. Compared to the FTIR spectra of untreated

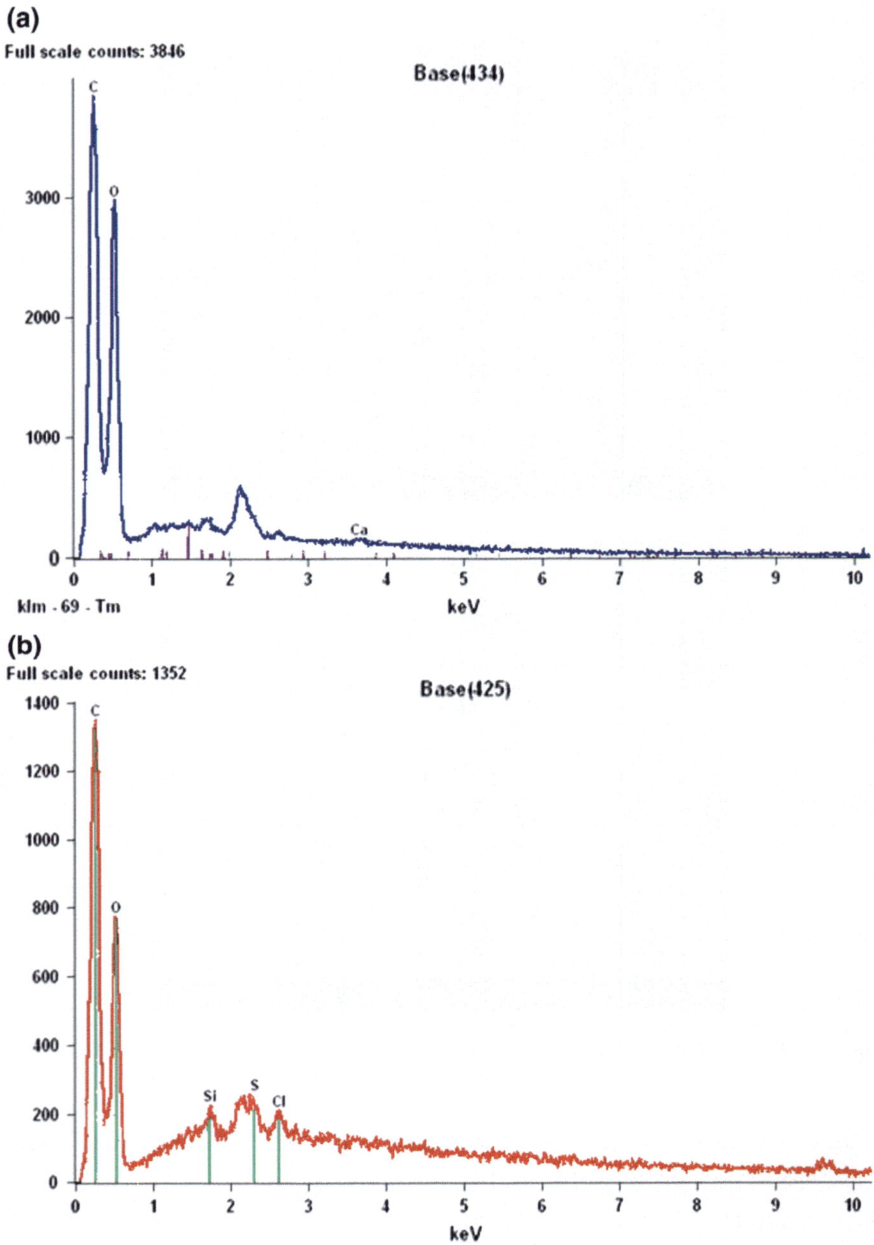

Fig. 2 EDX images of **a** untreated **b** lavender treated S/L 50:50 blended fabrics

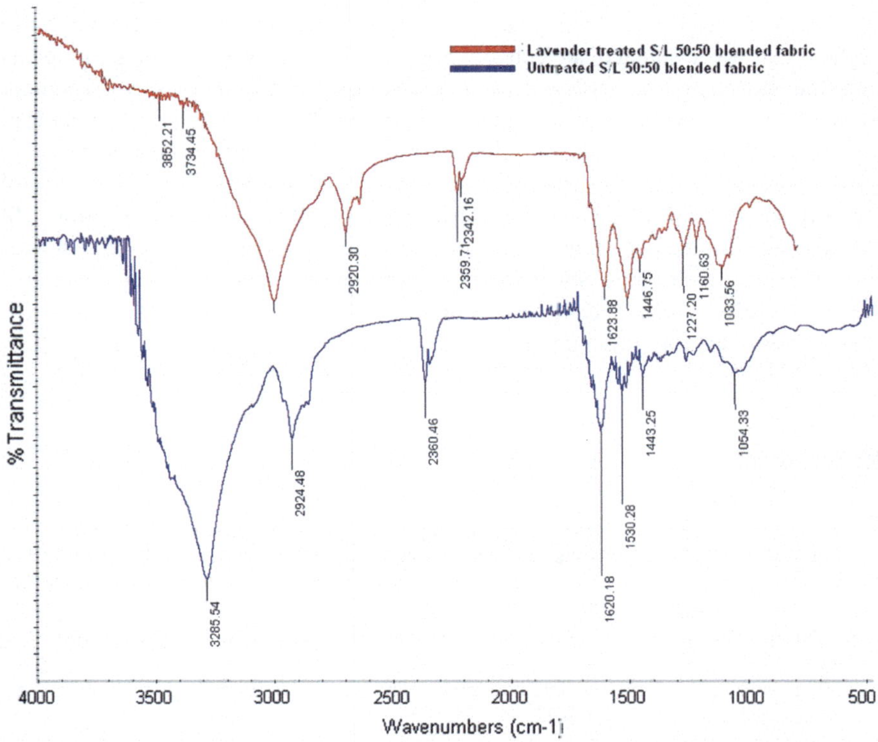

Fig. 3 FTIR spectra of untreated and lavender treated S/L 50:50 blended fabrics

S/L 50:50, the spectrum of lavender treated S/L 50:50 shows a peak at 1447 cm^{-1} for alkanes and 1034 cm^{-1} forester functional groups, indicating the presence of lavender in S/L 50:50 blended fabrics [8].

4 Conclusion

Scents can improve mood, promote optimism, reduce anxiety and facilitate creative thinking. Silk/lyocell 50:50 blended fabrics were treated with lavender microcapsules and its efficacy was evaluated by two methods, namely psychophysical assessment to test the fragrance properties and organoleptic evaluation of odour control for anti-odour assessment. Further, a panel of 25 judges was made to assess the wash durability of the finish by rating the intensity of the aroma after 5, 10 and 15 washes. Lavender treated silk/lyocell 50:50 blended fabrics performed well in terms of aroma finish. In the psychophysical assessment of the aroma finish, participants in the survey gave very good ratings for intensity and pleasantness of the aroma. The participants also gave positive ratings for their interest in purchasing scented products. Organoleptic evaluation of odour control also gave good anti-odour results. When wash durability of the finish was evaluated subjectively after 5, 15 and 25 washes, the participants rated the intensity of the finish as good up to 25 washes. Characterization of lavender treated fabrics with the untreated ones by SEM, EDX and FTIR analyses confirm the presence of lavender finish on the silk/lyocell 50:50 blended fabrics. Silk and lyocell possess similar properties in terms of lustre and strength. The cost of lyocell is nearly one-third of silk and lyocell is an eco-friendly fibre. Therefore, blending silk and lyocell will merge the richness of silk and softness of lyocell and give a techno-economic edge over other fabrics in the textile industry. Thus, it can be concluded that aroma infused silk/lyocell blended fabrics can be commercialized as there will be a niche market for aroma fabrics.

References

1. Chakraborty, A., Saha, P.K., Singha, K., Sengupta, A., Thakur, S.: Application of synthesized disperse-azo dyes on silk fabric-a new vista of silk dyeing. J. Text. Assoc. **72**, 229–241 (2011)
2. F2F Homepage: http://www.technicaltextile.net/articles/rawmaterial/detail.aspx?article_id=5100. Last accessed 21 Nov 2017
3. Gahlot, M., Pant, S.: Properties of oak tasar/viscose blended yarns. Indian J. Fiber Text. Res. **36**(2), 187–189 (2011)
4. F2F Homepage: http://www.fibre2fashion.com/industry-article/1929/fragrance-finishing-of-textiles. Last accessed 28 Nov 2017
5. Bradford, K.D., Desrochers, D.M.: The use of scents to influence consumers: the sense of using scents to make cents. J. Bus. Ethics **90**, 141–153 (2009)
6. Wang, J.M., Zheng, W., Song, Q.W., Zhu, H., Zhou, Y.: Preparation and characterization of natural fragrant microcapsules. Indian J. Fiber Bioeng. Inf. **1**(4), 293–300 (2009)
7. Liu, Y., Tovia, F., Balasubramian, K., Pierce, J.D.J., Dugan, J.: Scent Infused textiles to enhance consumer experiences. J. Ind. Text. **37**(3), 263–274 (2008)
8. Mulyani, W.E., Sunendar, B.: Synthesis and characterization of silica-lavender microencapsulation by sol gel-emulsion method for anti mosquito textile. Adv. Mater. Res. **789**, 215–218 (2013)

Part V
Garment and Accessory Design

Method Development for Modeling, Designing, and Digital Representation of Outdoor and Protective Clothing

Muhammad Awais and Sybille Krzywinski

Abstract Numerous researchers have focused on wear comfort for humans in the recent past. However, wear comfort is a complex and subjective issue, involving psychological and physiological effects in the form of clothing and the surrounding environment. Clothing, as immediate environment of the body, plays a vital role in achieving human comfort. Therefore, in addition to functional and aesthetic aspects, the interaction between garment design and human body including anatomical, physiological, and psychological requirements must be considered during the design process of comfortable and high-quality clothes. The aim of this research work was to develop a 3D simulation process based on geometrical, mechanical, and thermal characteristics, which gives insight into heat management through clothes and microclimate for the production of outdoor and protective clothes.

Keywords Thermophysiological comfort · Microclimate · Simulation

1 Introduction

Digitization of information and processes is currently an important topic in all areas. In terms of the clothing sector, researchers have been working intensively to make complex development processes more transparent and efficient. Considerable efforts have been directed toward virtual modeling in order to reduce the prototyping process for its efficient implementation in diversified applications. During the designing of outdoor and protective clothing, both constructive (fit, ergonomic wear comfort) and functional aspects (weather protection, thermal wear comfort) are very important. Therefore, the realization of holistic product design, which is predominantly based on empirical knowledge, is highly complex. Most of the simulations have been carried out regarding the ergonomic wear comfort of the garment, and virtually no computer-

M. Awais (✉) · S. Krzywinski
Fakultät Maschinenwesen, Institut Für Textilmaschinen Und Textile Hochleistungswerkstofftechnik, Technische Universität Dresden, Dresden, Germany
e-mail: muhammad.awais1@tu-dresden.de

A. Majumdar et al. (eds.), *Functional Textiles and Clothing*,
https://doi.org/10.1007/978-981-13-7721-1_17

assisted solutions exist for the realistic prediction of thermal comfort. Hence, the cross-linking of both approaches requires extensive research.

Clothes are an integral part of human life, and its main purpose is to protect us against environmental hazards [1]. It is easy to achieve this purpose when clothing comfort is neglected. Thermal comfort is one of the most important and experimentally investigated aspects of clothing comfort, which is affected by many factors, e.g., environment, characteristics of clothes, microclimate, and activity level. In addition to other factors, the thermal comfort of clothes considerably depends on the microclimate [2]. A microclimate is a small layer of air between the human body and clothing. This air layer is one of the most important factors for thermophysiological comfort as it extremely closes to the body. The properties of the microclimate (e.g., as insulator in winter and heat exchanger in summer) are affected by several aspects, for example, thermal and draping properties of the fabric, garment design, and environmental conditions.

The human body constantly interacts with the environment due to convection, conduction, radiation, and sweat evaporation. It is therefore continuously subjected to heat stresses, such as solar radiation, air temperature, humidity, and metabolic heat. A thermal regulation system helps maintain the heat balance between the human body and environment in order to achieve a body core temperature of 37 °C [3]. The heat exchange between the human body and environment during the human thermal regulation can be expressed with the following heat balance equation [4].

$$S = (M - W) - (E + C + R + B) \tag{1}$$

Here, heat storage (S) is the difference between heat production (metabolism M corrected for work W) and heat loss (evaporation E, convection and conduction C, radiation R and respiration B).

Many experiments have been conducted to understand the heat regulation process of the human body. Moreover, problems associated with clothing thermal comfort have been investigated, yet there is still a great comprehensive need for knowledge and research in order to resolve these problems. The use of thermal manikins or human is normal practice in the investigation of thermal insulation through clothing and evaporative resistance [5–7]. The only advantage of using thermal manikins is that the tests are repeatable with identical parameters and conditions; however, these tests are very expensive and time-consuming. The manikins have an average shape of a human body and are divided into different sections, which are equipped with heating sources as well as temperature and heat flux sensors. Modern thermal manikins can also sweat, walk, and even exhibit physiological responses similar to humans [8–11].

For the purpose of highly realistic investigations, subjective studies/tests involving humans can be conducted, which are, however, expensive and include ethical restrictions. In addition, such tests can exhibit widely varying test results due to intrasubjective and intersubjective variations [12].

Some efforts have also been directed toward simulating the thermoregulation during the interaction of the human body or a cylinder (which has the same temperature as the human body's surface) with the environment. The researchers have used computational fluid dynamics and investigated the thermal interaction of the nude human body in a naturally ventilated building [13], the heat transfer phenomena within microclimate [14], and the effect of slits in garment on airflow and temperature distributions [15]. Within these research projects, the influence of different body shapes (body surface, volume), clothing design, and fit has not been considered. 3D scanning techniques and an infrared thermal camera have also been used for investigating thermal insulation processes [16]; it has been reported that, if the microclimate is thicker than about 8 mm, internal convection will take place.

To avoid errors by simulation of thermal regulation of the human body, researchers have employed human thermophysiological models with varying accuracies. There is a large number of human thermoregulation models that were developed based on the first model of 1948 [17]. A multi-node model by Fiala [18] is the best performing model among others [19]. This model comprises 15 idealized spherical or cylindrical body elements: head, face, neck, shoulders, arms, hands, thorax, abdomen, legs, and feet. In accordance with former studies, a division was therefore made whenever a significant change of body tissue properties occurred. As a result, the present multilayer model consists of annular concentric tissue layers and seven different tissue materials: brain, lung, bone, muscle, viscera, fat, and skin.

The main focus of this research project is to introduce a method for the application of simulation techniques, which can not only simulate the thermal regulation of the human body in terms of microclimate, clothes, and environment, but also consider the fit of the clothes influencing the ergonomic comfort of outdoor clothes. This simulation process offers an enormous potential to sectors involving outdoor, working, and protective clothing as well as sportswear.

2 Simulation and Experimental Work

2.1 Thermophysiological Model Based on Individual Data

In order to simulate the heat regulation among human body, microclimate, clothes, and the environment, a digital anatomical surface model representing human thermoregulation was generated first. For this purpose, a human CAD model was used that had already been developed by 3D scanning and further processed with *Geomagic Studio* [20]. After repairing polygon structures, a NURBS[1] surface model was developed, which was then imported into *Lectra Design Concept* 3D [21] in IGES[2] format.

[1]NURBS—Non-uniform Rational B-Splines.

[2]IGES—Initial Graphics Exchange Specification.

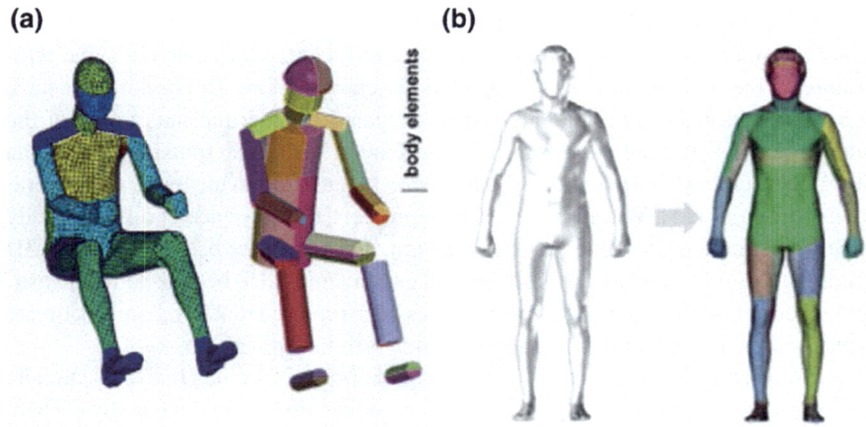

Fig. 1 **a** Thermophysiological model in *Theseus-FE* left: surface model, right: body elements [22], **b** thermophysiological model based on individual data

Subsequently, a thermophysiological model was developed by means of the Fiala Model and the framework for the software *Theseus-FE*[3] [23, 24]. For this purpose, a NASTRAN[4] format of the body model was imported into *Theseus-FE*, and each body part (body division was executed according to the Fiala Model) was assigned the Fiala thermophysiological properties. This generated human model (Fig. 1) that is based on individual scan data can now perform active and passive thermoregulation.

2.2 Selection of Material

The highly elastic material is selected, which has a smooth soft shell at the backside. It has also three layers and has a soft and thick knitted structure inside. It consists of polyamide, spandex, and laminated woven fabric. It is also very suitable for outdoor activities and sports. The physical, mechanical, and thermal characteristics of the fabric were tested (Table 1).

2.3 Virtual 3D Fit Modeling of Outdoor Jacket

According to the selected design, 2D patterns for the outdoor jacket were constructed using the 2D CAD software *GRAFIS* [25], as shown in Fig. 2. The software *GRAFIS*

[3]Special thanks to Dr.-Ing. Stefan Paulke, Group Manager Software Development THESEUS-FE for his technical support.

[4]NASTRAN—Nasa Structural Analysis System.

Table 1 Physical, mechanical, and thermal properties of the selected material

	Properties	Material	Instruments
Physical properties	Fabric thickness (mm)	0.74	Karl Schröder KG material tester
	Average mass per unit area of the fabrics (g/m^2)	301	GSM cutter
	Absolute water vapor permeability ($Pa\ m^2\ w^{-1}$)	5.76	Permetest
	Air permeability (mm/s)	21.736	Air permeability tester FX 3300 LABOTESTER III
Thermal properties	Thermal conductivity ($10^{-3} * W\ m^{-1}\ K^{-1}$)	69.33	Alambeta testing device
	Thermal diffusivity coefficient ($10^{-6} * m^2 s^{-1}$)	0.0641	
	Thermal absorptivity ($W\ ms^{-2}\ K^{-1}$)	273.20	
	Thermal resistivity ($10^{-3} * K\ m^2\ W^{-1}$)	10.49	
Mechanical properties	Bending stiffness (cN * cm)	1.580	Bending stiffness tester Cantilever ACPM 200
	Elongation (%) for 5, 20, 100 N/m (wale direction)	1; 10; 30	Zwick tensile strength tester
	Elongation (%) for 5, 20, 100 N/m (course direction)	0.5; 4; 15	
	Shear stiffness (N/m)	130	

offers the advantage that is quite easy to consider individual body measurements in pattern making and is also very user-friendly. The created pattern parts were exported as AAMA/DXF[5] file format and subsequently imported into *Modaris V8* [21] for the further 3D fit simulations.

These patterns were sewn together with *Modaris V8* and then assigned the properties of the material (Table 1). Besides the design factor, the draping behavior of garment also depends on fabric properties, such as elongation, bending and shearing stiffness, mass per unit, and fabric thickness. After sewing and assigning the fabric characteristics, the patterns were simulated on the 3D model (Fig. 3). The reliability of the draping simulation was checked with a sewn jacket. The jacket was scanned on the test person (compare Fig. 1) and the geometry data from the simulation and the sewn jacket were compared.

[5] AAMA—Standard of American Association of Medical Assistants/DXF—Drawing Interchange Format.

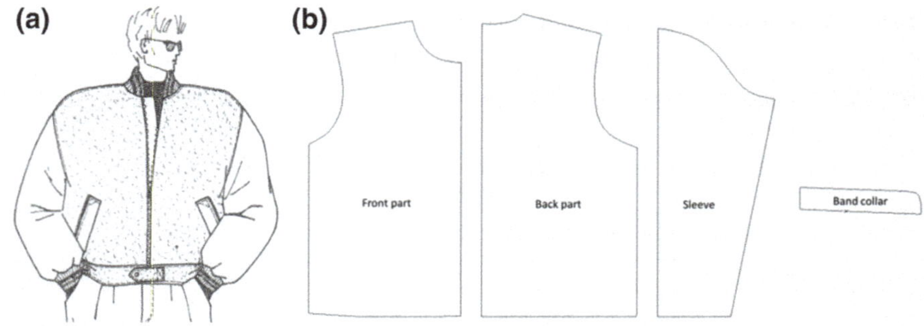

Fig. 2 **a** Selected design, **b** 2D patterns of jacket developed using *GRAFIS*

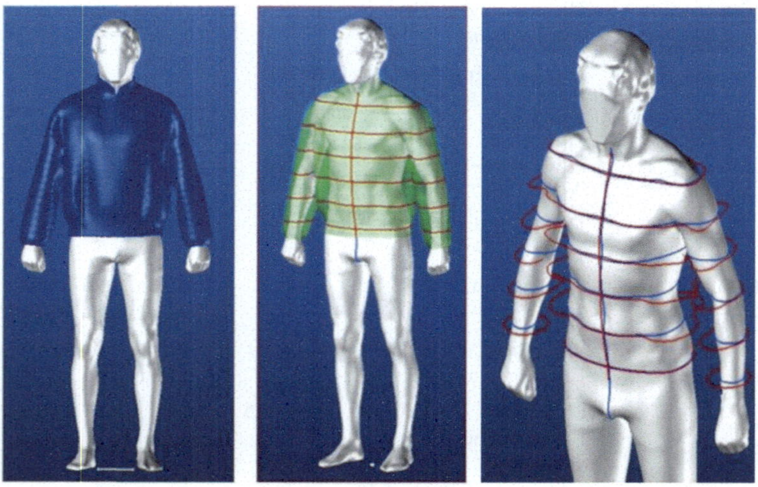

Fig. 3 Draping simulation, visualization of the distances between jacket and body

Shape and thickness of the microclimate depend on the draping properties of the clothing, which can be observed in Fig. 3. The red lines indicate the outer surface of the clothing, and the blue lines follow the body surface, revealing that the microclimate is not uniform across the entire body.

2.4 *Thermal Simulation of Human Body and Outdoor Jacket in a Time-Dependent Variable Environment*

Defining Air Zones in Microclimate In the first step, the 3D model of the jacket, which was generated using 3D fit simulation, was imported into *Theseus-FE* software via a NASTRAN interface. This 3D model of the jacket was an exact fit to the already developed thermophysiological model in *Theseus-FE* (Fig. 1). The thermal regulation system of the human body is strongly affected by the microclimate, which is developed by the influence of the body shape, cut patterns, and the draping properties of fabric. This microclimate, however, is not uniform across the entire body, so different air zones had to be defined within the microclimate in terms of thickness and their location on the body as shown in Fig. 4.

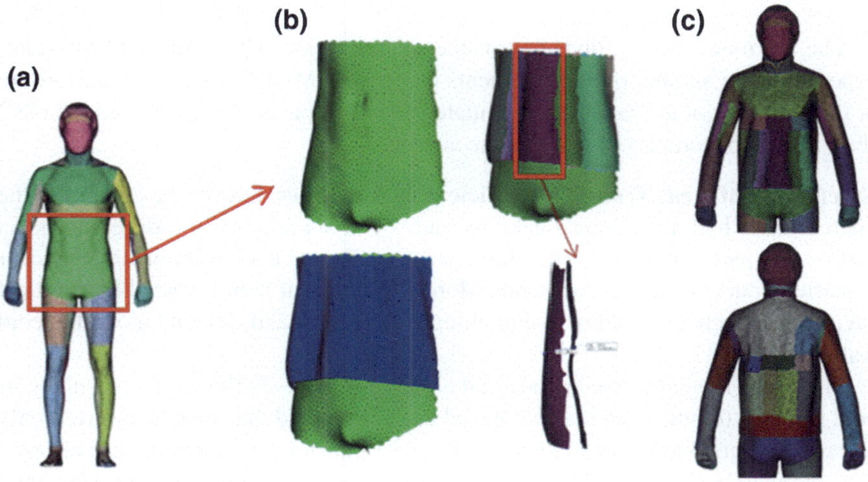

Fig. 4 **a** Thermophysiological model, **b** defining the air zones in microclimate, **c** different colors show different air zones in microclimate

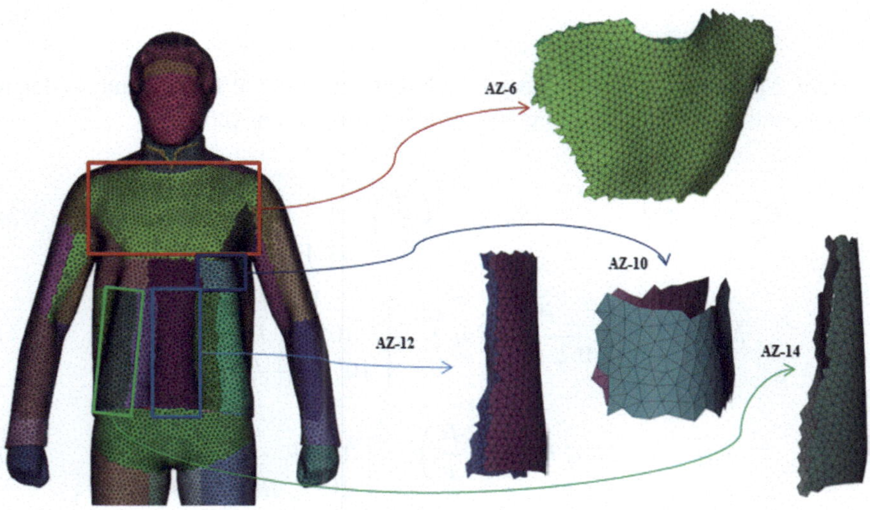

Fig. 5 Representation of some air zones within microclimate

Table 2 Characteristics of some specific air zones

Air zone	Volume (10^{-3} m^3)	Thickness (average) (cm)	Height (cm)
AZ-6	0.693	0.90	19.53
AZ-10	0.294	2.83	7.41
AZ-12	1.234	4.50	28
AZ-14	1.042	4.27	28

The air zones are the divisions of the microclimate. The number of air zones depends on the number of body elements in the Fiala Model and the non-uniformity of the thickness of air gapes/microclimate in such an element. Figure 5 and Table 2 show the characteristics of some of the air zones.

Calculation of Heat Transfer Coefficient of Air Zones Every air zone of the microclimate had a different thickness and was associated with a different jacket surface and part of the body. Therefore, heat transfer due to convection did not occur consistently across the microclimate. Moreover, these air zones reacted like vertical cavities in which air or fluid ascend along the hot wall and descend along the cold wall.

There was no gap between body and clothing surfaces in the area of shoulder, as both surfaces (clothes and body) touched each other, and thus heat transferred only as a result of conduction and radiation. The heat transfer coefficient of each air zone was calculated using the following correlation, which was suggested in Fundamental of Heat and Mass Transfer [26]:

$$h = \frac{\overline{Nu_L}\text{k}}{L}. \tag{2}$$

Here, Nu_L is the Nusselt number, which is the function of Prandtl and Rayleigh number, and can be calculated by the following correlations [26]:

$$\overline{Nu_L} = 0.22\left(\frac{P_r}{0.2+P_r}Ra_L\right)^{0.28}\left(\frac{H}{L}\right)^{-\frac{1}{4}}\begin{bmatrix} 2 \le \frac{H}{L} \le 10 \\ P_r \le 10^5 \\ 10^3 \le Ra_L \le 10^{10} \end{bmatrix} \tag{3}$$

$$\overline{Nu_L} = 0.18\left(\frac{P_r}{0.2+P_r}Ra_L\right)^{0.29}\begin{bmatrix} 1 \le \frac{H}{L} \le 2 \\ 10^{-3} \le P_r \le 10^5 \\ 10^3 \le \frac{Ra_L P_r}{0.2+P_r} \le 10^{10} \end{bmatrix} \tag{4}$$

$$\overline{Nu_L} = 0.42Ra_L^{\frac{1}{4}}Pr^{0.012}\left(\frac{H}{L}\right)^{-0.3}\begin{bmatrix} 10 \le \frac{H}{L} \le 40 \\ 1 \le P_r \le 2\times 10^4 \\ 10^4 \le Ra_L \le 10^7 \end{bmatrix} \tag{5}$$

$$\overline{Nu_L} = 0.046Ra_L^{\frac{1}{3}}\begin{bmatrix} 1 \le \frac{H}{L} \le 40 \\ 1 \le P_r \le 20 \\ 10^6 \le Ra_L \le 10^9 \end{bmatrix}. \tag{6}$$

Table 3 Used symbols and their definitions

Symbol	Definition	SI units
Nu_L	Nusselt number is the ratio of convective to conductive heat transfer across (normal to) the boundary	Dimensionless
Ra_L	Rayleigh number is the product of the Grashof and Prandtl numbers	Dimensionless
Pr	The Prandtl number is the ratio of momentum diffusivity to thermal diffusivity	Dimensionless
h	Heat transfer coefficient	W/m^2 K
k	Conductivity	W/m K
H	Height of cavity (air zone)	m
L	Characteristic length (thickness of air zone)	m
T_1	Temperature of jacket surface	°C
T_2	Temperature of body surface	°C
α	Thermal diffusivity	m^2/s
v	Kinematic viscosity	m^2/s
β	Thermal expansion coefficient (equals 1/T, for ideal gases, where T is absolute temperature)	K^{-1}
g	Acceleration due to gravity	m/s^2

In the above presented correlation, Ra_L is the Rayleigh number, which can be calculated using the following relations:

$$Ra_L = \frac{g\beta(T_1 + T_2)L^3}{\alpha v}. \tag{7}$$

The Prandtl number (P_r) depends on the fluid and fluid state and can therefore be taken from table of thermophysical properties of gases and fluid against the different temperature [26]. It has an almost constant value for the fluid over a wide temperature range (Table 3).

The calculated values for the heat transfer coefficient were later applied to each relevant air zone during defining the boundary conditions.

Defining Boundary Conditions The following boundary conditions were defined in *Theseus-FE* before starting the simulation:

- All thermal and physical properties—specific heat, conductivity, thickness, mass per unit area—were defined to the jacket surface. Scanned model was already

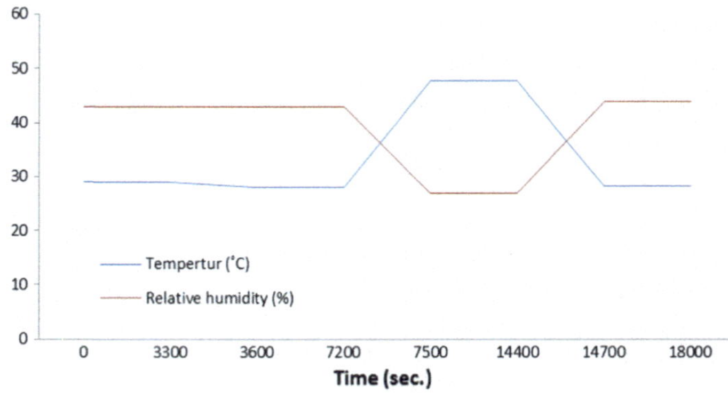

Fig. 6 Environmental temperature and relative humidity changes with respect to time

coupled with Fiala Model, therefore no need for definition of any other properties of the model.

- An environment was defined with a flow rate of 1 kg/s and a time-dependent temperature and relative humidity as shown in the following graph (Fig. 6).
- Linear convection was defined between outer shell (surface) of the jacket and the external environment (air zone), inner shell (surface) of the jacket and relevant microclimate (air zone), body shell (surface), and relevant microclimate. The boundary conditions for linear convection that are defined for microclimate were the function of the heat transfer coefficient, whereas the boundary conditions defined for the external environment and outer surface of the jacket depended on the velocity.
- Heat flux due to long-wave radiation was considered by defining the view factor cavity. For this purpose, radiation properties were defined to the jacket shell and human model.
- The initial temperature for all surfaces was defined as 27 °C.

3 Results and Discussion

The following results were obtained after running the simulation for 18,000 s. Figure 7 is a visual presentation of body and jacket surface temperatures at different times.

This graph (Fig. 8) reveals a trend according to which the temperatures of the body (thorax anterior), the relevant microclimate, and the surface of the jacket change in relation to a change in environmental temperature. It also indicates that the microclimate acts as an insulator between the body and external environment. It is evident that there is a difference between the temperature of the body surface and the microclimate. This is caused by the active system of the human body, trying to maintain body skin temperature at 35 °C (core temperature at 37 °C). The graph (Fig. 9) exhibits

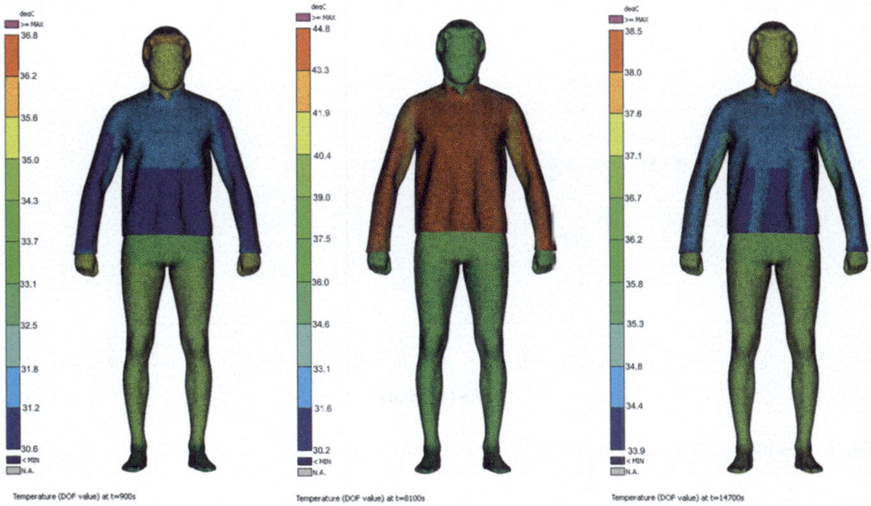

Fig. 7 Surface temperature of body and jacket during thermal simulation at different time steps (from left to right at time steps of 900, 8100, and 14,700 s)

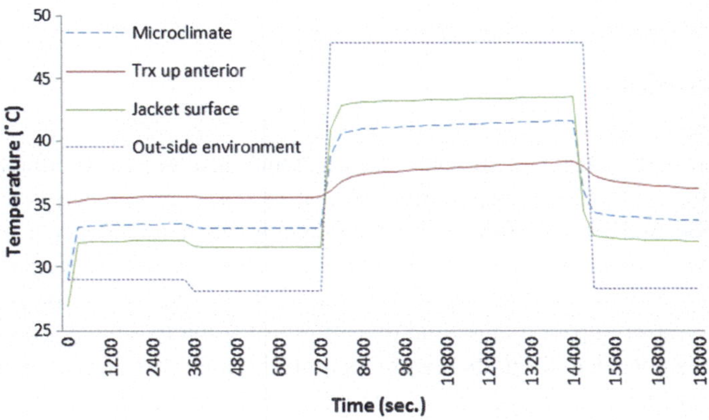

Fig. 8 Temperature behavior of surface at thorax up anterior with respect to time-dependent environmental conditions

the response of the active system due to a change in environmental temperature. The environmental temperature changes with respect to the time (28-29 °C, 47 °C and 28 °C in time interval 0–7200 sec, 7200–15000 sec and 15000–18000 sec respectively). The human model has heat stress in time interval 7200–15000 sec due to the high environment temperature of 47 °C. This heat stress activates the vasodilation and sweating processes in the human model so that the heat can be released to maintain body core temperature at 37 °C. It can also be seen in the graph that two other processes of active system; vasoconstriction and shivering do not start, because these

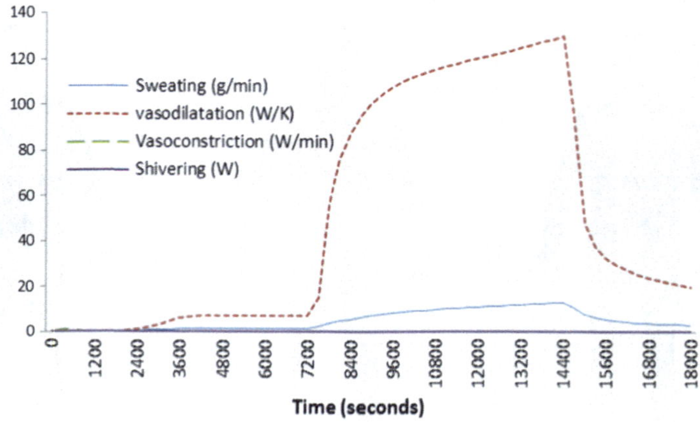

Fig. 9 Response of active system during the simulation

processes work only when body feels cool and there is a need to preserve the body heat in order to maintain body core temperature.

4 Conclusion

The human body releases heat continuously, which results in a microclimate surrounding the body. The transportation of heat in this microclimate is a complex phenomenon, and the mass flow within the microclimate mainly depends on the surrounding environment temperature, the specific body parts, and the properties of the garments a person wears. Most importantly, the velocity of air and its temperature also have a significant impact on the fluid dynamics of the microclimate. The design of the garments considerably influences mass transportation and thermophysiological comfort as well.

In the past, all thermal simulations involving the human body, its microclimate, clothes, and the environment have been performed based on the assumption that the microclimate is a uniform layer of air underneath the clothing. In reality, the thickness of the microclimate is not consistent all over the body, but it varies and depends on numerous factors, especially the draping properties of the fabric and the shape of the body. To simulate the actual scenario, a fit simulation of the jacket on individual scanned models was performed by giving the deformation properties of fabric, and the microclimate was subsequently divided into many small air zones according to thickness and location. Moreover, to realize the scanned model as actual human model, it was coupled with the Fiala Model.

This paper reveals the initial results of this research project. At this stage, the simulation does not support the transportation of wet heat from the clothing model.

However, upon completion of the project, a comprehensive research report will be published, which will be more realistic and will produce more authentic results.

References

1. Adams, P.S., Slocum, A.C., Keyserling, W.M.: A model for protective clothing effects on performance. Int. J. Cloth. Sci. Technol. **6**, 6–16 (1994). https://doi.org/10.1108/09556229410054495
2. Mert, E., Psikuta, A., Bueno, M.A., Rossi, R.M.: Effect of heterogenous and homogenous air gaps on dry heat loss through the garment. Int. J. Biometeorol. **59**, 1701–1710 (2015). https://doi.org/10.1007/s00484-015-0978-x
3. Chen, M.: The Tissue Energy Balance Equation Heat Transfer in Medicine and Biology, pp. 156–157. Plenum Press, New York (1985)
4. Parsons, K.: Human Thermal Environments (2003)
5. ASTM: Standard Test Method for Measuring the Thermal Insulation of Clothing Using a Heated Manikin, pp. 1–7 (2015). https://doi.org/10.1520/f1291-15.1
6. Holmer, I., Elnas, S.: Physiological evaluation of the resistance to evaporative heat transfer by clothing. Ergonomics **24**, 63–74 (1981). https://doi.org/10.1080/00140138108924831
7. McCullough, E.A., Jones, B.W.: A comprehensive data base for estimating clothing insulation. ASHRAE Trans. **2888**, 29 (1984)
8. Holmér, I.: Thermal manikin history and applications. Eur. J. Appl. Physiol. **92**, 614–618 (2004). https://doi.org/10.1007/s00421-004-1135-0
9. Fan, J., Chen, Y.S.: Measurement of clothing thermal insulation and moisture vapour resistance using a novel perspiring fabric thermal manikin. Meas. Sci. Technol. **13**, 1115–1123 (2002). https://doi.org/10.1088/0957-0233/13/7/320
10. Varheenmaa, M.: A new generation sweating thermal manikin for the evaluation of the thermoregulation properties of protective clothing. In: 19 Proceedings of the 10th International Meeting on Thermal Manikin and Modeling (10i3 m), pp. 1–5. Tampere, Finland (2014)
11. Richards, M.G.M., Matte, N.G.: A sweating agile thermal manikin (SAM) developed to test complete clothing systems under normal and extreme conditions. In: RTO HFM Symposium Blowing Hot Cold Protecting Against Climatic Extremes, vol. 4, pp. 1–7. Doi: ADP010487
12. Psikuta, A., Allegrini, J., Koelblen, B., Bogdan, A., Annaheim, S., Martínez, N., Derome, D., Carmeliet, J., Rossi, R.M.: Thermal manikins controlled by human thermoregulation models for energy efficiency and thermal comfort research—a review. Renew. Sustain. Energy Rev. **78**, 1315–1330 (2017). https://doi.org/10.1016/j.rser.2017.04.115
13. Yang, T., Cropper, P.C., Cook, M.J., Yousaf, R., Fiala, D.: A new simulation system to predict human-environment thermal interactions in naturally ventilated buildings. In: Proceedings of 10th International Conference on Building Simulation, pp. 751–756 (2007)
14. Mayor, T.S., Couto, S., Psikuta, A., Rossi, R.M.: Advanced modelling of the transport phenomena across horizontal clothing microclimates with natural convection. Int. J. Biometeorol. **59**, 1875–1889 (2015). https://doi.org/10.1007/s00484-015-0994-x
15. Lim, J., Choi, H., Roh, E.K., Yoo, H., Kim, E.: Assessment of airflow and microclimate for the running wear jacket with slits using CFD simulation. Fash. Text. **2**, 1 (2015). https://doi.org/10.1186/s40691-014-0025-2
16. Zhang, Z., Li, J., Wang, Y.: Improving garment thermal insulation property by combining two non-contact measuring tools. Indian J. Fibre Text. Res. **40**, 392–398 (2015)
17. Pennes, H.H.: Analysis of tissue and arterial blood temperatures in the resting human forearm. J. Appl. Physiol. **1**, 5–34 (1948)
18. Fiala, D.: Dynamic simulation of human heat transfer and thermal comfort (Thesis). Sustain. Dev. **45**, 1 (1998)

19. Psikuta, A., Fiala, D., Laschewski, G., Jendritzky, G., Richards, M., Blazejczyk, K., Mekjavic, I., Rintamaki, H., de Dear, R., Havenith, G.: Validation of the Fiala multi-node thermophysiological model for UTCI application. Int. J. Biometeorol. **56**, 443–460 (2012). https://doi.org/10.1007/s00484-011-0450-5
20. 3D Scanning, Design and Reverse Engineering Software From 3D Systems Geomagic. http://www.geomagic.com/en/
21. Lectra.: http://www.lectra.com/en
22. Paulke, S.: Finite element based implementation of Fiala's Thermal Manikin in Theseus-FE. http://www.theseus-fe.com/simulation-software/human-thermal-model
23. Fiala, D., Lomas, K.J., Stohrer, M.: Computer prediction of human themoregulatory and temperature responses to a wide range of environmental conditions. Int. J. Biometeorol. **45**, 143–159 (2001). https://doi.org/10.1007/s004840100099
24. Theseus, F.E.: http://www.theseus-fe.com/simulation-software/human-thermal-model
25. GRAFIS.: http://www.grafis.de/
26. Incropera, F.P., DeWitt, D.P., Bergman, T.L., Lavine, A.S.: Fundamentals of Heat and Mass Transfer. Wiley (2007)

Safer Custody Clothing: Designing for Female Prisoners at Risk of Self-Harm

Jane Ledbury, Nicholas Hall, Barbara Shepherd and Laura Parker

Abstract The 'Safer Custody Clothing' project offered by Her Majesty's Prison and Probation Service (HMPPS) to the Manchester Fashion Institute provided a unique opportunity for experiential learning in relation to functional clothing design and product development. Prisoners in custody may have difficulties coping with the prison environment. Women especially express distress through acts of self-harm or attempted suicide (Winters in Int. J. Art Des. Educ. 30(1):90–101, 2011 [1]). Specialist clothing is required that provides comfort, safety and eliminates the opportunity to use the garment for self-harm. MA Fashion Innovation students undertook a research project redesigning functional clothing for female prisoners at risk of suicide and self-harm. The project aimed to resolve the issue of female prisoners using their clothing to inflict harm upon themselves, through an experiential approach to problem solving and the design process for functional clothing. The pedagogical objective was to create a meta-learning environment (Watkins and Dunne in Functional clothing design: from sportswear to spacesuits, Bloomsbury Publishing, USA, 2015 [2]) in which, the students embarked upon an investigative journey that defined new ideas and generated a series of problem-oriented goals. A user-centred design process for functional clothing was adopted, with research and analysis in the early stages, followed by definition, idea generation, design development and evaluation (Lai et al. in Artif. Intell. Eng. Des. Anal. Manuf. 24(03):303–316, 2010 [3], Bulman in twofold rise in female prisoner suicides linked to 'inadequate' mental health support, 2017 [4]) was adopted, with research and analysis in the early stages, followed by definition, idea generation, design development and evaluation. Clothing solutions were deemed to have met the aim of the project, in providing cost-effective garments, which minimised risk of self-harm and suicide for female prisoners.

Keywords Functional clothing · Fashion education · Experiential learning

J. Ledbury (✉) · N. Hall · B. Shepherd
Manchester Fashion Institute, Manchester Metropolitan University, Righton Building, Cavendish Street, Manchester M15 6BG, UK
e-mail: j.ledbury@mmu.ac.uk

L. Parker
Her Majesty's Prison and Probation Service (HMPPS), 8th Floor 8.07, 102 Petty France, London SW1H 9AJ, UK

A. Majumdar et al. (eds.), *Functional Textiles and Clothing*,
https://doi.org/10.1007/978-981-13-7721-1_18

1 Introduction

A percentage of prisoners held in custody have problems coping with the environment or situation in which they find themselves. Women especially may express their distress through acts of self-harm or attempted suicide, and in some cases become disruptive, volatile and antagonistic towards both prisoners and staff. There are also instances where women will try to shock prison officers by removal of clothing. For their own protection, such prisoners are segregated in special cells, which have been adapted to remove all possible opportunities for inflicting harm upon themselves; however, for reasons of decency, a garment is required, specifically for women to wear in this environment.

Ministry of Justice data indicates that women represent 5% of the total population in custody, but account for 23% of occurrences of self-harm [1]. Lowering levels of self-harm and self-inflicted deaths is therefore a priority.

Her Majesty's Prison and Probation Service (HMPPS) approached the Manchester Fashion Institute with a sponsored project for MA Design Innovation students to design a garment or garments for this particular purpose. HMPPS developed a design brief and five students on the Masters in Product Innovation completed a twelve-week project that looked at redesigning the current clothing provided for female prisoners at risk of self-harm or in extreme distress situations.

The project aimed to resolve the issue of female prisoners using their clothing to inflict harm upon themselves, through an experiential approach to problem solving and the design process for functional clothing. The pedagogical objective was to create a meta-learning environment [2], in which the students embarked upon an investigative journey that defined new ideas and generated a series of problem-oriented goals.

2 Background

Her Majesty's Prison and Probation service (HMPPS), as part of the Ministry of Justice, is the agency responsible for delivering the British Government's vision and investment to make prisons places of safety and reform.

With a prison population in the UK exceeding 85,000 and expected to increase year on year, pressures within the service to offer a safe environment for both prisoners and staff are increasing. Interventions and new methods of operation are required to ensure maintenance of safety and that prisoners and staff are able to work in a safe and decent environment.

As in many situations, people in prisons express their feelings and emotions in a variety of ways. Prison is an especially alien and stressful environment, where some detainees have problems coping with the situation in which they find themselves. This appertains particularly to females, who may become disruptive, volatile and antagonistic towards both prisoners and staff and who are more inclined than males to express their distress through acts of self-harm or attempted suicide.

Government statistics state that although there has been a decrease in the number of self-inflicted deaths (reduction of 30% compared to the previous year), reports of self-harm and assaults continue to rise.

In 2016, incidents of self-harm alone increased by 12%, compared to the previous year. Although these incidents remain more prolific within the female estate, 2016 saw a decrease of 1% in numbers reported, compared to a 15% increase in the male estate, (413 per 1000 in male estate and 1914 per 1000 in the female estate). Importantly, reports of incidents resulting in hospital attendance increased by 9% [5].

Female occurrences predominantly involve cutting or scratching (57%), self-strangulation (29%), with the use of ligatures the lowest (8%) and consistently result in fewer hospital visits when compared to their male counterparts [5].

Across the prison estate, data suggests that more incidents occur amongst the 30–39-year old, white, British population and those prisoners awaiting sentence [5].

For their own protection and safety, prisoners identified as high risk from self-harm and are removed from normal locations on wings or landings and segregated in special safer custody cells, which have been adapted to remove all possible opportunities for harm and where staff can offer increased supervision and observation.

3 The Project

3.1 Prison Clothing Provision

Prison Industries support HMPPS through the provision of clothing, equipment and services. Items, predominantly manufactured by prisoners and designed specifically for the prison environment, utilise robust, hardwearing materials and components and are designed to reduce the opportunities for ligature points, damage or destruction.

The fabric, formed from quilting two layers of a densely woven, flame retardant polyester together, although heavy and durable, has proven difficult but not impossible to tear.

The gown, although not very flattering, had been in use for many years; designed to minimalise opportunities for ligatures, or self-harm. The garment had reinforced seams, and the armhole designed to prevent a head from fitting through.

With the changing demographics in prison, the gown is no longer deemed to meet the needs of the user. From a perspective of decency, it is easy to remove, is revealing and uncomfortable when worn for long periods and is especially unpopular within the female estate.

For many reasons, especially decency, a replacement garment was required, specifically for women to wear in this difficult environment.

Resources were limited and the project to enhance prisoner clothing had remained dormant for several years, until resources became available to undertake the complex redevelopment of better provision of clothing for female prisoners.

The chance to work with the Manchester Fashion Institute (MFI), Manchester Metropolitan University, recognised by HMPPS as an opportunity to explore alternative clothing solutions and to provide a durable garment, which optimised clothing comfort and which reduced opportunity for deconstruction for the purpose of self-harm.

3.2 Industry-Sponsored Projects

The HMPPS project formed part of a programme of postgraduate industry-sponsored projects at Manchester Fashion Institute, as part of an initiative, which developed a model of practice for industrial partners. The initiative offers an effective model for transferring research practice and knowledge into business outputs. This industry/university collaborative programme demonstrates a model of development and delivery that offers impactful outcomes for participating industry clients and valuable live project briefs for students. The programme offers a successful ongoing partnership development, based on industry needs and demonstrates the effective uses of interdisciplinary research into industry.

Enhancement for students includes improved employability skills in areas of organisation, communication and negotiation, working on real-life industry research problems against commercial constraints and timescales. In addition, students receive a financial sponsorship award. For the industrial partner, the scheme offers affordable access to research and blue-sky thinking through engagement with a university partner and the opportunity to challenge existing business reasoning and approaches. Collaborative projects also provide cost-effective graduate recruitment opportunities for the industrial partner as they source emerging talent. Additional benefits for the Fashion Institute are a contribution to development of research-informed teaching through valuable and contemporary updates on the needs of industry.

3.3 HMPPS Design Brief: Safer Custody Clothing

The HMPPS project, developed as a live sponsored brief for students on the MA Fashion Innovation course at Manchester Fashion Institute, focused on redeveloping a safer custody garment for female prisoners held on remand. Five students on the MA Fashion Innovation course completed a twelve-week project that looked at redesigning the current clothing provided for female prisoners at risk of self-harm or in extreme distress situations. The students as a team secured a £2 K award for their work on the project.

3.4 Project Aim

The aim of the project was to research and develop an original, functional garment that improved the user's likelihood of rehabilitation and reduced stress by enhancement of female prisoner's personal safety and well-being The prototypes created were to provide real-life solutions with the potential to have a genuine and lasting social impact, promoting positive rehabilitation and safety throughout the period in which the prisoner is segregated.

HMPPS provided a detailed design brief with specific criteria and allowed the group and key staff to visit their manufacturing workshops, to promote understanding of resources and constraints under which the garments are made. All clothing currently worn by prisoners is manufactured within the prison estate and quality affected by equipment and skill levels of the prison machinists.

The design brief dictated that materials were required that must be comfortable and cooling, whilst projecting an impression of warmth. Fabrics should also be strong, durable, impervious to soiling and washable at high temperatures. Stitches and seams must be resistant to deconstruction and aperture measurements controlled. The psychology of colour should be considered and garments should be easy to don by third parties and difficult to remove by the prisoner.

The HMPPS brief held significant challenges for design and development of the apparel product. A widely varied population of females in terms of age range, cultural diversity and body shape, necessitated particular considerations for pattern development in order to achieve satisfactory fit. Many common fastenings, such as buttons, zippers, ties, buckles and clips, eliminated as they presented possibilities for choke hazards or ligatures and innovative alternatives had to be found.

3.5 Design Criteria: Safer Custody Clothing

Due to specialised end use and design criteria generated from both a HMPPS and user needs perspectives, design criteria were specified by the provider (HMPPS) and included a number of key garment attributes, which must be addressed in the brief as seen below:

Health and safety:

- No metal fastenings such as zips or magnets (will activate metal detectors);
- No fastenings which can be easily removed (choking hazard);
- No elastic or draw cords (ligatures);
- No channels or large hems (to conceal contraband);
- Armholes and leg openings to be smaller than head circumference;
- Colour should be dark enough to hide everyday soiling, but light enough to show contaminants (i.e. blood or bodily fluids);
- All attachments such as Velcro must be secured on all four sides and securely finished.

Considerations of clothing comfort:

- Breathable fabric to aid temperature regulation, current provision is hot to wear;
- Fit: large size range, body shape variation, age range and ethnicities;
- Tactile comfort next to skin as the garment is worn without undergarments.

Fabric considerations:

- Material should be densely woven and robust enough to prevent abrasion and breaking of the yarns (deconstruction).
- Fabric required for laundering at high temperatures (in excess of 72 degrees centigrade) as garment is not for individual issue and reused for different prisoners after disinfectant washing.
- Must pass stringent tests appropriate to the prison environment.

In addition, the garment must be constructed in such a way as to allow ease of donning by officers on resistant prisoners and be difficult to remove once worn.

3.6 Design Process

The design process for functional clothing is a creative problem-solving process with in-depth research and analysis at the start of the process, followed by definition, idea generation, design development and evaluation was employed [3]. The model emphasises intensive research, which encourages scrutiny of the user and their needs, the use environment and activities engaged in by the wearer, as well as any hazards that they may encounter. Information gathered in the early stages of the process provides the designer with a sound understanding of the problems in need of solution. The HMPPS Safer Custody Clothing design brief further complicated the design process with additional consideration of the specific garment requirements of the organisation, and the clothing solution must meet the needs of the user and the needs of the prison service, in order to maximise safety and minimise opportunity for self-harm.

4 Pedagogy

The project aimed to build a meta-learning environment fostering students' control of their learning experience; encouraging self-regulation, self-motivation and reflective practice that enhanced their independence as a learner and confidence in their own creative endeavours [2]. The concept of meta-learning is to develop students' knowledge of how they learn—in this case—through the applied research, practical skills and acquisition of technical expertise implicit in completing the project [2]. Students are encouraged to embark on an experiential journey that defines new ideas and generates a series of problem-orientated goals. Goal-orientated projects are performance enhancing. The students' inclusivity and involvement in defining the problem and their control over the outcomes recognised as more rewarding than having pre-determined learning objectives and a planned, 'serialist' approach to learning as the key motivation [6].

The approach, designed to engage students in defining the boundaries and requirements of the project for themselves, built a comprehensive picture of its research contexts and topic areas, within which students could translate, reconcile and integrate discourses. This formed the 'scaffolding' students used to create and define experimental space for questioning and developing explanations around key issues they identified for consideration and evaluation of possible solutions. Meta-learning was further facilitated by opportunities to engage in cross-disciplinary thinking and the introduction of new knowledge perspectives to the students, so they can consider and construct new meanings through dialogue with experts in a dynamic and varied learning environment [6].

Manchester Metropolitan University's criminology department delivered sessions on prevention of self-injury and suicide in women's prisons [7] in order to gain better understanding of the psychological state of the intended end-user. This was important as direct access to female prisoners to evaluate their needs using conventional methods, such as interviews, was not possible.

The significance of this approach is to generate a learning environment where students gain personal interest in the project, fostering the intrinsic motivations to learn through the intention to understand problems in detail and the causes and effects of their resulting design decisions; ultimately aiming to deliver a useful and practicable product outcome to the client. The aim is that this becomes more important than gaining a mark or passing an assessment, the stronger cognitive engagement creating deeper learning [2]. This is important as deeper learning cannot be prompted simply by asking students to adopt it, and it must be generated and fostered in a carefully instigated, motivational learning environment.

To further explore student's connection with and understanding of meta-learning, they were also required to record the process of their experience; research decisions made, experiments that worked and failed, the artefacts they produce; reflecting upon their learning journey in critical contexts, so that they could visualise their progress. Reflective points were used to signpost key milestones within the project, as well as create a set of reference points to step back from progress towards the project deliverables and consider 'what has been learned, how have I responded, what should I do next?'. This process was further enhanced by the cooperative learning nature of the project, with the student's evaluation of the task forming regular informal and formal project discussions and critiques that involved presenting to themselves, teaching staff and project stakeholders when they reached milestones or simply wanted to discuss what they were doing. Key to this was presentation of preliminary prototypes to HMPPS staff during a formal 'milestone' presentation, where extensive critical discussion took place. Integrating suggestions and listening to perspectives of the clients further enhanced to final project results delivered a few weeks later, revisiting the design process and creating a series of adaption to the design to meet client expectations previously uncaptured, the prototypes formed a dialogue between the clients and students that was critical to the projects overall success.

The purpose of milestones was to encourage students to develop innovative solutions to the challenges within the HMPPS project through re-evaluating their effectiveness or relevance in a critical light and in reflection of their personal learning strategies. The students were encouraged to review and evaluate their learning in a project diary and project portfolio context, evaluated through presentation and discussion with project tutors. The diary took the approach of capturing and analysing student's personal design activities, asking them to evaluate the effectiveness of the research methods, their design activities and the project results, asking what they had learned during these processes, what might they do differently next time? The concept being that formative feedback and discussions around the reflecting on the process of their work would also allow a conversation to clarify their understanding of their meta-learning and its personal nature [8].

Concepts and prototypes were evaluated through group discussion with experts from a variety of disciplines (such as psychology, health; engineering) in order to address the complex needs of the user and the organisation, HMPPS. Meetings with HMPPS staff provided valuable feedback and viable prototypes then evaluated from both cost and manufacturing capability within Prison Industries. Students appreciated the requirement to satisfy design criteria on multiple levels and the need to rank conflicting criteria in order of importance in order to arrive at a garment solution in response to the design brief [9]. Importantly, the group gained understanding of the value of interdisciplinary approaches to a problem, in order to reach a credible solution.

5 Conclusion

Overall, it was clear that each of the students in the project ventured on a journey of discovery; engendering insightful research, innovative solutions to the prototype garments and generating high-quality work. Each of the students developed key roles and responsibilities as they collaborated towards the common goals set, which motivated their progress and overall learning. Students agreed that the concept of understanding how they learn and why helped them to progress more significantly than they may have done otherwise. Through a multi-disciplinary approach to common problems, a clothing solution was reached, which was deemed to meet the needs of the wearer within the constraints of the use environment and respond to multiple and complex design criteria as required by HMPPS. The resulting garment was deemed fit for purpose by the client. The live brief generated by HMPPS provided the student group with real-world problems requiring an innovative solution at a keen price point, which could be manufactured by Prison Industries. Each member of the group contributed towards achieving the aim of the project through teamwork, enhancing transferable skills of negotiation, organisation and communication in working towards a common goal. Students perceived the project aim as impactful and worked hard to find a solution towards safer custody clothing, which would reduce risk of self-harm in female prisoners. The ability to work in multidisciplinary teams contributing towards a common goal replicates industry practice and prepares students well for entry to the global apparel industry they seek to enter on graduation.

References

Winters, T.: Facilitating meta-learning in art and design education. Int. J. Art Des. Educ. **30**(1), 90–101 (2011). http://dx.doi.org/10.1111/j.1476-8070.2011.01685.x

Watkins, S.M., Dunne, L.: Functional Clothing Design: From Sportswear to Spacesuits. Bloomsbury Publishing, USA (2015)

Lai, J., Honda, T., Yang, M.: A study of the role of user-centered design methods in design team projects. Artif. Intell. Eng. Des. Anal. Manuf. **24**(03), 303–316 (2010). https://doi.org/10.1017/s0890060410000211

Bulman, M.: Twofold rise in female prisoner suicides linked to 'inadequate' mental health support. The Independent. Retrieved 22 March 2017 from http://www.independent.co.uk/news/uk/home-news/female-suicide-rate-inadequate-mental-health-support-womens-prisons-prison-and-probation-ombudsman-a7653111.html

https://www.gov.uk/government/statistics/safety-in-custody-quarterly-update-to-june-2017

Warburton, K.: Deep learning and education for sustainability. Int. J. Sustain. High. Educ. **4**(1), 44–56 (2003)

Walker, T., Towl, G., Harris, T.: Preventing Self-Injury and Suicide in Women's Prisons (2015)

Mezirow, J.: Transformative learning: theory to practice. New Dir. Adult Contin. Educ. **74**, 5–12 (1997)

Watkins, S.M.: Clothing the Portable Environment. The Iowa State University Press, Ames (1995)

Deeds Not Words, an Exploration into the Women's Suffrage Movement in Ulster Through the Fashion and Freedom Project

Alison Gault

Abstract This research project is the result of collaboration with the Textile Art, Design and Fashion Course students, residents and staff and the Belfast International Arts Festival. The 14-18 NOW Art Centenary Art Commissions set a student project for a number of fashion courses in England to produce work around the theme of Fashion and Freedom looking at Restriction and Release and the changing shape of women's garments pre-1914 and between the war years. The selected student work was exhibited over 2017 at a number of high profile exhibition events and galleries. The Textile Art, Design and Fashion Course at the request of Belfast International Arts Festival, were asked to bring a local dimension to the project. The focus was the women's suffrage movement in Ulster. Work produced by staff, residents and students was exhibited at the Belfast International Arts Festival. The focus of this paper is based on the discovery that my grandmother had been a part of the suffragette movement in Belfast during this period, and this research culminated in designing and knitting a dress as homage to her entitled *Lizzie*. This paper discusses the historical project, the creative process and the diverse outcomes in the context of the history of the suffragette movement in Ulster. It also references the use of fashion and textiles on that journey to emancipation, with the mantra Deeds not Words.

Keywords Textiles · Fashion · Education

1 Introduction

This research project began with the co-commissioned 14-18 NOW and Manchester Art Gallery, with the First World War Art Centenary Art Commissions Fashion and Freedom Exhibition [1]. The aim was to investigate and evaluate the impact of the First World War on the changing nature of women's dress in the interwar years. 'Students were given a brief called 'Restriction and Release' the results showcased the next generation of fashion talent from five universities who responded to the social

A. Gault (✉)
Ulster University, Belfast BT15, Northern Ireland
e-mail: ae.gault@ulster.ac.uk

A. Majumdar et al. (eds.), *Functional Textiles and Clothing*,
https://doi.org/10.1007/978-981-13-7721-1_19

and sartorial changes brought about by the First World War. The emerging creatives where students from Leeds College of Art, London College of Fashion, Manchester School of Art, the University of Salford and the University of Westminster, selected from numerous submissions. They responded to either the theme of pre-war *Restriction* or post-war *Release* and students from London College of Fashion were invited to respond to both' [2, 3]. Ulster University students were also invited to respond to both *Restriction and Release* in the brief and this time through textiles, fashion, ceramics and jewelry. The universities selected also represented areas across England and Northern Ireland that had large organized societies involved in the suffragette struggle and the 'votes for women' movement in the years running up to the First World War. This was seen to inform a number of the student designs. The exhibition also included the work of six designers, Vivienne Westwood, Holly Fulton, J JS Lee, Roksanda, Emilia Wickstead and Sadie Williams [4]. The Belfast International Arts Festival was the catalyst for Ulster students and staff to participate in the exhibition with an Irish/Ulster context to both the women's suffrage movement and their response to the First World War. The launch of the project to staff and students was followed by a co-curricular making extravaganza where 180 students from two programs, produced work for the exhibition. Inspiration was found in the textile banners, brooches and badges identifying their association with the suffragette movement and the propaganda statements that were emblazoned on the textiles and jewelry including the words of Millicent Fawcett 'courage calls to courage' [5] and our chosen mantra '*Deeds not Words*' [6]. The celebratory nature of the project was realized, personal approaches were developed through local history stories and responses to the period.

The historical context was very important and local historical research was introduced to the students at the launch of the project. An overview of the suffragette movement in 1832 and the timeline including the militant direction of the movement in 1909. The impact of the First World War in 1914 and then ultimately women being obtaining the vote as a result of the Representation of the Peoples Act in 1918. The ability to vote however was restricted to women over the age of 30 and either they or their husbands had to own property [7, 8].

2 Materials

The materials used by the students from the Ba (Hons) Textile Art, Design and Fashion Course were quite diverse and included cloth such as velvet, cotton, linen and silk; yarns such as wool, boucle and linen; beads; rhinestones; plastics; ceramics; kitchen implements such as whisks and beaters. The machinery used was also diverse and covered the technology available across the course. This technology included laser cutting, screen-printing, multi-needle embroidery and digital knitting technologies. The processes used also included slow and traditional ways of making work, hand embroidery, hand sewing, hand knitting and crochet. Students from the Ba (Hons) Ceramics, Silversmithing and Jewellery course used materials including clay, porcelain, found objects, plastics, metals and everyday household objects found in the kitchen or specifically for domestic use.

3 Methodology

Research methodologies used were video, photography, questionnaires and blogs. A public response to the exhibition was also captured through photography, written and digital feedback through social media. First-year students used blogs to reflect on their experience working on a project in the first week of Semester 1 and this will also form part of the critical enquiry. Research included local newspapers, historical societies, online resources and searching family history archives.

4 Historical Context

To launch the project, students were introduced to the context of the suffrage movement in Ulster with a lecture. The importance of the local history aspect was of particular interest and central to the research process. This introduction to the uniqueness of the political situation within Ireland allowed students to understand aspects pertinent to their local history. According to Diane Urquhart, Isabella Todd was a "self proclaimed chief pleader for women's suffrage in Ulster," this was also supported by Todd's writing at the time, "You know how deep is the conviction of the best women in Ulster … this claim has reached all parts of the Province, all grades of society, all creeds and classes … " [9].

The Ulster perspective was further complicated with the 'Home Rule' movement, and the differences between Unionism and Nationalism, an anonymous letter, published in *The Citizen* in February 1914, according to Urquhart ascertained that members of the Ulster Women's Unionist Council were making positive advances by invalidating:

> The cowardly desire to enjoy all the advantages of the State and leave to men all the drudgery of political life … The women of Ulster have left the Home … They realise that there is work in the political sphere for women as well as men … [10].

5 Personal History

The impact of women rallying together and demanding the vote can be seen in the numbers of women joining organizations such as the Irish Women's Federation (IWF) in various parts of Ulster. My own grandmother Lizzie Moore, born in 1894, lived during this time and is a source of inspiration in my own work. My memories are of her apron, and that she always seemed to be knitting. Feminine handicrafts were

Fig. 1 Lizzie Moore Circa 1918 my grandmother and the inspiration for the project image from the family archive. Photographer: unknown

also central to the suffrage movement. Badges, emblems and brooches identified the women who were aligning themselves with the suffragette movement. August 4, 1914, sees Britain declare war on Germany. During the war years, 1914–18, an estimated two million women replace men in traditionally male jobs Emmeline and Christabel Pankhurst (leaders of the Women's Social and Political Union) [11] cease campaigning, support recruitment of soldiers and urge women to join the war effort. Soon after the most popular of the Ulster suffragette groups closes its offices on College Square East in Belfast. NUWSS under Millicent Garrett Fawcett declares 'unconditional co-operation with the Government' [12] many women felt it their duty to knit for the war effort. This reference to knitting replacing the women's struggle for the vote and also the fact that aprons were sold to raise funds for the suffrage movement formed the rationale to make a knitted apron dress in homage to my grandmother. The knit design reflects the style of the period, in the colors of the suffragette mantra dignity (purple) purity (white) and hope (green). Made mostly from linen yarns retrieved from Ross's mill (closed in the 1990s) in Belfast. Belfast was world-renowned as the center of Linen, the Linenopolis. Other centers for yarns were Manchester, known as the Cottonopolis and Huddersfield known as the Woollenopolis. My grandmother and her two sisters worked in the Linen Mills. They had been successful at school and could have gone on to further education. However, this was not an acceptable option for women at that time. My great grandfather had decided, even though he was a man of some substance owning his own business that his daughters would take up employment at the local mill. Interestingly, this mill is on the same street that I now work. Imagine that, one hundred years later almost to the day, I have had the experience of being educated in a higher education environment and now work as a Senior Lecturer at Ulster University on the Belfast Campus, York Street. The York Street Mill not far from my current location was my grandmother and great aunts location for employment. The yarns used in the garment were the remnants from the linen spinning mills, it felt completely appropriate. My grandmother and Aunt Sarah were both a spinners and my Aunt Mary was an overseer all based in the linen mill in York Street Belfast (Fig. 1).

During this period, women in Ulster were able to go and work in the munitions factories. The linen industry had provided the opportunity up to the First World War for women to work as spinners, reelers, weavers and managers. As a result, more women had employment than in the rest of the UK. Their skills in knitting, weaving and spinning were transformed into jobs manufacturing shells and munitions [13].

The process used in the design of the aprons was 3-color jacquard, with birds-eye backing (Fig. 2). The garment was knitted in two pieces, with an overall swirling floral and foliage design combined with a small repetitive floral (Fig. 2). The integration of two different designs was an important interpretation of the Arts and Crafts movement from that era; Irish artists and designers such as Harry Clarke were the inspiration. I worked on a number of pieces for the exhibition including a neckpiece knitted in linen and cotton and made it into Suffolk puffs before sewing together (Fig. 3). A large rosette was created as an example for students relating to suffragette artefacts from that era (Fig. 4). All pieces used the three colors selected for the project, purple, white (in the garments off-white) and green. The designs were connected and assembled in a wrap style with the straps knitted in 100% linen yarns. One of the large knitted pieces was tied at the neck similar to a large apron and the other was tied round the waist, also like an apron. The memory of my grandmother was outworked through knit into an apron-style garment. The shape was also reflective of the shapes in fashion between 1910 and 1914 and the idea that they could be easily tied and untied and the stretch of the knit was also my response to the restriction and release theme and direction identified in the Fashion and Freedom Exhibition [4].

6 Artist/Designers in Residence

The Artist and Designers in Residence are recent graduates from the Textile Art, Design and Fashion Course. They are selected following graduation and subsequently offered studio space and access to the workshops and facilities. The residents in return support student learning by working with first- and second-year students in the workshops. This is a real opportunity for new graduates to make work in a supported environment. They can also (through supporting the students) gain a Higher Education Academy Fellowship Award (HEA). The HEA award is nationally accredited and will support future teaching in higher education, so is recognized as enhancing employability. The undergraduate students are able to work with recent graduates and feel more at ease to discuss the course, approaches to work and this then supports the new first-year students as they begin their journey into higher education. The residents decided during their research, that they would focus on a more recent history with the *Troubles*, they looked at the Peace Movement in Northern Ireland in the 1970s. The two main organizers at the time Mairead Corrigan and Betty Williams reached breaking point over the extreme violence during the troubles. They decided to unite, women from both the Protestant and Catholic communities to protest the continued violence in Northern Ireland. Recent televised documentary's called the *Peace makers* [14] were influential in their research and the students selected to make

Fig. 2 Image of Apron dress *Lizzie* designed by Alison Gault. Photography: Niall Murphy. Model: Melissa Elliot

Fig. 3 Knitted neckpiece and wrap designed by Alison Gault. Photography: Niall Murphy. Model: Melissa Elliott

their work focusing on these issues. Interestingly, it is noted on the *Peace People* Web site that 'the local parish priest was on holiday and was temporarily relieved by a missionary priest who had worked most of his mission in India and was familiar with the Gandhian movement there in the Forties and afterwards. Following a conversation with him, McKeown, himself committed to nonviolence and a community activist in his own neighbourhood of Ballynafeigh, wrote a feature that week on 'What would Gandhi do in Belfast?' [15] The residents' statement about the work

Fig. 4 Rosette designed by Alison Gault. Photography: Niall Murphy. Model: Melissa Elliott

Fig. 5 Artist in *Residence Dress The Movement of Freedom*. Photography Niall Murphy. Model: Melissa Elliott

that was selected and displayed alongside Vivienne Westwood exhibit *Camouflage Dress* was entitled *The Movement of Freedom.*

> The struggle for liberation is a never-ending story when it comes to women's emancipation. These pieces connect the struggles of the women who fought for the right to vote and those of the women of Ulster past and present. Our collection has been inspired by the cooperation and collaboration amongst female activists and agitators against the backdrop of years of civil war and The Troubles. Our own collaboration that of newly graduated women with diverse backgrounds and skills, has been undertaken in their spirit and their honour.
>
> We borrowed some of the language used by women to incite change – using the words of women of Ulster in combination with silhouettes from eighteenth Century Irish dress. Traditional fabrics in a monochromatic color palette contrasts the bold red text symbolizing the sometimes violent acts women felt driven to carry out against the places of male authority. The power and resonance of these words in our designs represents their struggle for release from societal and religious restrictions.
>
> Our collection is the outcome of the merging of individual visions into a single coherent one – inspired by debate and cooperation among women who brought about fundamental changes. The work is a love poem to the spirit of women everywhere. The TADF Artists/Designers in Residence (Fig. 5).

7 The One Week Project

The students in all years were introduced to the project with a lecture outlining the 14-18 NOW vision, with the Fashion and Freedom Project. The students then had to research how they were going to develop their own themes around Fashion and Freedom project. Students made a range of work, including brooches, badges, hairbands, bags, ribbons, hats, headpieces, sashes, and artifacts. As they worked together, there was an opportunity for the first-year students to familiarize themselves with other students on the course and the academic and technical staff that they will be working with for the while studying at university. Some of the comments from the student blogs showed the student response to this quick way of working to a brief within a defined deadline.

> I found that the Fashion and Freedom project was interesting as we learned about women's history and the suffragette movement. Before this I knew little about women's struggle to earn the right to vote, this is something we often take for granted and it was a motivating way to learn about our history as well as making designs [16].

Some of the students opted to help with the curation of the exhibition, and one of the first-year students reflected on this positively,

> Uni is so much better than I could have imagined…I don't even know where to start to explain. Maybe being art college it's a bit of a different experience but it's definitely the right one for me. Being thrown in at the deep end in week 1, into the 'Fashion and Freedom' Exhibition…was pretty cool to say the least but also seriously terrifying…I mean what if I just blanked and couldn't create anything…and they looked at my work and decided I wasn't right for this after all…but I think we were all in the same boat. It gave us or at least me, a lot more confidence. I realised you just have to go for it. Throw yourself into it…basically

> we had to come up with our own response to the exhibition…I didn't think I had any kind of opinion or response to it…I mean its about suffragettes…they seem pretty hard core feminists, right? And I had never considered feminism before. Its so easy to overthink and keep waiting for ideas to hit you but one thing I quickly realised is…I don't have time for that to happen not in Uni, not if I want to keep up…and we only had one week for this. But eventually I worked it out and I made the things… and they got displayed. I mean its pretty awesome for only being there for 1 week [17].

Another first-year student wrote in her blog the rationale for selecting to work the way she did,

> When coming up with the idea of what to make I was not sure what to choose, as we had learnt the colours of the suffragettes where green, white and purple. This meant that these were the colours that I was working with, I brainstormed my ideas, thinking of badges, or necklaces or slippers I could make. I then began to think of things which were not feminine at the time but are now acceptable for woman to wear. The main thing which I could think of was a suit, woman were not allowed to wear a suit and it was seen as unacceptable if they did. With this idea I chose to look at a suit jacket and its top pocket, thinking how I could make a masculine thing more feminine and how I could make a badge to be worn on it that would represent a suffragette. I was extremely happy with my outcome, especially the sequinned napkin and the medals [18].

8 Results

There is marked improvement in attendance and engagement in both subject areas. The one week project allowed students to focus and make design decisions quickly. There was an increased sense of belonging for 1st years, as a result of the co-curricular approach, working across courses and alongside artist and designers in residence and staff. The interdisciplinary nature developed cross-fertilization and students felt part of a 'community of practice' on the Belfast Campus. They were sharing materials and ideas, having conversations with different tutors. An increase in confidence was highlighted as a result of participating in an International Arts Exhibition (Table 1).

Attendance monitoring of the first-year students,

During the exhibition, I kept a book to register the attendance and the response, some of the comments were insightful,

> Great exhibition, I loved seeing the work of designers and students displayed together. Anon

> Lovely work, wish it was for sale, especially the small items, brooches and badges it would have been interesting to see how much they were, very unusual. Anon

Table 1 First-year attendance monitoring first semester 2017 weeks 1–12

Year	Weeks 1–4 (%)	Weeks 5–8 (%)	Weeks 9–12 (%)
2017	98	88	92
2016	94	87	90
2015	89	87	86

> Interesting exhibition, good variety of work displayed one of my favorites this year, well-done. Anon

The exhibition also attracted the press and Ulster Television (UTV) journalist, Sarah Clarke, reviewed the work. Twitter, Instagram and Facebook also commented and reviewed on the exhibition, and it was counted in one of the Belfast International Arts Exhibition 3 outstanding highlights.

> Welcome everyone to the official final day of the 2017 Belfast International Arts Festival. There's still plenty to see before it all ends. We have the FINAL days of three extraordinary Festival highlights in The Tempest: Ireland. Memory. Identity, Fashion and Freedom and Hard to be Soft [19].

A video was filmed and was jointly funded by Ulster University and Belfast International Arts Festival and was used as the Ulster University publicity for the project [20]. There was also a longer film created and this was viewed alongside the exhibition, explaining the project and included myself and representatives from the residents, Francesca Keaveney and Aoise Herron. Some of the feedback following the viewing of the video was also recorded in the comments book,

> I really enjoyed watching the video and how the students made their work. Visitor, Dublin

> It was interesting to listen to the local history stories and see how this was used as inspiration. Visitor, Ballymena

> Great video, impactful stories, I enjoyed seeing the process of making the work. Visitor, DownPatrick

The exhibition was well received with Richard Wakely director of Belfast International Arts Festival stating that the final numbers attending the exhibition were 3900.

9 Discussion

The Fashion and Freedom project presented opportunities for students to work together across all three years in Textile Art, Design and Fashion and across two years in Ceramics, Jewellery and SilverSmithing. The interdisciplinary nature of the project developed cross-fertilization in thinking and direction for the students, and encouraged the formation of a 'community of practice' on the Belfast Campus. They were sharing materials and ideas, having conversations with different tutors and becoming part of an International Exhibition with high profile designers and 27 students from five universities. The results showed that they were largely unaware of the Ulster Suffragette Movement and the fact that it was 'an articulate and definite cry for political freedom' [9]. The opportunity to showcase their work with outstanding designers was something of a highlight for their CVs and improving employability. The time constraints of the project brought focus and outputs that were often unexpected. The focus in my own research was to bring a personal dimension, for students to see how I approached this theme. The residents also experienced working as a team for the first time. They came from different parts of Ireland, North and South and they were working in different disciplines. This synergy was a liminal experience around a 'real life' scenario. Research and investigating their recent history provided diverse perspectives and discourse while making their work.

The students were surprised at the outcomes, realizing that they could produce work within a tight timeframe, get to know their cohort in a pressurized setting and learn to research, develop and manufacture to a brief. The first year feedback via the blogs shows that they have an understanding about their individual learning style, and how to learn and develop study skills in the early stages of their University Careers. They also experienced what it was to be part of an International Exhibition, and this in turn will develop confidence in their own work as identified in their blogs, being reviewed and selected for display. It is clear that they were astonished by their own ability to make work and be part of such a high profile so early in their University career. The attendance rate shows that they have been engaged in their activities across the semester.

10 Conclusion

The project was complex to organize and was multi-faceted; it included pedagogical themes such as induction, co-curricular, interdisciplinary and was also aligned to the suffragette movement and the subsequent impact and transformations for women with the outbreak of the First World War. The constraints of the one week project due to the International Exhibition brought a body of work from both staff and students that covered many aspects of making, craft, design and fashion. The diverse approaches supported the learning and allowed for students and staff to create work and learn from each other's tribes with creative results. For my own research, I discovered many aspects of my family's heritage not realized prior to the project. The networks that have been developed through this project will reap dividends into the future. These have included working the 14-18 NOW commission, the curator Jenna Rossi-Camus, Richard Wakely the director of Belfast International Arts Festival and the Ulster Museum. This coming year 2018 will celebrate 100 years since women were given the vote, I am sure there will be projects for the Textile Art, Design and Fashion staff team, residents and students to participate in and to raise the profile and role that textiles and fashion has had historically and the part it has to play in the future.

Acknowledgements Thanks to the Research Institute Art and Design, Ulster University for all their support with this project. Thanks also to the curator Jenna Rossi-Camus, Richard Wakely from the Belfast International Arts Festival and Staff and Students from Textile Art, Design and Fashion.

References

1. Fashion & Freedom: Fashion & Freedom [Internet] (2017) [cited 30 Sept 2017]. Available from: https://www.1418now.org.uk/fashion-freedom. Last accessed 2017/11/10
2. Festival Events 2017: Belfast International Arts Festival [Internet] (2017) [cited 30 Sept 2017]. Available from https://belfastinternationalartsfestival.com/festival-2017/. Last accessed 2017/11/10
3. Pidd, H.: How the first world war liberated women's wardrobes [online]. The Guardian (2018). Available at: https://www.theguardian.com/fashion/2016/may/13/how-the-first-world-war-liberated-womens-wardrobes. Accessed 14 Oct 2018
4. Designers Archive—Fashion & Freedom: Fashion & Freedom [Internet] (2017) [cited 15 Sept 2017]. Available from: https://www.1418now.org.uk/fashion-freedom/designers/. Last accessed 2017/11/10
5. Fawcett: Available from: https://pictorial.jezebel.com/suffragist-campaigner-millicent-fawcett-is-the-first-wo-181857758. Last accessed 2017/11/10
6. Deeds not Words: Oxford Words Blog [Internet] (2017) [cited 11 Sept 2017]. Available from: http://blog.oxforddictionaries.com/tag/deeds-not-words/. Last accessed 2017/11/10
7. NI Assembly: http://education.niassembly.gov.uk/sites/userfiles/files/suffragette_timeline.pdf. Last accessed 2017/11/10
8. Baker, J.: The Ulster Suffragettes [online]. Issuu (2018). Available at: https://issuu.com/glenravel/docs/suff. Accessed 14 Oct 2018
9. Urquart, D.: 'An articulate and definite cry for political freedom': the ulster suffrage movement. Womens's Hist. Rev. **11**(2), 273–292 (2002). http://dx.doi.org/10.1080/09612020200200321. Downloaded on the 15 Aug 2015, last accessed 2017/06/10
10. [ibid]
11. https://www.biography.com/people/emmeline-pankhurst-9432764. Last accessed 2017/11/10
12. http://education.niassembly.gov.uk/sites/userfiles/files/suffragette_timeline.pdf. Last accessed 2017/11/10
13. http://www.bbc.co.uk/programmes/p03f9k9h. Last accessed 2017/11/10
14. The Peacemakers: http://www.bbc.co.uk/programmes/b0956gws. Last accessed 2017/11/10
15. http://www.peacepeople.com/?page_id=8. Last accessed 2017/11/10
16. Yasmin Liggott: https://yasminliggett.wordpress.com/2017/11/01/learning-highlights/. Last accessed 2017/11/30
17. Katie Ireland, first year student TADF, https://katieirelanddesigns.wordpress.com/. Last accessed 2017/11/30
18. https://chloehalesuu.wordpress.com/. Last accessed 2017/11/10
19. https://en-gb.facebook.com/mybelfastfestival/. Last accessed 2017/11/10
20. https://vimeo.com/237115957. Last accessed 2017/11/10

Dynamic Anthropometry for Investigation of Body Movement Comfort in Protective Jacket

Inga Dabolina, Eva Lapkovska and Ausma Vilumsone

Abstract This paper focuses on analysis of human body measurements and movements—their maximum amplitudes—forced postures and characteristic poses for the performance of a particular work. Analysis of the work postures and PPE were performed on 155 males who discharge their duties in the Latvian National Armed Forces. From different posture tests, two were selected—bending and squat positions. Analytical results of both poses were compared with the human body measurements taken in the standard position. Measurements were made using a non-contact anthropometric method—human body 3D scanning. The results were analysed and classified for the entire sample. Scan results for the standard position were compared with traditional anthropometric techniques—primary measurements were taken with a tape measure and anthropometer. The most prominent implementation of the two poses was analysed in the paper by carrying out correlation analysis to evaluate mutuality of the results. However, no correlation has been found which would be described by the regression equations. To analyse a protective clothing jacket, a single test subject was scanned using physical markers for marking anthropometrical landmarks. Performance of different work positions dynamically identified various movements of anthropometric points. Such a result indicates necessity for further research in the field of anthropometry, dynamics and motion amplitude studies.

Keywords Garment fit · Anthropometrics · Human body 3D scanning

1 Introduction

1.1 Comfort in Clothing

Recent studies show that comfort in wear is the most important property of clothing demanded by users. Especially important comfort in wear is for athletes, physical

I. Dabolina (✉) · E. Lapkovska · A. Vilumsone
Faculty of Material Science and Applied Chemistry, Institute of de-Sign Technologies, Riga Technical University, Kipsalas Street 6-220, Riga 1048, Latvia
e-mail: inga.dabolina@rtu.lv

A. Majumdar et al. (eds.), *Functional Textiles and Clothing*,
https://doi.org/10.1007/978-981-13-7721-1_20

workers, rescuers, etc. Fundamental understanding of human comfort and knowledge of how to design textiles and garments to maximize comfort for the wearer is therefore essential in the clothing industry [1]. In addition, understanding the comfort of clothing can be described in several respects, for example, heat balance, energy metabolism, tactile comfort, body movement comfort, etc. This understanding reflects demands of a human body: anatomical (morphology, locomotor system, abnormalities), physiological (water regulation, thermal regulation, physical sensation), as well as psychological considerations (duration of use, wearability) [2].

1.2 Requirements for Protective Clothing Design for Wearer Comfort

Clothing as a constituent of personal protective equipment (PPE) system must comply with numerous and sometimes even incompatible or difficult to combine safety, comfort, performance and other conditions. Taking into consideration the said aspects, key design factors can be outlined: protection, comfort, mobility, connectivity and ease of use [3]. Along with the modern PPE, the look and the overall appearance are also important—protective clothing must be provided not just with PPE quality, but as well as with fashionable look and aesthetics [4]. Consequently, the task of designing becomes even more complex to provide contemporary materials and integrative consumables with the fit of workwear that not only allows productive execution of the task and use of PPE smart devices, but also offering an aesthetic appearance.

In addition, protective wear, defined as clothing containing protective covers covering or replacing individual clothing and designed to provide protection against one or more hazards, also must comply with safety and performance requirements (not affecting the health and hygiene of the user), size designations and clothing marking (measuring procedures and checking standards for compliance with standards ISO 3636, ISO 7250) and wear-out (changes in one or more initial qualities of protective clothing over time). Within all these requirements, users must be provided with a level of comfort that is in line with the level of protection against hazards, environmental conditions, user activity levels and expected wear time [5].

Standards already call for the ease of putting and removing the clothing, with a possibility of unrestricted bending of hands and knees, as well as leaning movements while at the same time leaving uncovered unprotected parts of the body during movements—there is sufficient jacket and trousers overlay [5].

1.3 Dynamic Anthropometry in Protective Clothing Design

Anthropometric standards (EN 13402, ISO 7250) characterizing types of human body sizes used in the design of clothing are based on data derived from measurements of male and female bodies in the standard posture. Such results may not always fully satisfy requirements of pattern makers who are to deliver clothing patterns for a man in motion, rather than to a tailor's dummy with "fixed posture". Due to movements, the size of the body shape and its parts vary, and positions differing from the basic static anthropometric posture are unlimited [6, 7].

Professionals must be familiar with human anatomy and functions to understand the body in its kinetic positions and to be able to design clothing that moves properly along with the body [8]. The adult body consists of 206 bones, most of which are joined by moving joints, creating many combinations of positions that are employed in everyday movements. In addition, their usability and frequency of use depend on the task to be performed and areas in which the person works. There are also a large number of variations for the potential volume/amplitude of movements between individuals in the population that are associated with the total morphological features as well as the differences between the members of the population: age, race, gender, health status, physical fitness, body proportions and temporary effects of exhaustion—these can all be significant factors that can affect movement [8]. For example, sports and active leisure design studies have shown that as a result of the movement, the skin of the human body significantly extends on the knee—35%, and elbow—in 45% of zones. Depending on the movements that can be made (movement segmentation of legs, lower back, thighs, shoulders, belly, hands, etc. in the human body), the wearer stretches in different ways, and the stretching of the skin in different areas ranges from 10 to 50% [9].

For motion description and research, it is initially worthwhile to study body size changes (increases and decreases of lengths and circumferences) at the extreme head, torso and limb states. With such an approach, the pattern maker will be able to apply not only the required mean values of measurements, but also the maximum increments obtained in the extreme position measurements [6]. However, the results of such dynamic indicators are only indicative, and their application in the development of the patterns will depend on the specific clothing type with the function assigned to it. When researching dynamic indicators, the following tasks must be addressed: the choice of dynamic indicators necessary for the development of patterns and the choice of measuring methods; determination of the statistical processing of data; determination of the direction of application of the obtained results [6, 10].

There are standards that define the movements for the development of special projects [11], but the ergonomic standards only describe different modes of motion, e.g. arm range, without any size references, just percentiles type. Due to this, the correlation between the two measuring systems is missing [12].

Different researches have been carried out at different times, differing both in terms of the size and gender of groups in question, the methods of measurement used, as well as the selected body postures and their amplitudes. Already a study carried out in the year 1968, including the CSR (Czechoslovak Socialist Republic) and the GDR (German Democratic Republic) population parts, as well as the USSR (Union of Soviet Socialist Republics) participating specialists [6], resulted in the compilation of various dynamic indicators. Large groups of people (340 individuals—170 females and 170 males) were studied using traditional anthropometric manual methods, and measurements were made in positions such as deep inhalation, back torsion, seated posture, chest torsion, bending, arms upwards (hands together), arms frontwards (wrists together), hands bent in the elbow at 90° angle, hands horizontally to the front at 90° angle, leg on a chair and squat [6]. In today's research, using contactless measurement methods, for example, changes in the measurements of 24 men's upper body in five movement positions were gained, which show both significant decreases (negative—22% chest width) as well as increases (positive—27% back width) [9]. For the purpose of exploring the dynamic suitability of diving suits, test persons were scanned at 5 positions specific to the diving process and testing of prototypes in real-world conditions was performed [13]. By analysing and classifying work positions to identify the most characteristic forms of motions, as well as to define poses so that they can be reproducible, even 10 different body poses were scanned [14]. For the study of the lower body, women's conditional mundane poses such as a standing position, 120° knee bending, one step climbing and sitting—knees bent at 90° angle were selected and the dynamic measurement deviations from the standard position measurements were analysed [15]. Another lower body study examines changes in values within the larger range of movements and with more detailed set of measurements when examining changes in the leg surface by comparing the results obtained in a standard posture and at 5 different heights in leg positions [16].

Considering the use of dynamic indicators in clothing pattern design, it should be noted that, if in the case of lingerie and sports clothing knitted fabrics are used to ensure freedom of movements, than in design of protective clothing outerwear made from non-rigid fabrics without the use of appropriate ease allowances, expansions/extensions may be allowed in certain garment areas [3]. The obtained dynamic indicators can be used to a reasonable extent on the ease allowances associated with significant body measurements for pattern making, such as waist and chest circumferences, back width, length and slope, the distance from the back armpit fold to the waist, the upper arm and elbow circumferences, the distance from the back armpit fold to the wrist and the hip/buttock circumference. [17]. In addition, there can be various constructive and technological solutions: pleats, anatomical seams, pre-shaped sleeves and trouser legs.

1.4 PPE Dynamic and Fit

The standard of general protective wear requirements states that the design must promote proper positioning of PPE on the user and ensuring that it stays in its place for the intended period of use, taking into account environmental factors along with movements and poses that the wearer can take during work or other activities. For this purpose, the appropriate means adequate size ranges and adequate adjusting systems, so that protective clothing would be adjustable to a user's morphology [5].

Mobility, or freedom of movements, is influenced by features of fabric used for the wear, as well as the design of the clothes [3], achieving that they are not too tight or too loose, and do not impair normal body movements [5]. Dynamic fit of workwear calls for maximum comfort of movements in response to minimum movements of clothing parts against the wearer's body, because, as defined by the standard [5], the design of clothing must ensure that body parts are not exposed during the foreseeable movements (e.g. the jacket should not rise above the waist by raising hands), thus failing to safeguard the body from surrounding hazards.

Weight and bulk of clothing can also encumber body movement. It has been found that in the case of bulkiness, each layer of the outfit may increase wearer's energy consumption by about 4%, caused by troubled pace and interlayer friction [3]. Interlayer friction is an essential factor in designing each of separate layers that are to be put atop one another so that outer layers do not sit too tight and do not impair body movement. In addition, friction may increase due to sweating—thus causing thermal discomfort, as well as limiting move of clothing over the surface of the body parts. For example, a scan experiment with fire proximity suits showed that protective layers can encumber wearer's movements by reducing lifting height of his or her hands by 32 mm (2056–2024 mm) [8].

In designing, feedbacks from end users are often used to evaluate the dynamic fit. Such a subjective assessment of product properties depends on the assessor's experience, as well as traditions and habits. In fact, ergonomic tests should be carried out by experts, as set by the standard [5], determining that the test or examination is to be carried out by one or more experienced evaluators or experts previously acquainted with manufacturer's information for further investigation of the protective wear manually and visually. Such an examination should give answers whether the wearer can stand, sit, walk and climb the stairs; is it possible to raise both arms above the head; can one easily lean down to pick up small items (e.g. a pencil). Also, significant requirements are set out for sleeves and trouser legs the length of which must not encumber movements of arms and legs; clothing must not be loose enough to flutter and move in a free and cumbersome manner; there must be no points for sudden and unexpected openings between or inside components of the clothing; no unreasonable movement limitation in any of the joints is permissible [5].

Methods for assessment of dynamic conformity can also be developed in directions regarding to assessment of the volume of deformation in strained parts of clothing, displacement of clothing parts against human body and degree of movement limitations.

As in traditional anthropometric studies, mobility tests may also reveal differences within different body types, e.g. data may vary significantly in assessment of changes in typical body poses of slim males or overweight respondents [3]. Studies on sport wear also address further investigation of interaction between clothing and human body, considering the fact that changes in dynamic body measurements will be affected by such factors as the kind of sport, flexibility and build of an athlete, as well as the type of the wear [9].

1.5 Aim of the Research

The target group of the performed research is soldiers of the Latvian National Armed Forces. This research is aimed at investigation of body movement comfort in protective jacket by looking through jackets wearability and compliance with human body movements. The fit evaluation is one of the main problems for uniforms. Choosing appropriate size uniform has large influence on protective clothing functionality. In case of wearing wrong size - too small or excessive clothing can cause inappropriate movement amplitude.

2 Materials and Methods

2.1 Target Group

Since workwear is an integral part of the PPE for fulfilment of military tasks in the field, soldiers, whose routine work include execution of these tasks, were studied as a target group. For testing and analysis of dynamic body changes, 162 soldiers were selected, initially including 7 females, in view of unisex division in the purchase of clothing. However, following recommendations of the standard [18], that in the case of both gender-group studies, both sexes should be represented in an equal number, only 155 males were discussed in the following section. The study group included soldiers aged 19–52 years, with a body height of 160–196 cm, a body weight of 62–142 kg and the following parameters: chest girth 91–143 cm, waist girth 73–138 cm and buttock girth 91–134 cm.

2.2 Methods

The two sections of the study employed a non-contact measuring method using 3D anthropometrical scanner Vitus Smart XXL® (© Human Solutions GmbH and VITRONIC GmbH) with Anthroscan software.

Each soldier was scanned in 3 positions: standard upright and 2 dynamic body positions. Oral examinations of the test subjects were described and performed before addressing body postures.

In the first posture—scanning in a standing position (see Fig. 1a)—an individual was standing in an upright position, without hunching shoulders, stretching out the spine and straining the abdomen. Not touching body with hands or palms. Palms slightly bent and elbows spread. Upper arm muscles unstrained. Palms fisted, thumbs pressed to other fingers. Legs positioned distant from one another, standing on footmarks. The look turned straight forward [19]. In the second position—bending (see Fig. 1b)—top of the body tilted forward to the maximum. Chin touches chest. Arms

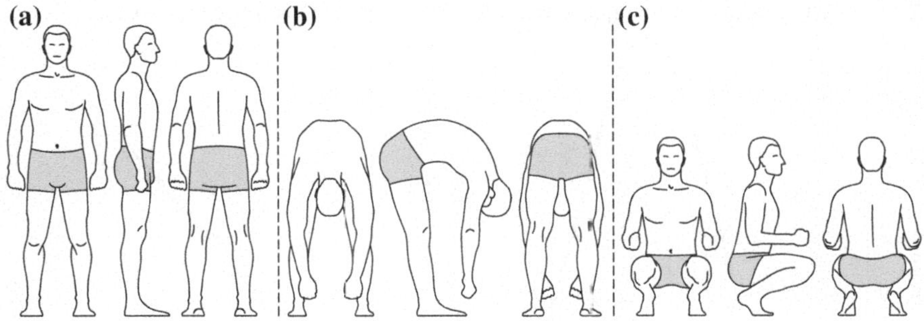

Fig. 1 Scanning positions. Standard (**a**), bend (**b**), squat (**c**)

held vertically downwards. Legs not bent in the knees. In the third pose—a squat (see Fig. 1c)—a full squat was made. The look directed straight forward. Hands bent in elbows at an angle of 90°. Forearms and wrists directed horizontally forward. Common conditions for all the postures—palms fisted, holding thumb close to other fingers; standing still during the scan process and avoiding taking deep breaths [19].

3 Results

3.1 Target Group Sample Analysis

Before detailed planning of dynamic measurements, applicable for measuring bodies of both undressed and dressed soldiers, and analysis of interactions, the entire target group was addressed covering general indicators focusing on the extent of a group of 155 males of the same profession showing represented the ability to effect certain movements within a given range. In order to group and to analyse the sample in question, information about the age of test subjects was collected (the division into age groups was chosen considering the division of the soldiers based on results of their annual physical fitness evaluation process [20], body control dimensions (body height, waist girth, hip girth), as well as body weight. Incidence of body height is shown in Fig. 2 and incidence of body weight—in Fig. 3.

Comparison of distribution of soldiers' body height with normal distribution according to the mean value and variation of sampling data shows that data comply well with the normal distribution curve—such a correspondence is also shown by QQ plot where distribution line of sample and theoretical quantiles is the line specifying data—the data points are arranged around the line in the biggest part of the interval. The differences are observed only at the ends of the sample group. However, they are not considered as significant. Therefore, it is considered that normal distribution density function approximates the selection data well and as such is suitable for data validation and analysis.

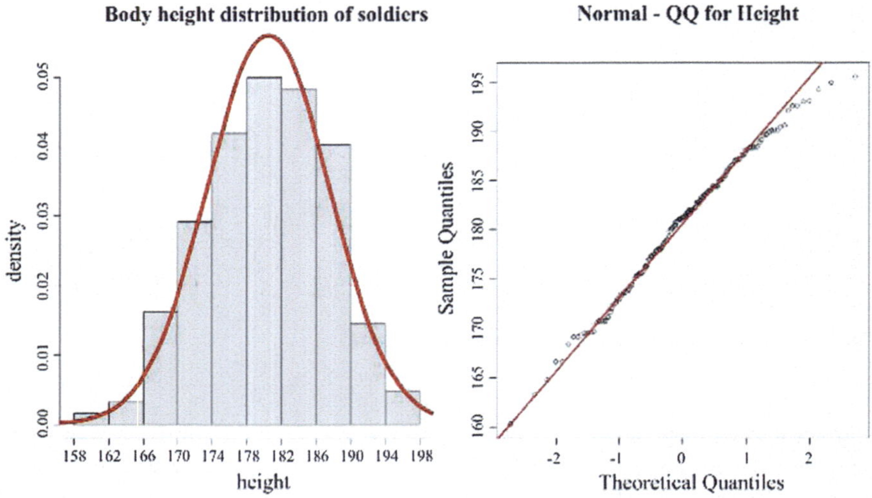

Fig. 2 Distribution of soldiers' body height in comparison with normal distribution

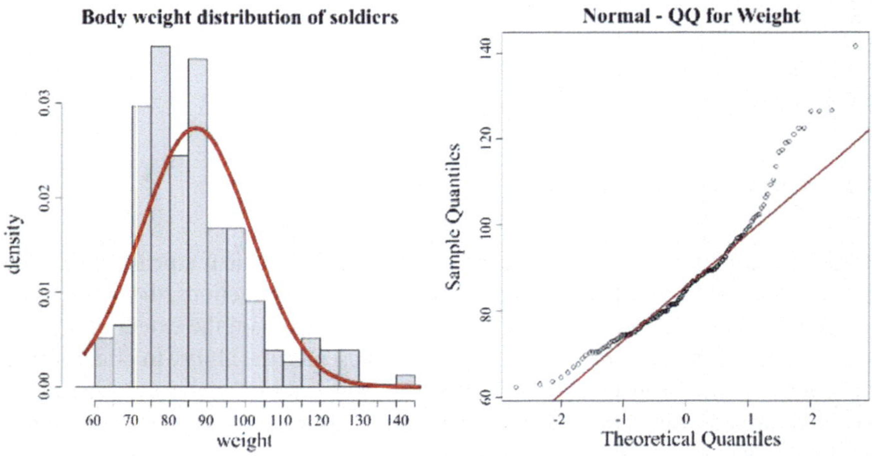

Fig. 3 Distribution of soldiers' body weight in comparison with normal distribution

Comparing distribution of soldiers' body weight with normal distribution according to the mean value and variation of sampling data it is evident that data partially meet the normal distribution curve—such a correspondence is also shown by QQ plot where sample and theoretical quantiles distribution line specifies data only in the middle of the interval. Examining the data comprehensively with the age of soldiers, coherence is observed showing that the older the soldier is, the bigger is his body weight. However, it cannot be considered as a correlating value. See boxplot in Fig. 4.

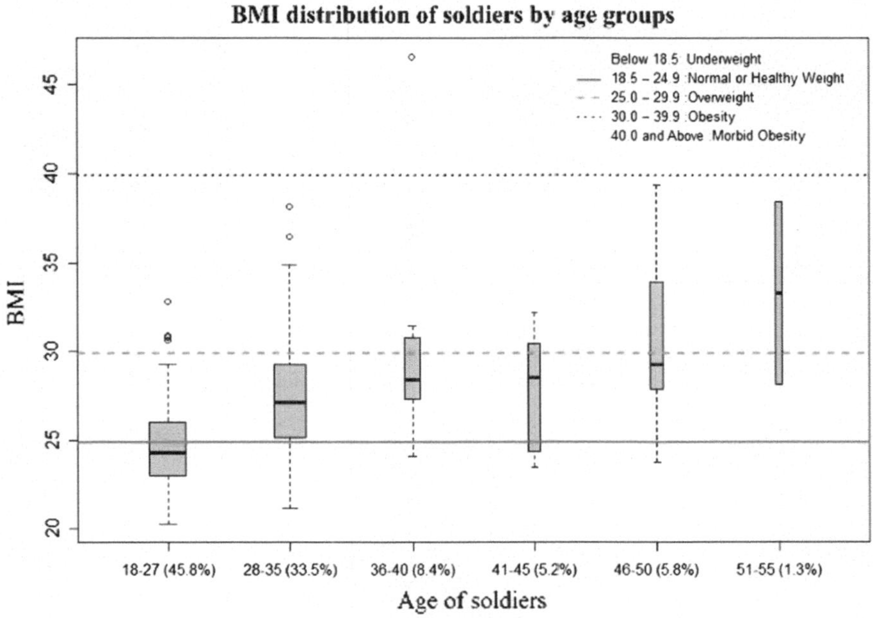

Fig. 4 Distribution of soldiers' BMI by age groups

Based on interactions observed in studies on the effect of physical fitness on possible amplitudes of motion, BMI was used as an additional indicator in the target group's analysis. This approach agrees with physical fitness requirements for soldiers and their control procedures [20], when, in line with rules, BMI is a tool to be used by commanders of relevant units to monitor physical fitness of soldiers.

Width of the boxes varies with the size of the data set, and outer points identify the outliers in the data. These data differ from the common age group tendency. Data analysis shows that the majority of soldiers (79.4%) is in the age group from 18 to 35. However, only less than a half of it has BMI indicating healthy weight. The general trend—the older the age is, the number of soldiers decreases in the age group and the BMI increases. Moreover, the diagram shows some very radical outliers.

For assessment of dynamical posture performance, measurements from the lowest palm position to the plantar plane were chosen for measuring the first dynamic posture (bending), while the second dynamic posture (squat) was measured from the lowest point of buttocks to the plantar plane, using distance plane-to-point measure used in Anthroscan system for both measures (see Fig. 5).

It was observed that the palm distance from plantar plane varies within the range from 0 when a soldier was able to touch the floor of platform with his fists, up to 41 cm, when the soldier's physical abilities do not allow to perform bending of such a large amplitude. See distribution in Fig. 6.

The "depth" indicators start at 12 cm, when, keeping the balance, trial subjects managed to perform high amplitude squat of up to 28 cm when physical fitness of

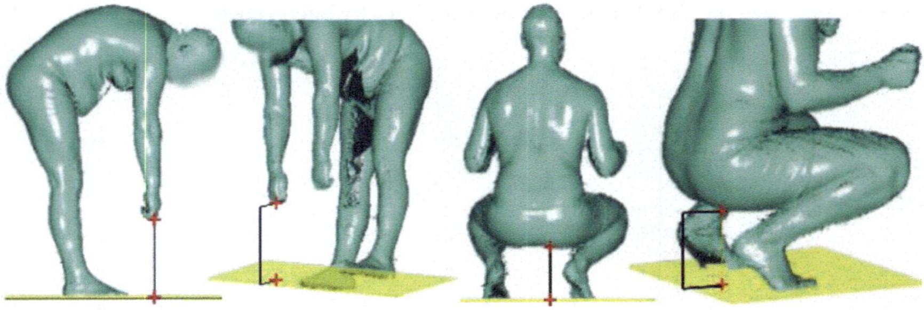

Fig. 5 Measurements from the lower palm position and the lower seat to the plantar plane

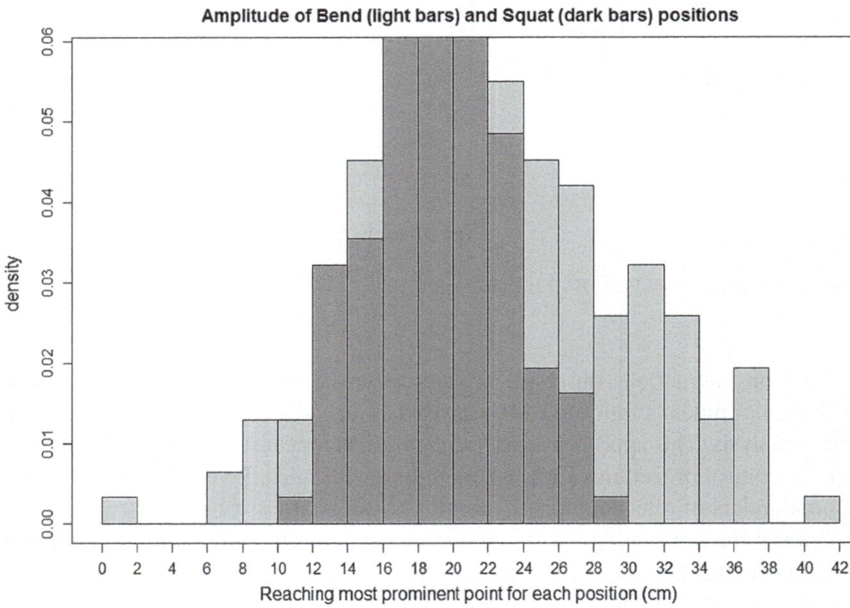

Fig. 6 Analysis of dynamical posture performance

the soldiers, compared with the others, did not allow them to make a deep squat and which often resulted in a loss of balance and an unsuccessful scanning procedure. See the distribution in Fig. 6.

As shown in the diagram (see Fig. 6), squat position results are more similar—they vary from 10 to 30 cm. However, bend position results are very diverse. It can be explained by diverse physical fitness of soldiers, indications set by anthropogenesis and side circumstances of civil life. Such analysis allows comprehension of thresholds of dynamic transfers which are significant in designing PPE (in defining ease allowances, overall dimensions of the product, gliding values of clothes and other parameters).

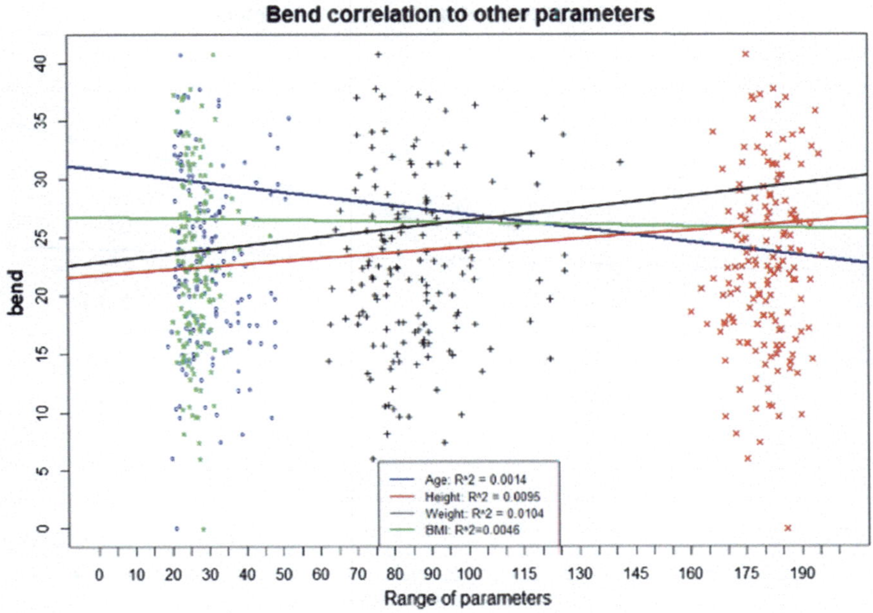

Fig. 7 Correlation analysis between the range of bend and other parameters

Neither in bending nor in squatting values correlation with indicators such as soldier age, body height, weight and BMI was observed. See analysis of parameters in Figs. 7 and 8.

Using matching of different scans available in the 3D system, eleven (11) soldiers referable to the group with body height of 184 cm were selected for visual comparison of bending posture performed by trial subjects compared to the group with a body height of 184 cm, are with normal weight (normal body weight according to BMI) and represent the most represented age group of 18–27 years, while for comparison of squatting posture seven (7) soldiers with a body height of 176 cm are of normal weight and also represent the most represented age group of 18–27 years.

Overprints of scans (see Figs. 10 and 11) show that, although soldiers have close height values (see Fig. 9), the bending amplitudes are different, which may also contribute to different dynamic measure results considering different degrees of skin surface stretching. Amplitude-influencing factors include various body features such as posture, body proportions, level of muscle growth, length of ligaments, in addition, the position of feet serves as the reference point for both postures, which, considering different body peculiarities, may be a hindering factor for deeper bending and squatting. Deviations from axial directions are also observed, for example, in case of bending, the body is redirected to the right or to the left, while in case of squatting, body weight is transmitted to one leg and the body inclines to the right or to the left.

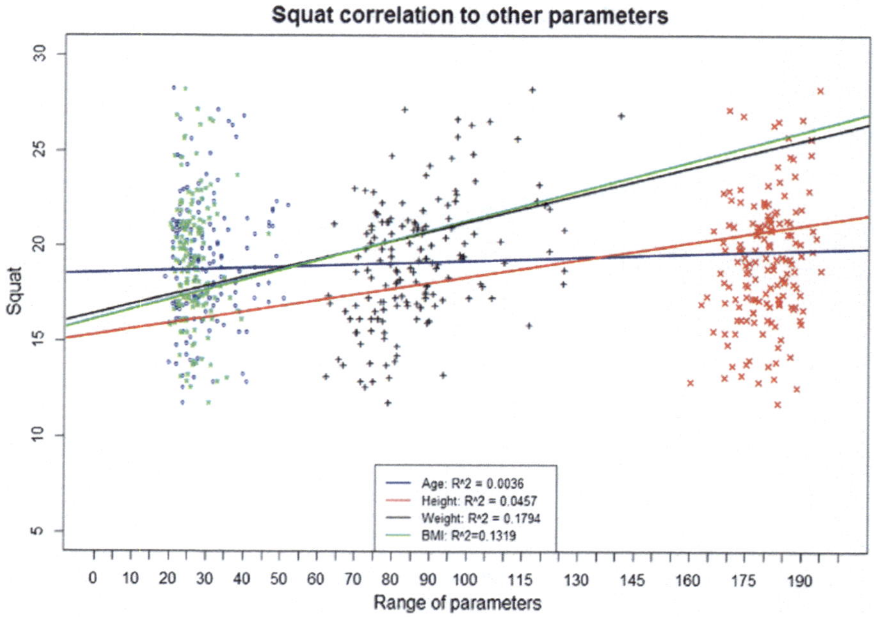

Fig. 8 Correlation analysis between the range of squat and other parameters

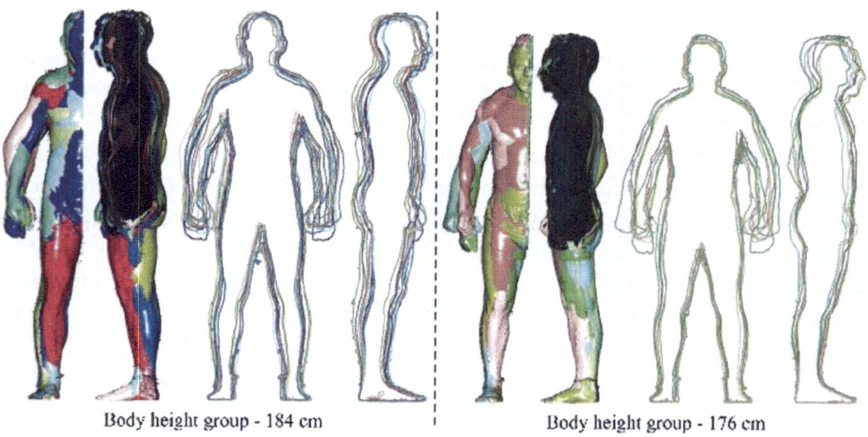

Fig. 9 Standard position visual analysis

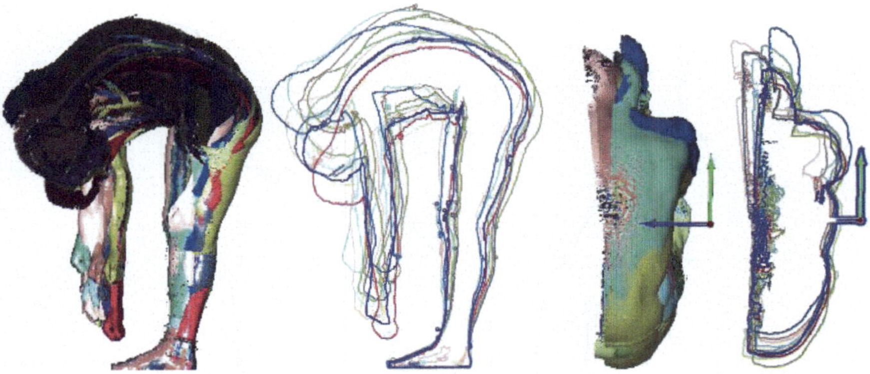

Fig. 10 Bend position visual analysis

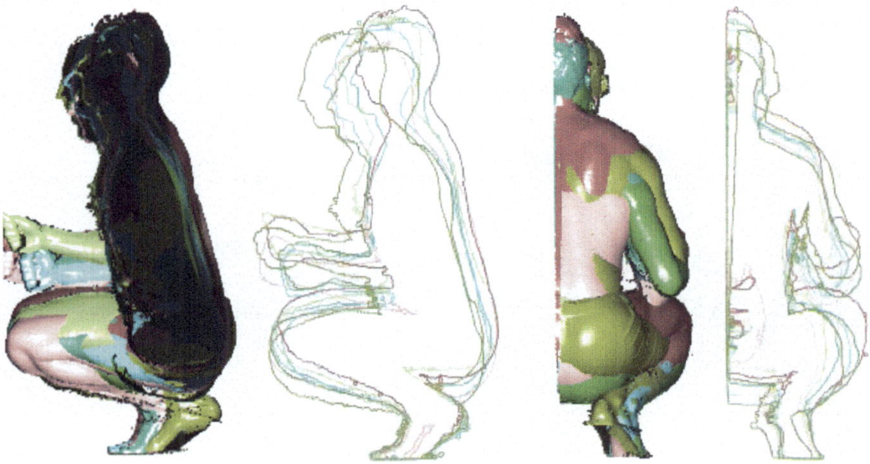

Fig. 11 Squat position visual analysis

3.2 Pilot Experiment

In a pilot experiment, the wearer of a field uniform with a jacket on and without a jacket was scanned in addition to using physical markers. The use of markers (coded for their use in the system) allows precise measurement of point movements, changes in distances and changes in body measurements in dynamics. An analysis of the transfer of characteristic points was made by developing a marker placement card according to the most important anthropometric points symmetrically on the human body (Fig. 12) for the naked body and the most characteristic nodes in jacket. Figure 13 shows an example of dynamic postures analysed.

For fixation of the posture, a light, durable tube with placement marks was used in the pilot experiment, or similarly, a so-called object for pose fixation was used in [21]

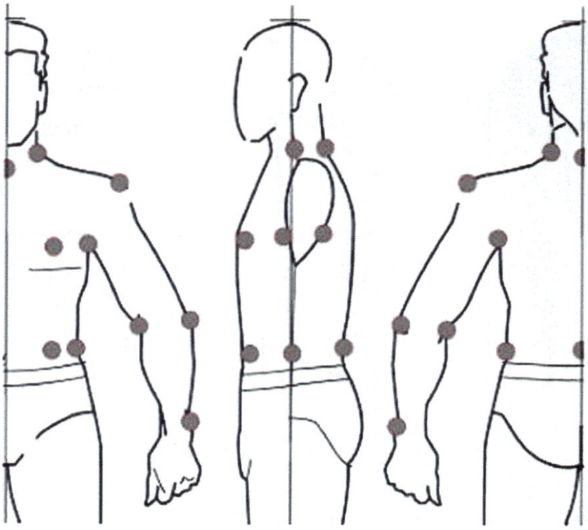

Fig. 12 Marker placement on human body anthropometrical points

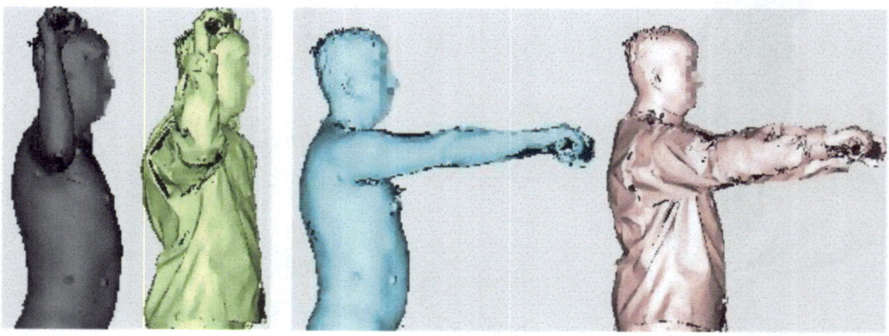

Fig. 13 Example of dynamic postures analysed (without and with a jacket)

the experiment. The standard [18] on scan techniques for internationally compatible databases also determines that one or more of so-called supporting devices can be used in case of an optimal posture which is optimally set, described and uniformly applied for all trial subjects depending on the need.

3D scanning has several advantages comparing to manual measurements—it is fast, sequential, the system stores scans for repeated measurements and complex analysis, measured values have higher precision level in terms of repeatability. Calculation of distance changes of anthropometric points and transfer of wearable items gives information about the interplay of clothing and the body in a moving person, performing their work duties and other activities. Comparing the obtained data with the standard tables of human body sizes [22] and the manufacturer's measurement table for the production of tested jacket, it can be concluded that the measurement

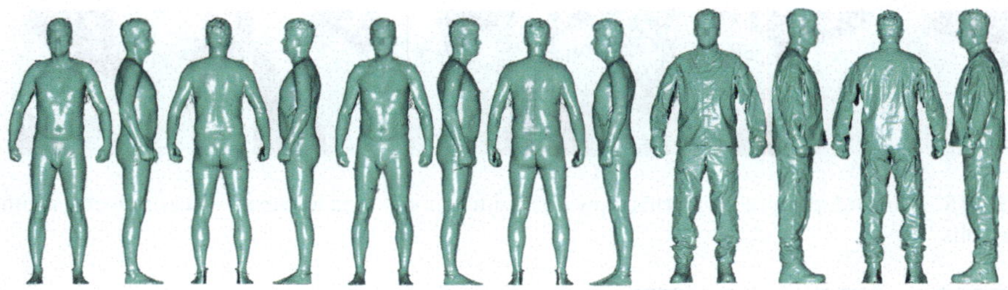

Fig. 14 Stand position. Four views (without and with markers and wearing protective clothing)

Table 1 Measurements for different postures

Posture	Arm length (cm)		Shoulder width (cm)		Width armpits (cm)	Side upper torso (cm)	
	Left	Right	Left	Right		Left	Right
Standard posture without markers (underwear)	70.9	69.8	14.5	14.7	43.6	28.0	27.9
Standard posture with markers (underwear)	70.3	71.4	16.1	14.9	42.9	29.1	29.4
$\|\Delta\|_1$	0.6	1.6	1.6	0.2	0.7	1.1	1.5
Arms upwards with markers (underwear)	71.5	71.5	9.9	8.9	34.3	43.3	41.0
$\|\Delta\|_2$	0.6	1.7	4.6	5.8	9.3	15.3	13.1
Standard posture with markers (wearing PPE)	72.2	72.2	14.5	15.1	43.3	–	–
$\|\Delta\|_3$	1.3	2.4	0	0.4	0.3	–	–

tables need to be processed—the dimensions and indifferent intervals given in them do not coincide with the results obtained in the dynamic transfer studies. Also, comparison of results by end users' measurements (155 persons tested) does not coincide to measurement tables used by manufacturers.

It is possible to use physical markers for identifying anthropometric landmarks correctly without impact of moves and dynamics of human body. These markers do not impact or restrict movements of a test subject. See Fig. 14 for scan visual evaluation with and without markers.

Body height of a test person is 194 cm and the weight is 94.6 kg. Positions for dynamic measurements are performed twice—while wearing PPE and for a person in underwear. Measurements after dynamics differ, see Table 1.

Delta absolute values ($|\Delta|$) show differences between measurements of standard posture without markers (underwear) and measurements in other postures. In some cases—for example, second delta, difference is large and thus it can show large

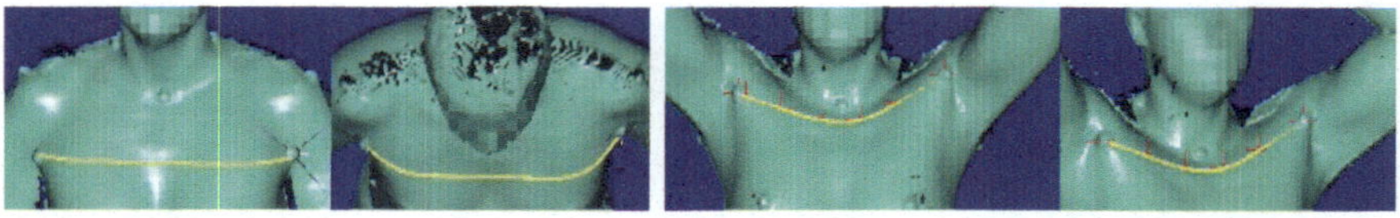

Fig. 15 Standard posture and arms upwards with markers (underwear)—measurement width armpits

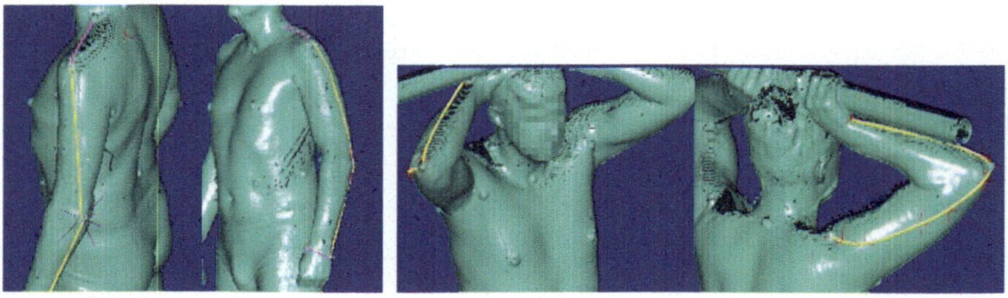

Fig. 16 Standard posture and arms upwards with markers (underwear)—measurement arm length

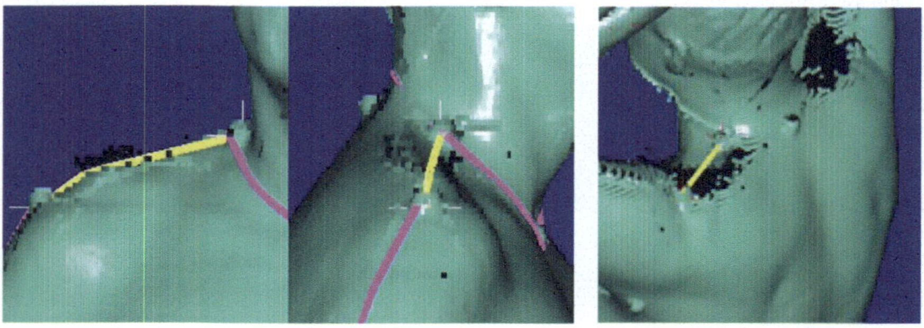

Fig. 17 Standard posture and arms upwards with markers (underwear)—measurement shoulder width

influence on ease allowance for movement comfort. See comparison in Figs. 15, 16, 17 and 18.

Fitting ease is the minimum amount of ease that needs to be added to the garment for unhindered movement. The constructive ease allowances are physiologically minimal (hygienic), movement comfort (dynamic) as well as silhouette allowances. Upper torso garment ease allowances define the basics of the chest, waist, hip, neck, shoulder, upper arm and palm, as well as adjust the length of the garment and its individual parts. A separated dynamic comfort ease is a component that is considered as particularly important in the design of workwear and special clothing (e.g. for army), or clothing that involves work or person's physical activity.

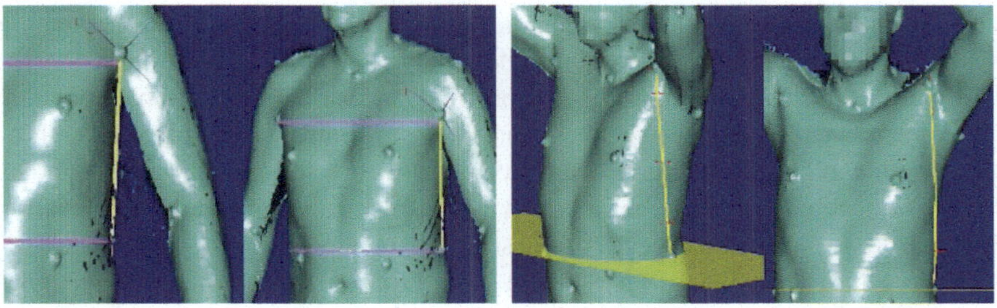

Fig. 18 Standard posture and arms upwards with markers (underwear)—measurement side upper torso

4 Discussion and Conclusions

One of the advantages of 3D scanning, in comparison with the manual method, is avoiding of physical contact between both the person to be measured and the measurer, which in some cases can cause embarrassing feelings, thus affecting the results. Particularly crucial it can be in measuring dynamic positions when a trail subject performs bending, deep squatting, etc. movements, as well as in measuring inside leg length and crotch length needed for the estimation.

The Vitus Smart XXL scan range is 1100 mm in width, 1000 in depth and 2100 mm in height. Such a limited range can narrow choices of body poses, for example, in case of hands raised in upright position, arms horizontally extended both to the front and to the side, as well as maximal stepping posture and similar active body positions. Some body poses may be included within the specified range by moving feet position away from feet marks shown on the scanner, however this may affect further work with the obtained scans, resulting in diverse, and thus incomparable distance between feet, as well as additional work for overlaying the scans.

Another problem is in missing areas of scans—body parts that cannot be reached by the laser light. In case of standing posture missing datas are expected in areas covered by arms and legs (armpits, crotch, etc.). However, in case of dynamic poses, they can be volumes significant for execution of measurements, such as bending, squatting postures—a part inside of thighs, crotch area, abdomen area, shoulder area, etc. In scanning of already dressed bodies, missing areas can be different as in naked ones—must be considered missing areas in clothing surface gathers.

While in various studies comparing measured results obtained by the two methods (contact and non-contact), no significant differences between the results regarding standard stand-up position were found, in some studies significant differences were observed in measurements of dynamic positions, for example, with inconsistencies of 6.8 up to 16 cm [21]. However, it should be taken into account that manual measurements must be confronted with effects such as the deformation of soft tissues by measuring instruments and the duration of the procedure during which the person to be measured may get tired of the posture either physically or emotionally thus leading to a change of poses and the extent of the pose-insuring muscle tension.

There is no single standard for dynamic measurements to handle data acquisition methods, define dynamic poses or provide reliable and sufficiently comprehensive information on values of dynamic indicators (body measurement allowances or cuts). If the precision of 3D measurements in upright poses compared to traditional anthropometric manual methods can be evaluated by means of standard [18], then studies of different volumes show contradictions in assessment of precision of dynamic measurements obtained by means of 3D scanning method.

Within the framework of the study, a large group of members of one occupation (soldiers) is addressed, and it is learned that physical fitness (stamina, strength, speed, agility) of soldiers is carried out during the entire period of service, and within the framework of comprehensive exercises, the aim is to strengthen the support and movement apparatus, to develop all muscle groups evenly, to build proper posture, as well as to improve sense of rhythm, movement precision and coherence necessary for performance. Physical fitness is regulated by regulations of the Cabinet of Ministers [20], however it cannot be said that there is a constant physical development within the group, and already initial execution of dynamic postures, or, more precisely, the posture amplitude for representatives of the same height, BMI and age group differs. It is influenced by both belonging to different age groups, daily routine conduct within the scope of service title, and individual approach to physical fitness, as well as by various physiological peculiarities regarding the range of body movement amplitudes. Even though the volume of the measured data is sufficient and corresponding to the normal distribution when looking on the body height, tendencies observable in other total morphological features imply that bigger amount of data would produce more believable measure results, However, the data obtained at the moment gives overall insight into the situation in the entire population (in the selection of soldiers).

In future studies, a research on the most characteristic work poses of the population group and their scan limitations should be carried out. It is necessary to classify dynamic parameters necessary for designing of protective clothing and to specify method of their measuring. It is necessary to continue assessment of their potential impact on the results of dynamic measurements, which can be affected by differences in posture amplitudes requiring expansion of comparable body parameters, such as the interlinked proportions between upper and lower body parts, as well as the ratio of limbs against body height.

Acknowledgements This research work was partially financed by the European Union European Regional Development Fund, through the INTERREG BSR Programme, which awarded a grant to the SWW project (#R006). The authors gratefully acknowledge the received financial support.

References

1. Song, G.: Improving Comfort in Clothing, p. 496. Woodhead Publishing Limited (2011). ISBN 1845695399

2. McCann, J., Bryson, D.: Smart Clothes and Wearable Technology. Woodhead Publishing Limited, p. 484 (2009). ISBN: 9781845693572
3. Goldman, F.R., Kampmann, B.: Handbook on clothing. In: Biomedical Effects of Military Clothing and Equipment Systems, 2nd edn, p. 321 (2007)
4. Podgórski, D.: A role of smart PPE and IoT technologies in future management. In: Report—Lecture, PPE Conference 2017, Brussels, 22–23 November 2017
5. Standard. ISO 13688:2013: Protective Clothing—General Requirements
6. Koblyakova, E.V., Kurshakova, Yu.S., Zenkevich, P.I., Dunaevskaya, T.N.: The Size Typology of the Population of the CMEA Member Countries, p. 440 (1974). Light industry, Moscow/Коблякова ЕВ, Куршакова ЮС, Зенкевич ПИ, Дунаевская ТН: Размерная типология населения стран—членов СЭВ. Легкая индустрия, Москва, с. 440 (1974)
7. Lim, H.W., Cassidy, T.: A comparative study of trouser pattern making methods. J. Text. Eng. Fashion Technol. **1**(5), 00031 (2017). https://doi.org/10.15406/jteft.2017.01.00031
8. Ashdown, S.P.: Improving Body Movement Comfort in Apparel. Woodhead Publishing Series in Textiles, Cambridge, UK (2011)
9. Hayes, G.S., Venkatraman, P.: Materials and Technology for Sportswear and Performance Apparel, p. 370. CRC Press, Taylor&Francis Group (2016)
10. Preedy, R.V.: Handbook of Anthropometry: Physical Measures of Human Form in Health and Disease, p. 3107. Springer Science & Business Media, (2012)
11. Standard. ISO/TR 7250-2:2010: Basic Human Body Measurements for Technological Design—Part 2: Statistical Summaries of Body Measurements from National Populations
12. Loercher, C., Morlock, S., Schenk, A.: Design of a motion-oriented size system for optimizing professional clothing and personal protective equipment. Hohenstein Institut für Textilinnovation gGmbH, Boennigheim, Germany. In: International Conference on Intelligent Textiles and Mass Customisation, Ghent, 15–18 October (2017)
13. Naglic, M.M., Petrak, S., Gersak, J., Rolich, T.: Analysis of dynamics and fit of diving suits. In: IOP Conference Series: Materials Science and Engineering, vol. 254, p. 152007 (2017). https://doi.org/10.1088/1757-899x/254/15/152007
14. Loercher, C., Morlock, S., Schenk, A.: Motion-oriented 3D analysis of body measurements. In: IOP Conference Series: Materials Science and Engineering, vol. 254, p. 172016 (2017). https://doi.org/10.1088/1757-899x/254/17/172016
15. Choi, S., Ashdown, S.P.: 3D body scan analysis of dimensional change in lower body measurements for active body positions. Text. Res. J. **81**(I), 81–93 (2010). https://doi.org/10.1177/0040517510377822
16. Xiao, P., Ashdown, S.P.: Analysis of lower body change in active body positions of varying degrees. In: Proceedings of the 4th International Conference on 3D Body Scanning Technologies, pp. 301–309, Long Beach CA, USA, 19–20 November 2013. http://dx.doi.org/10.15221/13.301
17. Sakharev, M.I., Boytsov, A.M.: Principles of engineering designing of clothes. Light and food industry, Moscow, p. 272 (1981)/ Сахарев МИ, Бойцов АМ: Принципы инженерного проектирования одежды. Легкая и пищевая промышленность, Москва, с. 272 (1981)
18. Standard. ISO 20685:2010: 3-D Scanning Methodologies for Internationally Compatible Anthropometric Databases
19. Human Solutions.: ScanWork User Guide Version 2.9, Human Solutions GmbH, Germany (2005)
20. Ministry of Defence, Republic of Latvia: Physical Preparation Provisions for Soldiers/Karavīru fiziskās sagatavošanas noteikumi (2008)
21. Chi, L., Kennon, W.: Body scanning of dynamic posture. Int. J. Cloth. Sci. Technol. **18**, 166–178 (2006). https://doi.org/10.1108/09556220610657934
22. Standard. EN 13402-3:2014: Size Designation of Clothes—Part 3: Body Measurement and Intervals

Development of Training Modules for Visually Impaired for Rehabilitation in Garment Manufacturing Units

Megha Gupta and Ritu Mathur

Abstract Disability is an inability to perform a normal process. It could be complete inability or partial inability to do something. Garment industry is highly labor-intensive and involves lots of routine, monotonous, and repetitive operations on which differently abled can easily be trained. The present study was conducted with the objectives of identifying the characteristic abilities of the identified subjects, to develop and administer training modules for training the visually impaired in the tasks of bar-tacking using bar-tack machine and lap seam using feed-off-the-arm machine. Two training modules were developed and conducted with stepwise demonstration of each task in "AMS Fashions" an export house. Training program was conducted under the constant supervision of the researcher. It was observed that visually impaired could perform the preliminary processes accurately. They could also perform the processes of bar-tacking and lap seam using feed-off-the-arm machine. To aid the threading of the machine, and accessory, viz. special threader was developed to make the process easier. Magnet guide was used to aid the visually impaired for the correct placement of the garment during the process of bar-tacking. It was observed that the speed and accuracy of the subjects improved significantly on practice.

1 Introduction

Happiness lies not in getting what you want but in making the most of what you get in any situation (Zen, www.nabindia.org). A physical or mental deficiency that prevents normal achievement is known as **disability**. About 600 million people live with various types of disabilities and reasons are chronic diseases, injuries, car crashes, falls, violence, and other causes such as aging. Of this total, 80% live in low- income

M. Gupta (✉)
IMS—Design and Innovation Academy, Noida, U.P, India
e-mail: gupta.megha1985@gmail.com

R. Mathur
Lady Irwin College, Delhi University, New Delhi, India

A. Majumdar et al. (eds.), *Functional Textiles and Clothing*,
https://doi.org/10.1007/978-981-13-7721-1_21

countries, most are poor and have limited or no access to basic services including rehabilitation facilities (www.who.int/nmh/).

The term "**Disability,**" "**Impairment,**" and "**Handicap**" all have different meanings. World Health Organization (WHO) which distinguishes between them as follows:

Impairment: It is defined as any damage or weakening of physiological, psychological, and anatomical function or structure.

Handicap: A handicap is a drawback for an individual which results from a disability or impairment and precludes an individual from being normal.

Disability: A disability is any restriction or lack of ability (resulting from an impairment) to perform an activity in the manner or within the range considered normal for a human being (www.disabilityhelper.com).

Defining visual impairment is complex. Any loss of ability to gather information by seeing might be considered a visual impairment. An attempt has been made to distinguish visual impairment from blindness for the sake of conceptual clarity.

Total Blindness: It is the inability to tell light from dark or the total inability to see.

Visual Impairment: Visual impairment or low vision is a severe reduction in vision that cannot be corrected with standard glasses or contact lenses and reduces a person's ability to function at certain or all tasks.

Legal Blindness: It is actually a severe visual impairment and refers to a best corrected central vision of 20/200 (inability to count fingers at a distance of 2 m) or worse in the better eye.

Tunnel Vision: The area of vision is restricted to a small central area and peripheral vision is limited.

Congenitally Blind: People born with a severe visual impairment are referred to as congenitally blind.

Adventitiously Blind: People who acquire a severe visual impairment after age two (www.disabilityhelper.com).

1.1 Visual Impairment

Visual impairment represents a continuum, from people with very poor vision or people who can see light but not shapes to people who have no perception of light at all. According to NSSO, "A person is treated as visually disabled if he/she may not be able to perceive any light with both eyes open, or if he/she had light perception but could not correctly count fingers of a hand (with or without use of spectacles/contact lens) from a distance of three meters in broad day light with both eyes."

Impairment can be as subtle as a mild astigmatism or as severe as total blindness. A person who possesses one or more of these traits may require aids for learning. Visual impairment can be divided into five sub-categories:

- Hyperopia—farsightedness
- Myopia—nearsightedness
- Astigmatism—blurred vision caused by uneven curvature of the cornea
- Strabismus—"crossed eyes"
- Blindness—condition of those whose visual acuity (sharpness of vision) is 20/200 or less in the better eye with the best possible corrections.

Magnitude of Visual Impairment

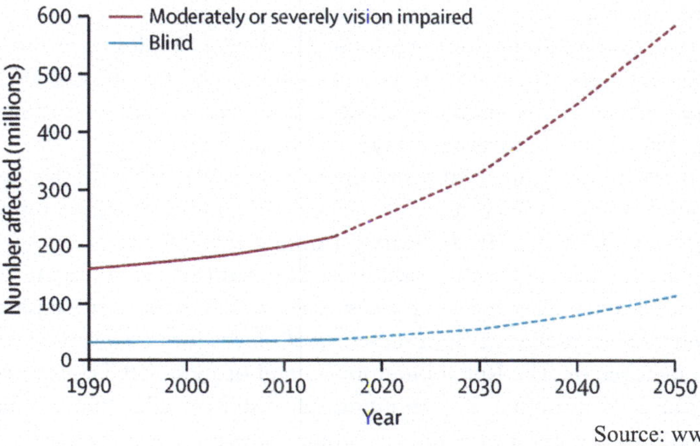

Source: www.ijo.in

Distribution of Visual Impairment

- **By age**: More than 82% of all people who are blind are 60 years of age and older, although they represent only 19% of the world's population. Childhood blindness remains a significant problem, with an estimated 1.4 million blind children below age 15. Thus, some training programs should be carried out which in turn will help them to earn their livelihood.
- **By gender**: Studies indicate that females have a significantly higher risk of being visually impaired than males.
- **Geographically**: More than 90% of the world's visually impaired live in developing countries [1].

Cause of Blindness and Visual Impairment

Visual impairment can be caused in many different ways ranging from birth defects to accidents and/or illness that inevitably affect vision. The main causes of visual impairment are cataract, glaucoma (*kala motia* or *neela motia),* and age-related macular degeneration (AMD). Other main causes of visual impairment are corneal opacities, diabetic retinopathy, childhood blindness, trachoma, and onchocerciasis.

1.2 Rehabilitation

Rehabilitation includes all measures aimed at reducing the impact of disability and handicapped condition and at enabling the disabled and handicapped to achieve social integration. Rehabilitation aims not only at training disabled persons to adapt to their environment, but also at intervening in their immediate environment and in society as a whole in order to facilitate their social integration. The concept of rehabilitation generally includes medical, social, psychological, educational, occupational, and economical measures aimed at securing for the individual the highest possible level of functional ability [2].

The Ministry of Labor through the Directorate General of Employment and Training (DGET) has extended its services to persons with disabilities through 17 vocational rehabilitation centers including seven skill development centers for the handicapped, located in 16 states, covering all types of disabilities.

Also, both the National Council of Vocational Training (NCVT) and the Apprenticeship Training Scheme (ATS) reserve an unspecified number of places for people with disabilities. In addition, 2000 Industrial Training Institutes have set apart 3% of their vacancies for persons with disabilities. The Ministry of Rural Areas and Employment runs a scheme known as Training of Rural Youth for Self-Employment (TRYSEM), which also guarantees to reserve 3% of places for persons with disabilities in government sector. The Rehabilitation Council of India (RCI) has also been trying to undertake the program of continuing rehabilitation education so that the country has a reservoir of trained people who could impart the best possible training to children and adults with disabilities.

1.3 Employment of Visually Impairment Persons in India

The concept of placement of the visually impaired in open employment was advanced by the formation of the Employment and Placement Committee of the National Association for the Blind in 1954. The rehabilitation program at National Association for the Blind (NAB) focuses on transforming the lives of the visually challenged men and women, by addressing the issue at the psychological, social, and vocational level. The employment department with aid from the various state and district branches strives to provide more and more blind individuals with diverse employment opportunities in the public and private sector. Since its inception in 1954 till date, more than 6500 blind individuals have emerged as successful professionals. (www.nabindia.org).

1.4 Objectives

- To identify the characteristic abilities of the visually impaired persons;
- To identify existing operations in different sections of a garment manufacturing unit;

- To develop and administer training modules for training of visually impaired individuals in the tasks of:

(a) Bar-tacking and
(b) Lap seam using feed-off-the-arm machine.

2 Materials and Methods

The present study was planned and conducted in the following two phases:

Phase I

- Selection of tasks for training the visually impaired persons;
- Identification of the subjects;
- Selection of the garment manufacturing unit for training program.

Phase II

- Formulation of training modules:
  - Module 1: Bar-tacking
  - Module 2: Lap seam using feed-off-the-arm machine
- Administration of training modules.
- Observation and recording of the performance of the subjects.
- Qualitative analysis of the data collected.

2.1 Phase I

Selection of tasks for training the visually impaired persons
The first step of the study was aimed at identifying appropriate tasks in the garment manufacturing process that can be carried out by the visually impaired persons. This was carried out by studying the production line of garment manufacturing units. On the basis of feedback received from garment manufacturers, the tasks of **bar-taking** and **lap seam using feed-off-the-arm machine** were identified for training the visually impaired persons in the present study.

Selection of Subjects
Two visually impaired subjects were selected through purposive sampling technique. Out of the two subjects, one visually impaired subject, Vishwanath was selected form Blind Relief Association (BRA) and the other subject namely Vijay was an employee of AMS Fashions, Noida. Although the subject was working in the garment manufacturing unit on collar turning machine, he did not have any knowledge about

bar-tacking and feed-off-the-arm machines. They had visual impairments of 50 and 10%, respectively.

Due to the unavoidable circumstances, the subject having 50% visual impairment had to discontinue the training in the middle of the training program. Hence, another subject namely, Ramakant was selected for training with 60% visual impairment. He was selected from Saburi, New Delhi. Like Vishwanath, he also did not have any previous knowledge about sewing machines.

Area Selection
The study was carried out in AMS Fashions, Noida, a garment manufacturing unit where both the machines were available for conducting the training program. The area selection was facilitated with the assistance of the Program Coordinator of Saburi.

2.2 *Phase II*

Formulation of the Training Modules
The second phase of the study began with the formulation of the training modules for both the above-mentioned tasks. A separate module was made for each process. Keeping in mind the capabilities and limitations of the visually impaired subjects, each module was broken down into smaller tasks as the processes were very complex in the nature. Each task was further divided into sequential steps to facilitate better understanding.

In the garment manufacturing units, bar-tacking is used on moderate and heavy weight clothing such as jeans, trousers, shirts, and jackets. Bar-tacking can also be done at different places as per the buyer's requirements. This is usually done to reinforce the stitching. Initially, the subjects were familiarized with the task of bar-taking on flat fabric pieces. Subsequently, the training was carried out on pocket flaps of shirts as bar-tacking is easier to carry out on the shirt as compared to the trouser.

Also, in the garment manufacturing process, Fee-off-the-arm sewing machine is used wherever lap seam is required. It is usually done on trouser crotch, shirt side seam, trouser inner seam, underarm seam, etc. Therefore, initially the training was carried out on flat fabric pieces to familiarize the subjects with the task, and subsequently, training was carried out on the side seam of the shirt. Hence, Module I for bar-tacking and Module II for lap seam were formulated. The detailed modules are given in Appendix-I and Appendix-II.

Note: The speed and tack length of bar-tacking machine are controlled with the computer. Although the basic principles were explained to the visually impaired, the setting of the computer was done by a normal-sighted person.

Application of the Training Modules: The training program was conducted on the basis of the training modules. Both the modules were taken up simultaneously. Training was conducted in the task of bar-tacking to the first subject, and simultaneously, the second subject was trained in the other task, i.e., lap seam using Brother feed-off-the-arm machine. Only after the trainee got thorough with one module, the next module was introduced. Initially, the training was facilitated with the help of a trainer for visually impaired persons. The trainer helped in rapport formation with subjects. All the steps involved in the process of bar-tacking and lap seam followed a sequence. Therefore, the tasks were explained to the visually impaired in the same sequence. Both the trainees were paid individual attention. One step was explained at a time. After the explanation, adequate time was given to the subjects for practice of each step to gain competency. The next step was only introduced after the trainees got thorough with the previous one. The training of visually impaired individuals had to be conducted in the presence of researcher as the tasks of bar-tacking and lap seam were carried out on high-speed machine and therefore required constant supervision.

During the training program, the subjects faced some problems. Hence, slight modifications were made during the application of modules on the basis of the problems faces by the visually impaired persons. Accordingly, an accessory, viz., a special threader was designed and developed in stainless steel to aid the threading of feed-off-the-arm machine. Also, a three-inch magnet guide was used to aid the visually impaired for the correct placement of the garment during the process of bar-tacking.

Observation and Recording of the Performance of the Subjects: All the trainees were observed while performing each task, and their performance in each task was recorded. The accuracy and speed of performing the tasks by each subject was noted at the beginning, once the subjects attained the basic skills of performing a task, as well as after gaining competency in each task. The problems faced by the visually impaired individuals in performing the tasks of bar-taking and lap seam were also recorded. The time taken by the visually impaired subjects and a normal-sighted person in executing the task were compared.

Qualitative Analysis of the Data: The data collected were tabulated and qualitatively analyzed by the researcher. Time taken by subjects in performing each task was evaluated. Performance of completely blind and partially blind was then compared. The problems were analyzed. Also, the observation and performance of visually impaired persons were compared with normal-sighted person.

3 Result and Discussion

3.1 Task Evaluation

It was observed that the visually impaired individuals can be trained to perform complex tasks efficiently if proper guidance is provided. Depending upon their learning speeds, each trainee was trained individually in each task. A new training module was initiated only after each subject reached level of proficiency. Thus, time taken to perform a task was recorded at intervals after the trainees had developed the basic skills of performing the tasks and also at the end of the training program.

MODULE-1

Bar-tacking on Brother Bar-tack machine

Bar-tacking is an activity of the production department of garment manufacturing unit. It is usually done on pocket flaps, trouser loops, etc. (Fig. 1). Sometimes it is used for decoration purposes.

The task of bar-tacking was divided into eight steps. Each task was explained individually to both the trainees. It was observed that both the trainees were very keen to learn the task of bar-tacking and had good memory power. They quickly understood all the tasks and retained the steps properly. They participated positively during the entire training program. They often gave suggestions for executing the tasks in an easier manner. Thus, these were incorporated into the training. The observation recorded during the administration of the training modules and their analysis is as follows:

Preliminary process involved in the task of bar-tacking

Both of the subjects were able to grasp all the preliminary processes for setting up of the machine quickly. These included the process of winding of the bobbin, insertion of bobbin in bobbin case, and threading of bar-tacking machine. Initially, they faced problems in a few tasks like winding of the bobbin on the machine and threading of the machine. But with practice they were able to overcome these problems. The subjects had very focused attention. Initially, they took some time to do the task but were able to perform well after practicing for 3-4 sessions. Later, they were capable of performing all the tasks on their own without any supervision (Tables 1, 2 and 3).

During the insertion of the bobbin in bobbin case, it was observed that both the trainees took lots of time while passing the thread through the hole of the lever of the bobbin case. Also, they were unable to pass the thread through the needle hole while threading of the machine. Hence, in order to overcome these problems, a needle threader was used (Fig. 2). A needle threader is easily available in the market. They improved upon their performances gradually and reduced upon their time taken to complete the tasks.

Fig. 1 A trouser with bar-tacking

Table 1 Observation chart for winding of the bobbin on the Brother bar-tacking machine

Vijay		Vishwanath	
Performance	Problem faced	Performance	Problem faced
He understood the task quickly but took time to achieve proficiency. Initially, he took **2 min 9 s** to complete the task but after giving competency he was able to complete it in **53 s**	He faced problem in passing the thread between the disks of the tension post. Sometimes he forgot to press tension/wind key on computer before starting bobbing winding	He understood the task quickly. He was able to perform the task efficiently after the practice. Initially, he took **1 min 5 s** but after gaining competency he was able to complete the task in **59 s**	While threading the machine for winding the bobbin, sometimes thread got entangled in the tension post. Sometimes he was unable to pass the thread between the two disks of the tension post

Table 2 Observation chart for insertion of bobbin in bobbin case of Brother bar-tacking machine

Vijay		Vishwanath	
Performance	Problem faced	Performance	Problem faced
He took time in learning the task. Initially, he completed the task in **1 min 58 s** but after gaining competency he performed the task in **56 s**	The thread broke sometimes while passing the thread through the hole of the bobbin case. Also he faced a problem in passing the thread through the hole of the lever which was solved by the use of a needle threader	He was able to understand the task quickly. He improved with practice and was performing the task satisfactorily. He took **2 min 15 s** to complete the task initially but after gaining competency he completed the task in **1 min 20 s**	Sometimes he forgot to pass the thread through the hole of lever

Table 3 Threading of the Brother bar-tacking machine

Vijay		Vishwanath	
Performance	Problem faced	Performance	Problem faced
He was able to understand the task quickly but took time to achieve efficiency. He did the job in **12 min 55 s** initially but after gaining the competency he was able to complete the task in **3 min 6 s**	The subject faced problems in passing the thread between the disks of the tension posts. Sometimes the thread got entangled in the tension. He took time in threading the needle	The subject understood the task but could not perform efficiently. Initially, he completed the task in **13 min 05 s** but after gaining competency he performed the task in **6 min 07 s**	Initially, the subject took a lot of time in passing the thread (wound on the metal clip). Sometime he forgot to pass the thread through the tension spring before passing through the L-shaped guide. Also he took a lot of time in threading the needle

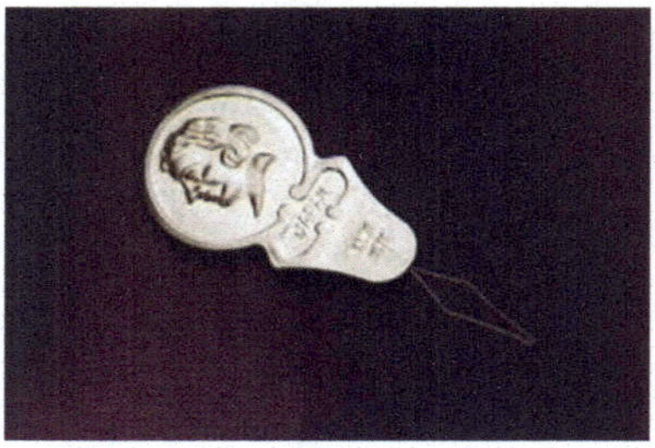

Fig. 2 Needle threader

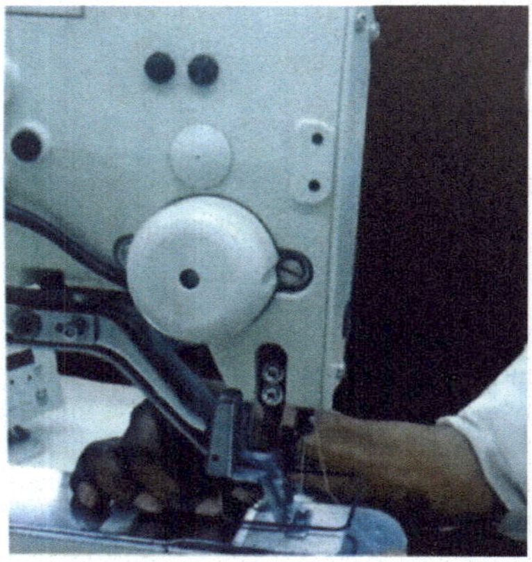

Fig. 3 Threading of bar-tacking machine

It was observed that with practice, both the subjects improved their performance greatly. However, there was a remarkable difference in the time taken by the subjects in threading of the machine (Fig. 3).

In case of Vijay, the time taken for threading of the machine was reduced to **3 min 6 s** from 12 min 55 s by the end of the training. On the other hand, time taken by Vishwanath was reduced to **6 min 7 s** from 13 min 5 s. After two and half months, Vishwanath had to discontinue the training. Even though Vishwanath was partially sighted, he took more time than Vijay for threading of the machine. Vijay took **5 min 30 s** after completing equal number of sessions and had reduced it further by the end of the training program. This might be because he had experience of working on a collar turning machine. Thus, he was able to understand the task quickly and performed better.

The process of bar-tacking

It was observed that Vijay was able to perform the task of bar-tacking efficiently (Fig. 4). Initially, the training was carried out on flat fabric pieces, and the subject was able to do the task competently.

However, subsequently, when the training was carried out on pocket flaps of shirts, he was unable to keep the garment position correctly on the machine with the help

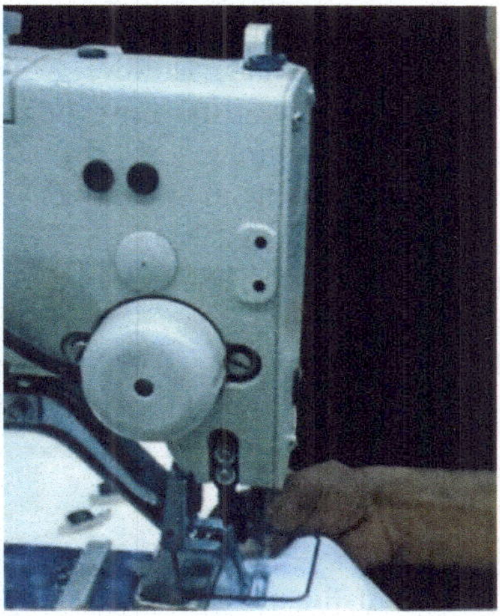

Fig. 4 Bar-tacking on pocket flap

Table 4 Observation chart for bar-tacking on pocket flap using Brother bar-tacking machine

Performance	Problem faced
He was able to understand the task but could not perform it efficiently. He took **1 min 10 s** to do the task initially. After gaining competency he did the task in **50 s**	Sometimes, he was unable to place the sticker edge under the needle point. He could not keep the pocket flap straight while bar-tacking

Fig. 5 Magnet guide on bar-tacking machine

of sticker. However, such marking has no significance if the task has to be carried out by the visually impaired individuals (Table 4).

Hence, in order to overcome this problem, an accessory viz. magnet guide was introduced to facilitate correct positioning of the garment by the visually impaired person. Magnet guides were placed in line with the throat plate of the machine which helped to keep the garment in position during bar-tacking. Magnet guides (Fig. 5) of different sized were tried out, and finally, three-inch magnet guide was found to be most appropriate.

Sometimes, the subject also faced problems in bar-tacking on the pocket flap even with the help of sticker and the magnet guide. However, with constant guidance and practice he was able to perform the task satisfactorily. He took 50 s to complete the task whereas the normal-sighted person took 20 s as he took more time in correct placement of the garment. However, with constant practice, the time taken by the visually impaired for bar-tacking can be reduced further.

Also, the setting of machine through the computer was done by a normal-sighted person. Although the process of setting up of machine with computer was explained to the subject and he understood and comprehended the same, due to the limitations, visually impaired person was not able to do the settings of the machine with the computer.

MODULE-2

Lap seam using Brother feed-off-the-arm machine

Lap seam is usually done on trouser crotch, inseam of trouser, side seam of the shirt, underarm area, and sleeve. It can be done on light weight to medium and heavy weight clothing. Machine could be of double needle or triple needle type. Mainly double needle machine is used in garment manufacturing units; therefore, the training was conducted on the double needle machine. All the trainees were trained in performing lap seam on the feed-off-the-arm machine. Vijay and Vishwanath had been trained in all the processes involved in executing the lap seam using the feed-off-the-arm machine. Due to the limitation of time and course, Ramakant was only trained in the preliminary processes involved in the task of lap seam using the feed-off-the-arm machine. His training is still in progress. The process of the lap seam was divided into six steps. Each task was explained individually to all the trainees.

Preliminary process involved in the task of lap seam using feed-off-the-arm machine

Preliminary processes involved threading of the double needle machine and threading of loopers. Threading of the machine and loopers was very complex. Although all the trainees understood the tasks well for completing the preliminary processes in setting up of the machine, they took a lot of time for threading of the loopers. The subjects performed all the tasks stepwise in the correct sequence.

The machine was of double needle type; hence, they faced some problems initially during threading the machine (Figs. 6 and 7).

A needle threader was used for the threading of needles. Sometimes, during the threading of the second needle, threads of both the needles got entangled. It was observed that they improved with the practice and performed satisfactorily (Table 5).

All the subjects were able to understand the task. However, they took a lot of time in learning the task. It was seen that in spite of a great difference in the level of visual impairments of all the subjects, there was no significant difference in their performance. This might be due to the fact that all three subjects relied more on their sense of touch while performing the task rather than their vision (Table 6).

The main problem was faced in threading of the levers during the threading of the loopers (Fig. 8). There were two tiny holes present in each of the lever. Thus, it was

Table 5 Threading of the Brother feed-off-the-arm machine

Vijay		Vishwanath		Ramakant	
Performance	Problem faced	Performance	Problem faced	Performance	Problem faced
He quickly understood the task improved with practice and was performing the task satisfactorily. Initially, he took **22 min 15 s** to perform the task but after gaining competency he completed the same in **13 min 27 s**	He took a lot of time during the threading of needles	He was able to understand but could not perform it efficiently. He did the task in **21 min 58 s** initially but after gaining competency he was able to complete it in **15 min 16 s**	Sometimes the thread got entangled in the tension post	He understood the task quickly and improved with the practice. He took **20 min 13 s** in completing the task with guidance. After gaining competency, he was able to complete it in **15 min 25 s**	He was unable to pass the thread from the hook. Sometimes the thread got entangled in the tension post

Table 6 Threading of the loopers of the Brother feed-off-the-arm machine

Vijay		Vishwanath		Ramakant	
Performance	Problem faced	Performance	Problem faced	Performance	Problem faced
He was able to understand the task but took time to achieve efficiency. Initially, he took **40 min 12 s** to do the task. After gaining competency, he did the same job in **28 min 47 s**	He took a lot of time in feeling each part of the loopers during the process of threading. He faced a problem during the threading of looper 2	Initially, he did perform the task satisfactorily. He took **41 min 50 s** to complete the task. However, he improved with practice and was able to complete the task in **31 min 05 s** after gaining competency	Sometimes he forgot to pass the thread from the groove	He was able to understand the task but took time to achieve efficiency. He took **38 min 40 s** to do the task initially. After gaining competency, he did the job in **29 min 50 s**	The subject faced problem in passing the thread from the groove. Sometimes thread got entangled while passing the thread through the looper

Fig. 6 Threading of the needle of the feed-off-the-arm machine

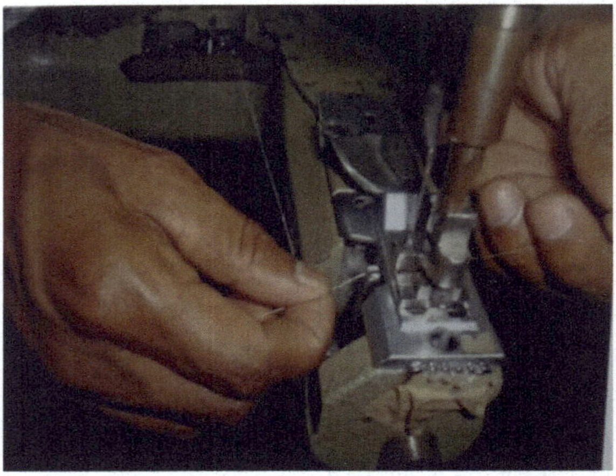

Fig. 7 Threading of the looper of the feed-off-the-arm machine

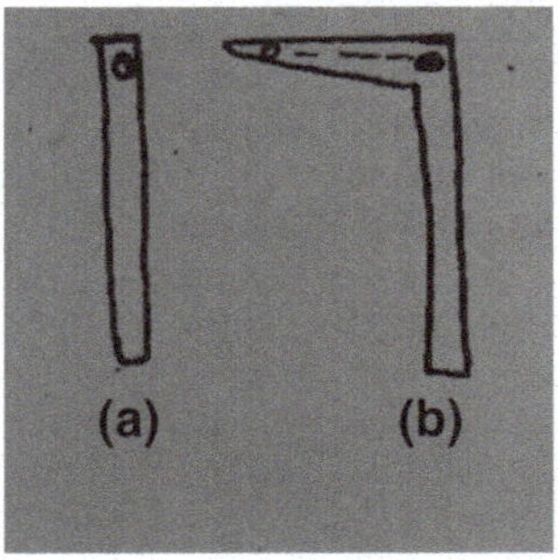

Fig. 8 Lever in the looper, **a** front view, **b** side view

very difficult for visually impaired person to pass the thread through the lever. The process involved a lot of time.

Hence, in order to overcome this problem, and accessory, viz. a special **threader** was made by the researcher during the training program (Fig. 9). It was made in

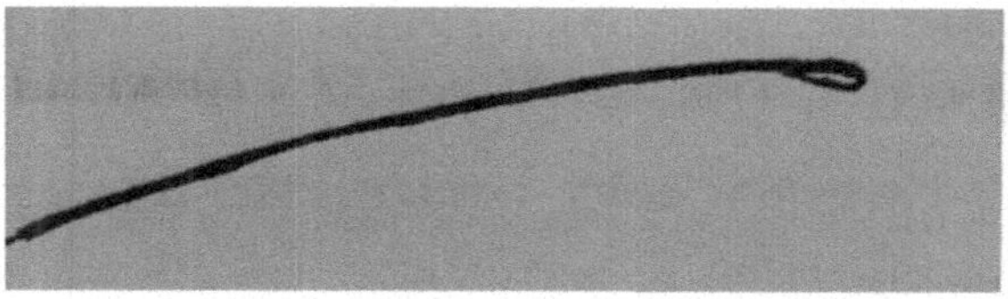

Fig. 9 Special threader

stainless steel metal to assist the visually impaired subjects in the threading of the levers during the threading of the loopers of the machine (Refer Fig. 7).

The threader made was a straight wire with a hook on one side. The threader worked in a following way:

- The threader was passed through the hole in the lever keeping the hook facing outside.
- The thread was passed through the hook of the threader and the thread was held in that position with one hand.
- The threader was pulled from the other side through the hole of the lever. Hence, the lever is threaded.

Process of performing the lap seam on feed-off-the-arm machine

Vijay and Vishwanath were trained in the process of the lap seam using feed-off-the-arm machine. Initially, the training was carried out on flat fabric pieces to familiarize the subjects with the task, and subsequently, training was carried out on the side seam of the shirt.

The subjects initially faced problems in executing the task. For executing lap seam, the fabric has to be inserted through the folder in feed-off-the-arm machine. The folder is an attachment to the machine. It was attached in the front of feed-off-the-arm machine before the pressure foot. It helped to keep the seam allowance of the garment in a folded position.

The problem that the subjects faced was that they were not able to set the fabric in the folder. However, with constant guidance and practice they were able to perform the task. Vijay was able to perform the task of lap seam on a shirt in **1 min 4 s** whereas Vishwanath took 58 s to complete the same task (Table 7).

It might be because partial vision helped in guiding the fabric through the folder of the machine. Time taken by a normal-sighted person was also observed. A normal-sighted person took 40 s to complete the task. Hence, it could be concluded that they can improve even further with the practice.

To conclude, it can be stated that the tasks of bar-tacking and lap seam using feed-off-the-arm machine can be efficiently performed by the visually impaired person. The problems faced by the visually impaired during the training were overcome by the accessories viz. magnet guide and a special threader. The processes involved in setting and threading of feed-off-the-arm machine were very complex but visually impaired were able to do it satisfactorily with the aid of the accessory.

Table 7 Observation chart for lap seam using Brother feed-off-the-arm machine

Vijay		Vishwanath	
Performance	Problem faced	Performance	Problem faced
He was able to understand the task quickly but took time to achieve efficiency. He did the job in **1 min 13 s** initially but after giving the competency he was able to complete the task in **1 min 6 s**	The subject faced problems in passing the fabric through the folder	The subject understood the task and performed satisfactorily. Initially, he completed the task in **1 min 10 s** but after giving competency he performed the task in **58 s**	He did not face any problem in performing the task

4 Summary and Conclusions

Visually impaired individuals are considered suitable for working in the industry because they have a remarkable capacity for learning and they overcome their natural shortcomings by heightened concentration, sensory powers, and a sense of commitment to the job.

The study was carried out in two phases. The **first phase** of the study aimed at identification of the subjects and garment manufacturing unit to carry out the training program. Two visually impaired subjects with substantial difference in their visual impairments were selected. They had visual impairments of 100% and 50%, respectively. Only two subjects were selected for the training because the training required that individual attention be given to the subjects as the tasks of bar-tacking using bar-tack machine and lap seam using feed-off-the-arm machine were very complex.

In the **second phase,** training modules were formulated and administered in a stepwise manner to the subjects. Training was carried out on Brother bar-tacking and feed-off-the-arm machines in AMS Fashions, Noida, for six months. During the first two sessions of training program, the training was facilitated with the help of a trainer for visually impaired persons. As the tasks of bar-tacking and lap seam using feed-off-the-arm machine were very complex, each step of the training program was conducted in the presence of researcher. The performance of the subjects was observed and recorded. The data obtained were qualitatively analyzed. The performance of the subjects was compared on the basis of the visual impairment and working experience. Also, the time taken by the visually impaired subjects and normal-sighted person in executing the task was compared.

During the training program, problems faced in the execution of the tasks were as follows:

- Difficulty in threading the loopers of feed-off-the-arm machine;
- Inability of the visually impaired persons to set the correct position of garment with the help of sticker on the bar-tacking machine;

- Inability to set the program in the computer attached with bar-tacking machine.

In order to overcome these problems, slight modifications were made during the application of the modules.

- Designing of an accessory viz. a special threader, in order to facilitate the threading of feed-off-the-arm machine. It was observed that the efficiency and accuracy of the visually impaired subjects improved significantly.
- Three-inch magnet guide was used to keep the garment in straight position during the process of bar-tacking.
- Due to the limitation of capabilities of the visually impaired person, setting of the computer program for the bar-tacking machine had to be done by a normal-sighted person.

To conclude, it can be stated that visually impaired persons belong to a special category that requires sensitivity and proper guidance. There are many jobs, which can be easily performed by the disabled if some special training is given. With practice and guidance, the visually impaired persons can gain competency in repetitive tasks. With the aid of the specific tools and accessories, they can perform the task as well as any other person.

Acknowledgements I would like to express my deep sense of gratitude and indebtedness to my guide Dr. Ritu Mathur, Reader, Department of Textiles and Clothing, Lady Irwin College. Her valuable guidance, constant encouragement at all stages of the study enabled me to achieve the desired goals. My sincere thanks extend to Dr. Anupa Sidhu, Director, Lady Irwin College; Mrs. Nitasha Kachroo, Program Coordinator, Suburi Disability Division, New Delhi; Mr. K. C. Pandey, Executive Secretary and Mr. A. David, workshop manager and Placement Officer, The Blind Relief Association, Delhi, Mr. Sanjay Gupta, Director and Mr. Sanjay Anand, Production Head, AMS Fashions, Noida, for his support during the training program. I am thankful to all the trainees for their patience, cooperation, and sparing their valuable time without which the research would not have been possible. Last but not the least I deeply value the help and support given to me by all those whom I could not acknowledge individually.

Appendix I

Module 1

Bar-Tacking on Brother Bar-Tack Machine (KE-430D)

Bar-tacking is done for making reinforced stitching. Usually, it is done on moderate and heavy weight clothing like suits, Jeans, trousers, shirts, etc. For the normal-sighted person, the position of the bar-tack is marked. This is done for exact positioning of the bar-tack. However, such marking has no significance if the task has to be carried out by the visually impaired individuals. Therefore, magnet guides of different sizes were used to assist the visually impaired trainees. The magnet guides help to guide the correct positioning of the garment during the bar-tacking. Thus, the process of bar-tacking with the help of "magnet guides" was broken into smaller tacks which were further divided into sequential steps as follows:

Orientation to Parts of the Machine

The process of bar-tacking was carried out on Brother bar-tack machine, Model no. KE-430D. The first step toward working on the bar-tack machine was to introduce the task to the subjects. The application and use of bar-tacking was explained to the subject. The next step was to familiarizing the subjects with various parts of the machine.

Each part of the machine was introduced to the trainees through touching and feeling, in the sequence in which it was threaded and subsequently operated, so as to fully orient them to the machine. It helps them to create an image of machine in their mind and thus facilitate better understanding. The major components of the machine with their functions were explained to the subjects.

Orientation to Computer and Its Parts

Computer is placed between the cotton stand and the machine. Each part of the computer was introduced to the trainees through touching and feeling. It helps them to create an image of computer in their mind and thus facilitate better understanding. Although the basic principles were explained to the visually impaired, the setting of the computer was done by a normal-sighted person. The subjects were explained the parts of the computer with their functions.

Attachment of Magnet Guide

Magnet guide is a rectangular piece of a magnet having stainless steel covering on the three sides. These are available in different sizes in the market (Fig. 4). For bar-tacking machine, magnet guide of three inches in size was found to be most appropriately used in the training program. It helps to keep the fabric straight during bar-tacking.

Magnet guide is placed on the working surface of the machine, next to the finger guard (toward the back side of the pressure foot). Placement of magnet guide is done in a manner so that it does not cause any hindrance during the operation.

Preliminary Processes for Setting up the Machine

Winding the bobbin on the machine

Attaching the bobbing to the bobbin case

Installation/Removal of the bobbin case from the machine

Threading of the machine and needle—The threading can be done easily by using the needle threader. The needle of this machine is threaded from front to back, but the thread guide is inserted from back to front.

Process of Bar-Tacking on the Brother Bar-Tack Machine

The needle bar must be at its highest position before starting. To achieve that turn the machine pulley in clockwise direction so that the index mark present on the pulley get inside the mark on the mark on the back cover.

Note: If the machine is started while the index mark is not inside the mark, error code "E 110" will be displayed on the menu bar in the computer with a beep. The error will be cleared if you turn the machine pulley to set the needle to needle up stop position.

Appendix II

Module II

Lap Seam Using Brother Feed-off-the-Arm Machine (DA-9270)

This machine is used to do a lap seam. Lap seam is usually done on trouser crotch, inseam of trouser, side seam of the shirt, underarm area, and sleeve. It can be done on light weight to medium and heavy weight clothing. Machine could be of double needle or triple needle type. Mainly double needle machine is used in garment manufacturing units; therefore, the training was conducted on the double needle machine. The process of the lap seam was broken into smaller tasks, and each task was broken into sub-steps as follows:

Orientation to Parts of the Machine

The first step toward working on the feed-off-the-arm machine is familiarizing the subjects with various parts of machine. The process of lap seam was carried out on Brother feed-off-the-arm machine, Model no. DA-9270.

Each part of the machine was introduced to the trainees through touching and feeling, in the sequence in which it is threaded and subsequently operated, so as to fully orient them to the machine. This also helps them to create an image of machine in their mind and thus facilitate better understanding. The major components of the machine with their functions were explained to the subjects.

Preliminary Processes for Setting up the Machine

Threading of the Loopers—Since machine is double needle machine, it has two loopers. The second one is threaded after the first one has been threaded.

Threading of the machine and needles for needle 1 and 2—The threading can be done easily by using the needle threader.

Note: The needle of this machine is threaded from front to back, but the thread guide is inserted from back to front.

Process of Lap Seam on Brother Feed-off-the-Arm Machine

Step 1 Press "ON" button on right-hand side of the machine below the table.
Step 2 Hold the garment pieces in hand and pass the seam allowance through the folder.
Step 3 Hold one piece in one hand and another piece in the other hand in a manner such that the seam allowance of both the pieces stands parallel to each other (Fig. 6).
Step 4 Press the foot pedal (F.P.1) to make a lap seam.
Step 5 Remove the garment from the machine by cutting the thread with the help of cutter.

References[1]

1. Feldman, R.S.: Development Across the Life Span, pp. 299–300. Prentice Hall Inc., New Jersey (1997)

[1] *Original not seen.

2. Pandey, R.S., Advani, Lal, Perspectives in Disability And Rehabilitation, Vikas Publishing House Ltd., pp. 1–14, 100–107, New Delhi (1995)

Part VI
Testing and Characterization

Thermo-Physiological Comfort and Microbial Properties of Different Textile Raw Materials and Structures

Dragana Kopitar, Beti Rogina-Car and Zenun Skenderi

Abstract Textile materials used in healthcare sector, food processing industry, nursing homes, hotels, restaurants and similar areas of application often present a problem when need to provide comfort and microbial barrier properties at the same time. Impact of different fabric structures made of different raw materials commonly used for hospital bed linen, healthcare professionals' uniforms and surgical drapes on thermo-physiological comfort and microbial barrier properties was investigated. Investigated textile fabrics are woven satin fabric made of 100% cotton, twill woven fabric made of 100% Tencel®, plain woven fabric made of 50% PES/50% cotton, nonwoven fabric made of 100% polypropylene fibres and PES/PU/PES three-layered fabric laminate. Thermal and water vapour resistance was determined using the Sweating Guarded Hot Plate while microbial barrier properties were tested using bacterial spores of the Bacillus genus *Geobacillus stearothermophilus* and *Bacillus atrophaeus*. The woven and nonwoven fabrics have very good thermo-physiological comfort, respectively, extremely breathable and comfortable at higher level of activity while PES/PU/PES three-layered fabric laminate has satisfactory comfort where fabric is breathable but uncomfortable at higher rate of activity. The woven fabrics allow easier penetration of bacterial colonies related to the polypropylene nonwoven fabric which is almost impermeable to microorganism due to layered structure with densely arranged polypropylene fibres. The laminate, due to polyurethane membranes between polyester knitted fabrics, is impermeable for microorganism but has highest water vapour resistance.

Keywords Microbial barrier permeability · Sweating guarded hot plate · Thermal and water vapour resistance · *Geobacillus stearothermophilus* · *Bacillus atrophaeus*

D. Kopitar (✉) · Z. Skenderi
Faculty of Textile Technology, Department of Textile Design and Management, University of Zagreb, Prilaz Baruna Filipovica 28a, 10000 Zagreb, Croatia
e-mail: dragana.kopitar@ttf.hr

B. Rogina-Car
Faculty of Textile Technology, Department of Clothing Technology, University of Zagreb, Prilaz Baruna Filipovica 28a, 10000 Zagreb, Croatia

A. Majumdar et al. (eds.), *Functional Textiles and Clothing*,
https://doi.org/10.1007/978-981-13-7721-1_22

1 Introduction

Raw material and fabric structure have a strong effect on sweating and thus bacterial development. High humidity represents a favourable condition for bacteria growth where sweat is an ideal bacterial breeding ground. Textile materials that are used in healthcare sector, food processing industry, nursing homes, hotels, restaurants and similar areas of application often present a problem when need to provide comfort and microbial barrier properties at the same time [1, 2]. Textile products are used in various forms such as gowns, caps, masks, uniforms drapes, covers, beddings bed sheets, blankets and pillow cases. Today the medical textiles sector represents one of the fastest growing areas in the textile sector with wide range of fabrics with different properties [2–5].

Hospital bed linens are used as a protection barrier against cooling environment between human body and the place where the human body takes rest (sleep). The comfort provided by a fabric during sleeping depends on various factors. When a patient lies on the resin hospital bed covered by bed sheets for a long time, unbearable heat is generated on the patient's body. A patient's body temperature rises from 39 °C up to 40 °C creating patient restlessness. The thermal comfort of fabrics is associated with the thermal balance of the human body and its thermal responses to the dynamic interactions with the clothing and environment systems [6].

Both heat and moisture transmission behaviour of a fabric play a very important role in order to maintain thermo-physiological properties, respectively, comfort. A fabric should allow moisture, in the form of sensible and insensible perspiration, to be transmitted from the body to the environment. A fabric which is in contact with the skin should be dry to the touch. If a fabric is wet, heat which is flowing from the body will increase causing unwanted body heat loss and a sweaty feeling [7, 8].

Thermal transport properties of polyester, cotton and polyester/cotton blended fabric in an effort to understand the physical basis of clothing comfort were investigated [9, 10].

The lyocell fibres have very high moisture absorption rate, where at 65% of relative humidity lyocell fibres still have unused capacity to absorb moisture from the human skin comparing to the cotton fibres. Lyocell fibre has an extremely smooth surface and feels soft and pleasant on the skin [11]. The combination of smooth fibre surface and excellent moisture absorption creates a positive environment for healthy skin. According to recent dermatological studies, wearing clothing made of lyocell fibres significantly improves comfort and promotes a feeling of well-being [12].

Disposable nonwoven polypropylene (PP) fibres do not absorb water or moisture and are resistant to microorganisms and mildew. Despite the low hydrophilicity, they are not prone to charging static electricity. Due to low prices of polypropylene fibres, disposable clothes made of polypropylene fibres are inexpensive [13].

The three-layer textile laminate PET/PU/PET, which meets the European standards for surgical drapes EN 13795, was used as the reference material in studies [14].

Impact of different fabrics with different raw material composition commonly used for hospital bed linen, healthcare professionals' uniforms and surgical drapes on thermo-physiological comfort and microbial barrier properties was investigated.

2 Materials and Methods

The thermo-physiological and microbial barrier properties of three woven fabric are as follows: 100% cotton woven fabric in satin 4/1(2) weave with warp density of 40 threads/cm and weft density of 28 threads/cm, 100% Tencel® woven fabric in twill 2/1 weave with warp density of 50 threads/cm and weft density of 27 threads/cm as well 50% PES/50% cotton woven fabric in plain 1/1 weave with warp density of 34 threads/cm and weft density of 25 threads/cm were investigated. The thermo-physiological and microbial barrier properties of nonwoven fabric made of 100% polypropylene fibres and PES/PU/PES three-layered fabric laminate were carried out as well.

The constructional fabrics parameters were tested according to the standards ISO 3801 (fabric mass per unit area), ISO 5084 (fabric thickness) and ISO 7211-2 (fabric density). Thermal and water vapour resistance was determined using the Sweating Guarded Hot Plate according to ISO 11092:2014. Microbial barrier properties were tested using bacterial spores of the Bacillus genus *Geobacillus stearothermophilus* and *Bacillus atrophaeus*.

2.1 Thermal and Water Vapour Resistance

The thermo-physiological comfort was determined by measuring thermal and water vapour resistance under steady-state conditions according to ISO 11092:2014 using the Sweating Guarded Hot Plate device (Fig. 1). The data is used to measure the thermal and water vapour resistance of the fabrics and to determine the level of comfort.

The sample to be tested is placed on the heated plate with the conditioned air ducted to flow along and parallel to its upper surface. The test conditions for measuring thermal resistance are as follows: temperature of the plate is 35 °C, air temperature at 20 °C, relative humidity of 65% and airspeed at 1 m s^{-1}. After reaching the test conditions and steady state, the recording of values can be started.

During determination of water vapour resistance, the surface of the porous plate is kept constantly moist by means of a water-dosing device. A smooth, water vapour permeable, but liquid-water impermeable cellophane membrane shall be fitted over the porous plate. For the determination of water vapour resistance (R_{et}), the temperature of the measuring unit needs to be set at 35 °C, air temperature at 35 °C, relative humidity of 40% and airspeed at 1 m s^{-1}. After reaching the test conditions and steady state, the recording of values can be started. Thermal and water vapour

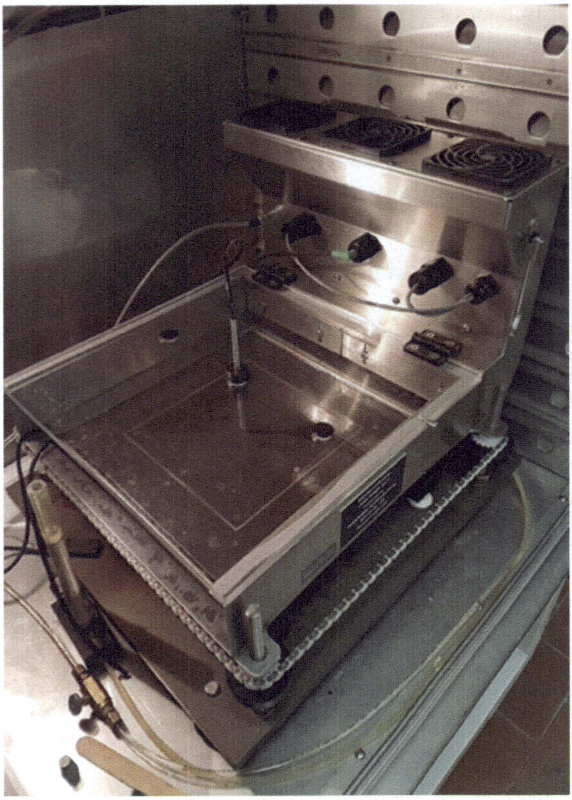

Fig. 1 Sweating guarded hot plate, measurement technology north-west

resistance of the tested material is determined as the arithmetic mean of the values of three individual specimens. The thermal and water vapour resistance of the fabric is calculated using calculations by the ThermDac software listed below [15, 16]:

$$R_{ct} = \frac{(T_s - T_a)}{Q/A} - R_{ct,0} \quad (1)$$

where R_{ct} is the thermal resistance in m^2 °C W^{-1}, T_s is the hotplate surface temperature in °C, T_a is ambient temperature in °C, Q/A is zone heat flux W m^{-2} and $R_{ct,0}$ is the bare plate thermal resistance in m^2 °C W^{-1}.

$$R_{et} = \frac{(P_s - P_a)}{Q/A} - R_{et,0} \quad (2)$$

where R_{et} is the water vapour resistance in m^2 Pa W^{-1}, P_s is the saturation vapour pressure at hotplate surface in Pa, P_a is ambient partial vapour pressure in Pa, Q/A is zone heat flux W m^{-2} and $R_{et,0}$ is the bare plate evaporative resistance in m^2 Pa W^{-1}.

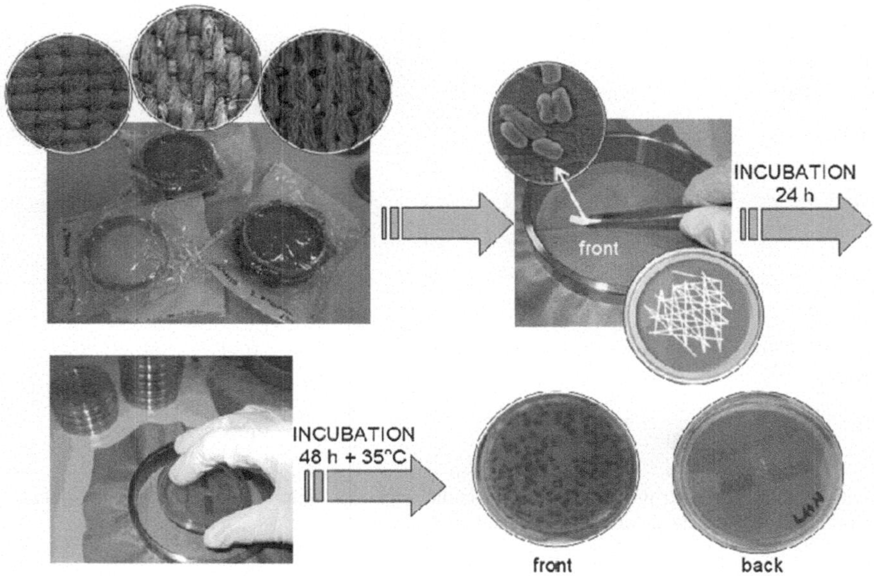

Fig. 2 Schematic overview of microbial barrier permeability testing [19]

2.2 *Microbial Barrier Properties*

Microbial barrier properties were tested using bacterial spores of the Bacillus genus *Geobacillus stearothermophilus* 10^5 (ATCC 12980, DSM 22) and *Bacillus atrophaeus* 10^6 (ATCC 49337, DSM 7264) in dry state (Fig. 2). Fabric was fastened in a ring device (diameter of 55 mm). The method involves directly rubbing the microorganisms onto the sterilized samples. Incubation of 24 h followed. After 24 h, the print was taken using a CT3P agar print plate (bioMe´rieux SA, Marcy I'Etoile, France). The first print was taken from the back side and then with new plate from the front side of the fabric. Agar plates were incubated for 72 h at 35 °C. After incubation, colony forming units (CFU) were counted [17, 18].

3 Results and Discussion

The fabric raw material, fabric type and weave as well as basic fabric parameters are presented in Table 1. All tested fabrics have application in healthcare sector, nursing homes and hotels where they need to provide comfort and microbial barrier permeability. Woven fabrics are woven in three different weaves (satin, twill and plain) with similar weft density (from 25 threads/cm to 28 threads/cm) and different warp density (from 34 threads/cm to 50 threads/cm) Three-layer laminate consists

Table 1 Basic fabric parameters

Samples		Statistic	m ($g\,m^{-2}$)	t (mm)	d (cm^{-1})	
Raw material	Fabric type				Warp	Weft
100% cotton	Woven fabric (satin 4/1(2))	M	195.9	0.40	40	28
		SD	0.6	0.00		
100% PP	Nonwoven	M	36.17	0.15	/	/
		SD	0.6	0.01		
100% Tencel®	Woven fabric (twill 2/1)	M	193.7	0.34	50	27
		SD	1.5	0.01		
50% PES/50% cotton	Woven fabric (plain 1/1)	M	178.6	0.31	34	25
		SD	2.1	0.01		
PES/PU/PES	Three-layer laminate	M	216.0	0.65	/	/
		SD	0.4	0.01		

where m is the mass per unit area, g m^{-2}; t is the thickness, mm; d is the woven fabric density, threads per cm; M is mean value; SD is standard deviation

of two polyester knits layers with a polyurethane membrane between. Nonwoven polypropylene fabric is made by melt blowing process.

Mass per unit area of the woven fabrics is similar and ranges from 178.6 to 195.9 g m^{-2}. Three-layered laminate mass per unit area is 216.0 g m^{-2}. Nonwoven fabric has considerably lower mass per unit area (36.2 g m^{-2}) compared with the woven fabrics and three-layered laminate. Thickness of the fabrics is in the range of 0.15 up to 0.65 mm. The woven fabrics thickness is similar ranging from 0.31 to 0.40 mm. The three-layered laminate has greatest thickness of 0.65 mm, while nonwoven PP fabric has lowest thickness of 0.15 mm compared to woven fabrics.

Results of thermal and water vapour resistance are shown in Table 2. Thermal resistance of tested fabrics is in the range from 0.0125 to 0.0755 m^2 °C W^{-1}. The highest thermal resistance has PES/PU/PES laminate (0.0755 m^2 °C W^{-1}) while Tencel® woven fabric has lowest (0.0125 m^2 °C W^{-1}). Highest laminate thermal resistance is a result of polyurethane membrane within three-layered PES/PU/PES laminate. The lowest thermal resistance of Tencel® woven fabric could be the result of weave and/or raw material. The warp density of tested woven fabric did not effect on thermal resistance since cotton and 50% PES/50% cotton woven fabrics have lower warp density than Tencel® woven fabric. Comparing to woven fabrics, nonwoven has higher thermal resistance due to own structure filled with trapped air within the pores.

The water vapour resistance ranged from 3.65 to 14.99 m^2 Pa W^{-1}. According to the Hohenstein Institute from Germany, all three woven fabrics and nonwoven fabric have very good comfort, respectively, extremely breathable and comfortable at higher level of activity [20].

According to the obtained results of the woven fabrics water vapour resistance, it can be concluded that different weave and fabrics weft density do not significantly

Table 2 Thermal and water vapour resistance

Samples		Statistic	R_{ct} (m^2°C W^{-1})	R_{et} (m^2 Pa W^{-1})
Raw material	Fabric type			
100% cotton	Woven fabric (satin 4/1(2))	M	0.0194	3.81
		SD	0.0001	0.16
100% PP	Nonwoven	M	0.0220	4.84
		SD	0.0003	0.31
100% Tencel®	Woven fabric (twill 2/1)	M	0.0125	3.64
		SD	0.0002	0.19
50% PES/50% cotton	Woven fabric (plain 1/1)	M	0.0201	3.65
		SD	0.0002	0.09
PES/PU/PES	Three-layer laminate	M	0.0755	14.99
		SD	0.0029	0.06

where R_{ct} is the thermal resistance, m^2 °C W^{-1}, R_{et} is the water vapour resistance, m^2 Pa W^{-1}; *M* is mean value; SD is standard deviation

effect on the water vapour resistance. The nonwoven fabric with lowest mass per unit area (36.2 g m^{-2}) and thickness (0.15 mm) has higher water vapour resistance compared to the woven fabrics. The higher water vapour resistance is due to specific nonwoven structure. Nonwoven fabric is composed of fibres which cross each other, creating structure with numerous voids and pores filled with air. Highest water vapour resistance compared to the woven fabrics is a result of air volume within pores of nonwoven fabric.

The water vapour resistance of PES/PU/PES three-layered fabric laminate is 14.99 m^2 Pa W^{-1} or having satisfactory comfort where fabric is breathable but uncomfortable at higher rate of activity. The highest water vapour resistance compared to the woven and nonwoven fabrics is the result of polyurethane membrane within three-layered PES/PU/PES laminate.

The results of microbial barrier permeability testing of the fabrics show that three-layered textile laminate provided full protection against the penetration of microorganisms (Table 3). Laminate is an efficient microbial barrier since even a single bacterial colony was not able to penetrate through fabric. Due to the polyurethane membrane between PES knit fabrics, laminate is completely impermeable to microorganism's penetration. Although PES/PU/PES laminate is an efficient microbial barrier, the fabric under higher rate of activity is uncomfortable.

The results of the woven fabrics microbial barrier permeability made by different raw materials, weave and warp density after extreme contamination with bacterial spores *G. stearothermophilus* and *B. atrophaeus* showed that 100% Tencel® fabric (57:1) provided slightly better microbe barrier than 100% cotton fabric (60:1) and noticeably better microbial barrier permeability compared to PES/cotton blend (32:1). Based on fabrics diversity, difference of microbial barrier permeability between the fabrics is the result of warp density, weave and raw material. Which of

Table 3 Microbial barrier permeability

Samples	*CFU*—colony forming unit			
	The average number of bacterial colonies on the face side (CFU)	The average number of bacterial colonies on the back side (CFU)	Ratio (CFU)	ΔCFU (%)
100% cotton satin 4/1(2)	800	14	57:1	
100% PP nonwoven	407	2	204:1	−258
100% Tencel® twill 2/1	419	7	60:1	−5
50% PES/50% cotton plain 1/1	356	11	32:1	44
PES/PU/PES laminate	155	0	/	/

the mention components, i.e. warp density, weave and raw material, influence more on woven fabrics microbial barrier permeability, should be additionally explored.

The polypropylene nonwoven fabrics, although have the lowest mass per unit area and thickness, is providing very good protection against microorganism penetration (of 204 microbial colonies on the fabric face side only one penetrated through fabric) and very good thermo-physiological comfort (4.84 m^2 Pa W^{-1}). Because of specific porous nonwoven structure, passage of bacteria is difficult. Additional advantage of nonwoven fabrics comparing to woven fabrics and especially laminate is high economic efficiency of production, e.g. low price of nonwoven fabric related to woven and laminate. Disadvantage of the nonwoven fabric with application in healthcare sector and nursing homes is the fact that they are disposable which create ecological problems after usage.

Comparing microbial barrier permeability of fabrics regarding to 100% cotton woven fabric (*ΔCFU,* bacterial reduction with regard to 100% cotton woven fabric), the greatest reduction is visible considering the polypropylene nonwoven fabrics (−258%) which is the result of the specific nonwoven fabric structure itself. The 100% Tencel® woven fabric has only 5% lower bacterial reduction compared to 100% cotton woven fabric which could be result of warp density, weave and raw material.

4 Conclusion

The tested fabrics with application for hospital bed linen, healthcare professionals' uniforms and surgical drapes showing different thermo-physiological and microbial barrier properties, respectively, impact of raw material and fabric structure is evident.

The woven and nonwoven fabrics have very good thermo-physiological comfort while three-layered PES/PU/PES laminate due to polyurethane membrane within fabric shows satisfactory comfort, but at higher rate of activity fabric is breathable but uncomfortable. The highest thermal resistance has PES/PU/PES laminate because of polyurethane membrane within own structure. The thermal and water vapour resistance of the woven fabrics showed that different weave and fabrics weft density do not significantly effect the thermo-physiological comfort. Comparing to woven fabrics, nonwoven has higher thermal and water vapour resistance due to specific nonwoven structure with numerous pores and voids. The air volume within nonwoven fabric pores provides higher resistance to heat and water vapour to pass.

Laminate is an efficient microbial barrier since even a single bacterial colony was not able to penetrate through fabric due to the polyurethane membrane between PES knit fabrics. Tencel® fabric provided slightly better microbe barrier than cotton fabric and noticeably better microbial barrier permeability compared to PES/cotton blend. Result of poor microbial barrier permeability of PES/cotton blend fabric could be result of lowest warp density. Influence of weave and raw material on woven fabrics microbial barrier permeability should be additionally explored.

Although the polypropylene nonwoven fabric has the lowest mass per unit area and thickness regard to the woven fabrics, fabric is providing very good protection against microorganism penetration. The reason could be found in specific porous nonwoven structure where the passage of bacteria is more difficult.

Taking cotton woven fabric as a reference fabric, the greatest reduction in bacterial passage through the tested fabrics is visible for the polypropylene nonwoven fabrics (−258%) as a result of the specific nonwoven fabric structure. The bacterial reduction in Tencel® woven fabric is only 5% lower.

Acknowledgements This work has been fully supported by the Croatian Science Foundation under the project number IP-2016-06-5278.

References

1. Bloomfield, S.F. et al.: The Infection Risks Associated with Clothing and Household Linens in Home and Everyday Life Settings, and the Role of Laundry. http://www.ifh-homehygiene.org/best-practice-review/infection-risks-associated-clothing-dhouseholdlinens-home-and-everyday-life. Accessed Apr 2011
2. Fijan, S., Turk, S.S.: Hospital textiles are they a possible vehicle for healthcare—associated infections? Int. J. Environ. Res. Public Health **9**(9), 3330–3343 (2012)
3. Belkin, N.L.: A historical gowns and drapers. AORN J. **76**(4), 648–653 (2002)
4. Pissiotis, C.A., Komborozos, V., Paputsi, C., Skrekas, G.: Factors that influence the effectiveness of surgical gowns in the operating theatre. Eur. J. Surg. **163**, 597–604 (1997)
5. Stephens-Borg, K.: Surgical dressings and turbulent years of cotton industry. J. Perioper. Pract. **18**(7), 305–306 (2008)
6. Kandavadivu, P., Vigneswaran, C., Ramachandran, T., Geetha, M.: A comparative study on the performance of commercial hospital bedlinen and multilayered bedlinen developed. J. Text. Assoc. **72**(3), 183–191 (2011)

7. Das, B., Das, A., Kothari, V.K., Fangueiro, R., DeAraujo, M.: Moisture transmission through textiles part II: evaluation methods and mathematical modeling. AUTEX Res. J. **7**(3), 194–216 (2007)
8. Chellamani, K.P., Veerasubramanian, D., Panner Selvam, G.: Thermal characteristics of hospital bed linens—a sensitive science. J. Acad. Ind. Res. **2**(6), 374–377 (2013)
9. Diepgen, T.L., Christian Schuster, K.: Dermatological examinations on the skin compatibility of textiles made from Tencel fibers. Lenzinger Berichte **85**, 61–67 (2006)
10. Yoon, H.N., Buckley, A.: Improved comfort polyester part I: transport properties and thermal comfort of polyester/cotton blend fabrics. Text. Res. J. **54**(9), 602–613 (1984)
11. Schurz, J.: What is new about new fibres of the Lyocell type. Lenzinger Berichte **9**(94), 37 (2003)
12. Sundaresan, S., Ramesh, M., Sabitha, V., Ramesh, M., Ramesh, V.: A detailed analysis on physical and comfort properties of bed linen woven fabrics. IJARIIE **2**(2), 1649–1658 (2016)
13. Cunko, R., Andrassy, M.: Vlakna. Zrinski, Cakovec (2005)
14. EN 13795-2: Surgical Drapes, Gowns and Clean air Suits, used as Medical 2: Surgical Drapes, Gowns and Clean Air Suits, used as Medical Devices for Patients, Clinical Staff and Equipment
15. ISO 11092 Textiles—Physiological Effects—Measurements of Thermal and Water-Vapour Resistance Under Steady-State Conditions (sweating guarded-hotplate test)
16. Operators Manual for Material Evaluation Hotplate Model SGHP-8.2 S/N 223-xx, Measurement Technology Northwest (2006)
17. Rogina-Car, B., Budimir, A., Turcic, V., Katovic, D.: Do multi-use cellulosic textiles provide safe protection against contamination of sterilized items? Cellulose **21**(3), 2101–2109 (2014)
18. Rogina-Car, B., Budimir, A., Turcic, V., Katovic, D.: Cellulose and cellulose composites: modification, characterization and applications. In: Mondal, I.H. (ed.). Nova Science Publishers, Inc. New York, pp. 415–444 (2015)
19. Rogina-Car, B., Budimir, A., Katovic, D.: Microbial barrier properties of healthcare professional uniforms. Text. Res. J. **87**(15), 1860–1868 (2017)
20. Reljic, M., Stepanovic, J., Lazic, B., Cirkovic, N., Cerovic, D.: The change of water vapour resistance of materials used for the clothing production during exploitation. Adv. Tech. **5**(2), 73–78 (2016)

Handle Assessment of Knitted Mattress Fabrics Treated with Flame Retardant Finishes Using Fabric Touch Tester Device

Atiyyah Binti Haji Musa, Benny Malengier, Simona Vasile and Lieva Van Langenhove

Abstract Flame retardant finishes are often imparted to the mattress fabrics nowadays as to inhibit the ignition of fire that is commonly dispersed rapidly by textile materials. As consumers buy their mattress fabrics partially after making some touch evaluation, hence the handle assessment of these materials is significant to be quantified. In this study, 12 types of double jersey mattress ticking fabrics were analyzed by the fabric touch tester (FTT), a relatively recently commercialized instrument for fabric handle measurement. The fabrics were differentiated by three production parameters, i.e., mass per unit area, fiber composition of the upper layer (i.e., closest to the body), and concentration of flame retardant finish. Based on the generated models, strong correlations were found between FTT fabric indices and mass per unit area which is indicated as the dominant factor that influenced most of the handle properties, while other factors are less pronounced. This suggests the capability of the FTT to assess the variation in these types of fabric.

Keywords Fabric touch tester · Fabric handle · Fabric comfort

1 Introduction

Comfort is a feel of being stress-free that is always a goal for humans. Being in a state of comfort relates to satisfaction which leads to quality of life. In the context of sleeping quality, one of the many aspects that contribute to it is the handle properties

A. Binti Haji Musa (✉) · B. Malengier · L. Van Langenhove
Faculty of Engineering and Architecture, Department of Materials, Textiles and Chemical Engineering, Centre for Textile Science and Engineering, Ghent University, Technologiepark 70A, 9052 Zwijnaarde (Gent), Belgium
e-mail: Atiyyah.BintiHajiMusa@UGent.be

S. Vasile
Faculty of Science and Technology, Department of Fashion, Textiles and Wood Technology/FTI Lab, University College Gent, Buchtenstraat 11, 9051 Ghent, Belgium

A. Binti Haji Musa
Universiti Teknologi MARA Cawangan Negeri Sembilan, Kuala Pilah, Negeri Sembilan, Malaysia

A. Majumdar et al. (eds.), *Functional Textiles and Clothing*,
https://doi.org/10.1007/978-981-13-7721-1_23

of the mattress fabrics. More often than not, the handle properties of the fabrics are an important and decisive criterion for purchasing it, as the consumers touch and squeeze the fabrics before they buy them. The touch motion by the palm and fingers gives information about some handle attributes which are commonly used to describe fabric comforts such as smoothness, softness, and warmth. A number of studies dealing with thermal comfort of the mattresses were reported in the literature where the researchers found that thickness of the mattress ticking fabrics is the most important parameter for a comfortable sleep environment [1, 2]. Moreover, the addition of phase-change materials also contributed to a good thermal comfort for the bedding system [3]. These studies are mainly focused on the thermal comfort of the mattress. Some studies about tactile perception can also be found [4, 5] but they are limited in scope.

As reported by the United States National Fire Protection Association, two-thirds (66%) of the home smoking-material fire fatalities resulted from fires originating with mattresses or bedding other than upholstered furniture [6]. Hence, due to the awareness that the common cause for fire in the house is due to cigarettes lit in the bedroom, these mattresses and beddings nowadays are imparted with some flame retardant finishes, other than having self-extinguish fibers or yarn in the fabric structure itself. Having the ability to retard or resist the flame, the manufacturers then need to compensate that attribute with some deterioration in the handle of the fabrics [7]. Nevertheless, not all flame retardant finishes add a harsher hand since researchers also found that some flame retardant finishes used in combination with certain weave structures can make the fabric softer [8]. The effects of flame retardant finishes on fabric hand also vary depending on the finish type and fabric conditions (dry and wet) as the fabrics treated with Indura gives better handle compared to those treated with Pyrovatex and Flamex II [9]. Wet flame retardant-treated fabric was rougher as it required more force due to a larger contact area which resulted in higher frictional resistance of the fabrics when the fabrics were pulled through the nozzle of an extraction method [9]. The studies of the effect of flame retardant finishes on fabric handle mainly address common fabrics. None has measured the effect on mattress fabrics such as mattress ticking, pad, or protector.

In the previously mentioned study [9], the extraction method was used in which the force needed to pull the fabrics through a nozzle is measured and then the correlation between the measured force with other hand-related properties such as drapability, flexural rigidity, and static friction coefficient was determined in order to measure the handle properties of the fabrics [8–10]. There are many other objective measurements developed to assess the tactile perception of fabrics such as Kawabata Evaluation System (KES) [11], Fabric Assurance by Simple Testing (FAST), [12] Tissue Softness Analyzer (TSA) [13], and the recently commercialized equipment known as Fabric Touch Tester (FTT) [14]. In this study, the handle of mattress ticking fabrics is evaluated by using the FTT. The capability of the FTT to assess the variation in fabric handle was verified in a number of publications [4, 15–17]; thus, this study will further acknowledge the capability of this new instrument to distinguish the changes in production parameters related to flame retardant treatment on mattress ticking fabrics.

2 Materials and Methods

2.1 Materials

Twelve types of mattress ticking fabrics, encompassing three types of production settings which are commonly used to alter the properties of the mattress ticking materials, were investigated. Those fabrics were differentiated by mass per unit area, fiber composition, and flame retardant finish concentration. The fabrics consist of three layers where the top and bottom layers are double jersey knitted fabrics and in between the layers, thick bulk continuous filament (BCF) polyester yarns were filled. For a mattress ticking, the top layer or the face side is our main concern as it is the closest to the body. Hence, the variation in production parameters was made only to the top layer of the knitted fabrics, while the bottom layer was equal for all samples. Mass per unit area of the fabrics ranges from 186 to 619 g/m^2, the fiber composition of the top layer of the fabrics is either polyester (PES) or viscose (CV) and the concentration of the flame retardant finish were 5, 20, or 100 g/L. The specification of each fabric is shown in Table 1.

2.2 Methods

The tactile behavior of the mattress ticking fabric was measured using the FTT. It is a relatively recent tool in the market and introduces a new way to quantify the han-

Table 1 Specification of the materials

Fabric ID	Composition of the top layer		Mass per unit area m (g/m^2) (SD)		Flame retardant concentration (g/L)	Thickness (mm) at 4.14 kPa (SD)
	Fiber	Code	Projected	Actual		
1	PES	0	560	619.17 (10.48)	100	3.62 (0.10)
2	CV	1	560	604.03 (13.06)	100	3.16 (0.25)
3	PES	0	560	591.43 (5.97)	20	3.53 (0.32)
4	CV	1	185	214.93 (5.18)	5	1.09 (0.06)
5	CV	1	275	283.40 (2.65)	20	2.01(0.03)
6	CV	1	185	225.37 (3.25)	100	1.29 (0.02)
7	PES	0	185	194.13 (1.62)	20	1.16 (0.07)
8	PES	0	560	575.90 (11.17)	5	3.49 (0.16)
9	PES	0	185	203.73 (2.39)	100	1.20 (0.03)
10	PES	0	185	186.70 (0.44)	5	1.09 (0.04)
11	CV	1	560	563.83 (19.73)	5	3.55 (0.09)
12	PES	0	275	275.67 (4.63)	5	1.97 (0.04)

dle properties of fabrics. The device was manufactured by SDL Atlas as a result of research done at The Hong Kong Polytechnique University. Within the four modules of the FTT, bending, compression, surface, and thermal properties are concomitantly measured, resulting in 13 fabric indices as in Table 2. The measurements are comprehensive and cover both sides (inside and outside of the fabric) and both directions (warp/wale and weft/course). Based on obtained values of the 13 fabric indices, the FTT software will subsequently compute three primary comfort indices, i.e., smoothness, softness, and warmness, as well as two global comfort indices, i.e., total hand and total touch. Details about the modules of the instrument and calculation of the fabric indices have been reported elsewhere [14, 18, 19]. Prior testing, all samples were placed for at least 24 h in a conditioning room, controlled at 21 °C ± 2 °C and relative humidity of 65% ± 4% [20].

3 Results and Discussion

The mean and standard deviation (SD) of the 13 FTT fabric indices corresponding to the face side of the 12 mattress ticking fabrics are given in Table 3, both for wale and course direction. A small letter a or e following an index refers to the value of that particular index in wale or course direction while a letter m stands for the average value of both directions. The highest value for bending properties, i.e., bending average rigidity (BAR) and bending work (BW) is noticed for the heaviest fabrics (ID 1, 2, 3, 8, and 11), while the least is for the lightest (ID 4, 6, 7, 9 and 10). Since higher BW and BAR relates to a hasher handle, based on the results, it can be said that the heavier fabrics are harsher than the lighter ones as more forces and work are needed to bend the heavier specimen. The same trend can also be seen with the compression property (T and CW), surface roughness (SRA and SRW), and thermal (Qmax, TCC, and TCR). The composition of viscose (CV) can be noted in the surface friction property (SFC) as the viscose (CV) fabrics showed the highest friction coefficient as compared to polyester (PES). The trend for concentration of flame retardant is barely visible in the FTT fabric indices in which a further statistical analysis was performed to quantify the influence of it and the other investigated production parameters toward fabric handle.

The python statsmodel package was employed in which the composition of the top layer of the mattress ticking fabrics was treated as categorical value, i.e., 0 is for PES and 1 is for viscose. A stepwise regression was then executed, considering linear and quadratic terms with $p < 0.05$ for the coefficients, which indicate the relevance of the respective coefficient of the model. This analysis procedure enables an automatic selection of the predictive variables based on all data from the FTT measurements. The term or variable that is added in every step is the one that most increased the adjusted R^2 value of the ordinary least squares (OLS), while satisfying the requirement $p < 0.05$. Thus, a valid model is generated which comprises all the significant terms and the optimal adjusted R^2 value. The influence of the mass per unit area (m), flame retardant concentration (FR), and viscose composition (CV)

Table 2 Interpretations of the FTT fabric indices

Item	Fabric property	Index	Description	Unit given by FTT software	SI unit	Usual interpretations
1	Bending	BAR	Bending average rigidity	gf mm rad^{-1}	N m rad^{-1}	Force needed to bend per radian
2		BW	Bending work	gf mm rad	N m rad	Work needed to bend
3	Surface friction	SFC	Surface friction coefficient	–	–	Friction coefficient on surface with ribbed plate
4	Surface roughness	SRA	Surface roughness amplitude	μm	m	Roughness irregular wave amplitude
5		SRW	Surface roughness wavelength	mm	m	Roughness irregular wave wavelength
6	Compression	CW	Compression work	gf mm	N m	Work needed to compress the specimen
7		CRR	Compression recovery rate	–	–	Percentage of thickness changes after compressed
8		CAR	Compression average rigidity	gf cm^{-2} mm^{-1}	N m^{-3}	Forces needed to compress per mm
9		RAR	Recovery average rigidity	gf cm^{-2} mm^{-1}	N m^{-3}	Forces reflected when recovery per mm
10		T	Thickness	mm	m	Thickness of the materials

(continued)

Table 2 (continued)

Item	Fabric property	Index	Description	Unit given by FTT software	SI unit	Usual interpretations
11	Thermal conductivity	TCC	Thermal conductivity under compression	10^{-3} W m^{-1} °C^{-1}	W m^{-1} °C^{-1}	Energy transmitted per degree per m per second under specimen compression
12		TCR	Thermal conductivity under recovery	10^{-3} W m^{-1} °C^{-1}	W m^{-1} °C^{-1}	Energy transmitted per degree per m per second under specimen recovery
13		Qmax	Thermal maximum flux	W mm^{-2}	W m^{-2}	Maximum energy transmitted during compression

of the measured mattress ticking fabrics can be reflected in the statistical models shown in Table 4. The sign of each coefficient in the models indicates a positive (+) or negative (−) influence of the respective variable on the FTT fabric indices considered and high R^2-values (i.e., above 0.8) designate a strong FTT indices–input variable relationship. The results discussed here are all based on the 13 measured properties in FTT, whereas the computed models from it, i.e., smoothness, softness, warmth, total hand, and total touch are neglected since the values are out of the index range (0–1). This is because the tested fabrics are not the typical clothing fabrics which were used to develop the models; hence, the generated models by the FTT software are not suitable for them.

Referring to Table 4, mass per unit area (m) were exposed in all FTT indices, either in linear or quadratic or both, except surface friction coefficient in wale direction (SFCa). These findings are in agreement with another study [4] that investigated the effect of softener concentration on tactile properties of mattress ticking fabrics of similar weight range. Hence, it can be said that mass of the fabrics is the dominant factor which governs the handle properties of the fabrics, in comparison with flame retardant concentration and viscose composition with an exceptional denotation to RAR and SRAa where the R^2 is less than 0.5, meaning a poor relationship with the input variable, though the trend is statistically significant. Composition CV also tends to influence bending properties (BAR and BW) and surface friction (SFC) in both linear and quadratic relationships, along with mass per unit area. However, the impact is more pronounced in SFCa as composition is the only variable that appears

Table 3 Mean (standard deviation) for 13 FTT fabric indices for the face side of 12 mattress ticking fabrics

	Fabric ID											
	1	2	3	4	5	6	7	8	9	10	11	12
BARa	4016.74	3524.72	3444.55	184.11	512.66	206.84	221.97	3677.73	190.79	193.38	3417.19	461.70
	(546.89)	(535.34)	(473.39)	(18.03)	(53.83)	(23.27)	(24.28)	(298.66)	15.25	(13.17)	(893.83)	(66.83)
BARe	5644.69	4910.45	4694.60	202.08	769.85	198.27	194.21	4199.49	195.65	183.59	4890.95	712.12
	(988.70)	(739.12)	(771.98)	(35.75)	(119.19)	(32.31)	(20.06)	(740.27)	(32.42)	(19.74)	(950.78)	(117.38)
BWa	17550.14	15146.83	16521.27	903.44	2548.11	1034.35	1134.59	16384.81	1019.65	1017.51	16416.55	2301.19
	(2754.93)	(2341.43)	(1085.82)	(77.23)	(87.19)	(104.92)	(54.59)	(1056.95)	(75.83)	(83.00)	(3149.16)	(121.24)
BWe	27438.24	26086.81	24809.53	1084.75	3928.12	1112.98	1112.57	22492.23	1022.95	1023.51	24965.13	3670.76
	(5598.83)	(2725.95)	(3771.00)	(100.24)	(429.47)	(107.43)	(116.06)	(2669.63)	(72.61)	(97.32)	(4220.21)	(402.50)
CW	2204.26	2501.00	2170.57	1259.64	1715.09	1298.12	1333.29	2115.15	1225.34	1199.21	2243.00	1647.88
	(165.71)	(335.24)	(173.17)	(111.71)	(189.06)	(151.49)	(142.89)	(169.88)	(170.12)	(136.85)	(122.55)	(131.29)
CRR	0.53	0.48	0.55	0.81	0.64	0.78	0.82	0.56	0.84	0.83	0.46	0.62
	(0.11)	(0.03)	(0.12)	(0.08)	(0.08)	(0.11)	(0.08)	(0.10)	(0.06)	(0.09)	(0.02)	(0.03)
CAR	127.26	111.69	129.96	185.88	151.85	182.37	174.00	128.01	196.41	191.03	117.55	164.96
	(16.43)	(17.82)	(17.89)	(25.73)	(23.78)	(23.37)	(28.56)	(16.07)	(30.02)	(28.45)	(11.55)	(20.87)
RAR	212.58	193.23	213.84	234.41	231.24	240.34	216.06	221.36	228.66	231.57	215.90	235.14
	(35.42)	(37.67)	(32.46)	(36.46)	(43.07)	(46.58)	(38.59)	(32.99)	(30.63)	(35.71)	(32.81)	(19.81)
T	3.61	3.32	3.44	1.19	1.93	1.17	1.18	3.41	1.10	1.10	3.45	1.82

(continued)

Table 3 (continued)

	Fabric ID											
	1	2	3	4	5	6	7	8	9	10	11	12
	(0.20)	(0.16)	(0.09)	(0.03)	(0.02)	(0.06)	(0.02)	(0.12)	(0.01)	(0.03)	(0.15)	(0.04)
TCC	83.35	81.96	64.86	43.10	53.16	44.84	41.77	63.52	42.46	40.97	73.52	48.38
	(5.01)	(5.61)	(1.79)	(0.77)	(1.17)	(1.63)	(0.32)	(1.47)	(1.10)	(0.85)	(4.71)	(0.71)
TCR	66.36	66.71	60.19	43.79	50.12	44.53	43.48	58.49	43.25	42.48	61.20	48.61
	(3.06)	(2.78)	(1.97)	(0.48)	(1.64)	(1.45)	(0.96)	(1.76)	(1.03)	(1.03)	(3.40)	(1.49)
Qmax	658.29	779.18	644.55	728.67	781.37	777.48	541.09	650.58	628.63	510.22	762.27	565.61
	(147.28)	(166.64)	(98.81)	(37.33)	(59.79)	(35.25)	(29.21)	(63.48)	(22.35)	(28.45)	(72.46)	(34.96)
SFCa	0.26	0.30	0.27	0.35	0.36	0.38	0.27	0.26	0.25	0.23	0.31	0.30
	(0.02)	(0.03)	(0.02)	(0.06)	(0.03)	(0.02)	(0.07)	(0.01)	(0.05)	(0.01)	(0.04)	(0.03)
SFCe	0.35	0.42	0.38	0.65	0.63	0.66	0.74	0.40	0.69	0.74	0.44	0.49
	(0.03)	(0.04)	(0.04)	(0.15)	(0.09)	(0.12)	(0.04)	(0.04)	(0.12)	(0.03)	(0.04)	(0.02)
SRAa	254.02	249.83	167.82	127.22	154.08	95.00	65.29	151.73	72.23	95.69	269.71	153.91
	(119.67)	(101.35)	(124.05)	(50.29)	(61.45)	(27.39)	(9.45)	(22.82)	(35.53)	(25.21)	(178.28)	(87.92)
SRAe	301.73	260.86	266.91	105.73	202.27	161.14	97.30	298.63	98.72	97.52	305.45	144.10
	(40.12)	(47.49)	(66.48)	(39.69)	(71.77)	(99.61)	(15.06)	(55.65)	(20.72)	(12.79)	(56.06)	(35.69)
SRWa	10.76	11.67	7.95	3.63	7.42	3.54	3.16	9.60	3.45	3.48	9.26	4.52
	(3.38)	(2.71)	(5.82)	(0.62)	(3.79)	(1.46)	(0.97)	(3.28)	(0.95)	(0.87)	(1.13)	(2.09)
SRWe	5.47	6.02	5.99	3.60	3.44	2.81	1.68	5.69	1.45	1.33	5.37	4.39
	(0.69)	(0.83)	(0.58)	(2.06)	(1.02)	(1.44)	(0.48)	(0.71)	(0.30)	(0.29)	(0.12)	(2.87)

Table 4 Influence of mass per unit area (m), flame retardant concentration (FR) and viscose composition (CV) on FTT fabric indices

FTT fabric indices	Statistical model	Adjusted R^2
BARa	$-291.16 + 0.01\ \text{m}^2$	0.992
BARe	$-346.85 + 0.01\ \text{m}^2 - 5573.06\ \text{CV}^2 + 1825.57\ \text{CV}$	0.999
BARm	$-296.30 + 0.01\ \text{m} - 310.17\ \text{CV}$	0.999
BWa	$-754.87 + 0.05\ \text{m}^2 - 11.73\ \text{FR}$	0.990
BWe	$503.92 + 0.09\ \text{m}^2 - 26614.01\ \text{CV}^2 + 8810.87\ \text{CV} - 15.07\ \text{m}$	0.999
BWm	$-1349.73 + 0.06\ \text{m}^2 - 1655.73\ \text{CV}$	0.999
CW	$805.10 + 2.49\ \text{m}$	0.915
CRR	$1.50 - (4.25 \times 10^{-3})\ \text{m} + (4.0 \times 10^{-6})\ \text{m}^2$	0.937
CAR	$283.18 - 0.58\ \text{m} + (5.26 \times 10^{-4})\ \text{m}^2$	0.909
RAR	$235.74 - (7.5 \times 10^{-5})\ \text{m}^2$	0.488
T	$-1.27 + 0.01\ \text{m} - (1.1 \times 10^{-5})\ \text{m}^2 - 1.20\ \text{CV}^2$	0.991
TCC	$39.53 + (9.9 \times 10^{-5})\ \text{m}^2$	0.883
TCR	$34.00 + 0.05\ \text{m}$	0.951
Qmax	$227.84 + 478.37\ \text{CV} + 1.91\ \text{m} + 0.67\ \text{FR} - (2.09 \times 10^{-3})\ \text{m}^2$	0.928
SFCa	$0.26 + 0.33\ \text{CV}$	0.839
SFCe	$1.25 - (3.34 \times 10^{-3})\ \text{m} + (3 \times 10^{-6})\ \text{m}^2 + 0.19\ \text{CV}$	0.929
SFCm	$(4.87 \times 10^{-1}) - (4.84 \times 10^{-7})\ \text{m}^2 + 0.22\ \text{CV}$	0.949
SRAa	$-512.21 + 3.93\ \text{m} - (4.5 \times 10^{-3})\ \text{m}^2$	0.330
SRAe	$-145.27 + 1.50\ \text{m} - (1.3 \times 10^{-3})\ \text{m}^2$	0.947
SRAm	$-328.74 + 2.71\ \text{m} - (2.9 \times 10^{-3})\ \text{m}^2$	0.764
SRWa	$-10.86 + 0.09\ \text{m} - (9.5 \times 10^{-5})\ \text{m}^2$	0.724
SRWe	$-4.57 + 0.04\ \text{m} - (3.90 \times 10^{-5})\ \text{m}^2$	0.878
SRWm	$-7.72 + 0.07\ \text{m} - (6.7 \times 10^{-5})\ \text{m}^2$	0.838

in the model for SFCa. Flame retardant concentration (FR) only came out twice, namely in the model for BWa and Qmax, and then as the third term or higher after mass and/or CV. This indicates a less significant term which exhibits less influence to the fabric property.

4 Conclusion

The influence of three production settings towards the handle of the mattress ticking fabrics was assessed by FTT, a relatively recent device in the market meant for clothing fabrics. For that, 12 mattress ticking fabrics were produced which varied in their mass per unit area, concentration of flame retardant finish and composition of the top layer for the assembled fabric. Among the considered variables, it was found that fabric mass per unit area has the greatest contribution toward FTT fabric indices. The generated statistical models also suggest that almost all properties are influenced by mass of the fabrics in linear and quadratic ways, with statistically strong correlation, except for RAR and SRAa indices. Nonetheless, surface friction property tends to be impacted by the presence of viscose in the top layer of the fabric, which increased the coefficient of friction, thus generally associated with a rougher handle. In addition, the concentration of flame retardant does not give a major change in the handle of the considered fabrics, so we can conclude that the concentration should not be a limiting factor when considering the handle of the mattress ticking.

Although the mass seems to be dominant in most properties, note that it is distributed in a large range (186–619 g/m^2). Hence, a narrow range could be further explored not to hinder the effects of other parameters. Through the results, it shows that FTT can differentiate the fabrics with the said production settings, proving its potential to assess the changes caused by these variations. The involvement of expert panels also can be a good measure to confirm the results. Nonetheless, the variation of values of the fabric indices for the fabrics with the selected production parameters will give the fabric manufacturers a guideline to optimize the use of the resources (i.e., polyester/viscose fibers and flame retardant concentration) as to achieve the intended output.

Acknowledgements We would like to acknowledge Ministry of Education, Malaysia and Universiti Teknologi MARA, Malaysia for the sponsorship given for the study. The materials were received from BekaertDeslee Textiles, Belgium and the results are obtained within an international ERANET-CORNET research project—IWT 140387 (2015–2016) named Touché: Boosting innovation through application of basic understanding of the process and testing of textile touch and fabric feel. The authors also acknowledge the support of Myréne Vanderhoeven in performing part of the tests.

References

1. Onal, L., Yildirim, M.: Comfort properties of functional three-dimensional knitted spacer fabrics for home-textile applications. Text. Res. J. **82**, 1751–1764 (2012)
2. Terliksiz, S., Kalaoglu, F., Eryürük, S.H.: Analysis of thermal comfort properties of jacquard knitted mattress ticking fabrics. Int. J. Cloth Sci. Technol. **28**, 105–114 (2016)
3. De Mey, G., Özçelik, M., Schwarz, A.: Designing of Thermal Comfortable Bed Systems. Ambience 14&10i3m, Tampere, Finland (2014)
4. Vasile, S., Malengier, B., Deruyck, F., De Raeve, A.: Influence of selected production parameters on the hand of mattress knitted fabrics assessed by the Fabric Touch Tester. Text. Res. J. **89**, 98 (2017)
5. De Vilder, I., Hör, I., Hertleer, C.: Integrated solutions for improve the rest quality. Aachen Dresden Int. Text. Conf, Aachen, Germany (2013)
6. National Fire Protection Association. Fact Sheet: An Overview of the U.S. Fire Problem. Natl Fire Prot Assoc 2017

7. Mamalis, P., Andreopoulos, A., Spyrellis, N.: The effect of a durable flame-retardant finishing on the mechanical properties of cotton knitted fabrics. Int. J. Cloth Sci. Technol. **13**, 132–144 (2001)
8. Kim, J.O., Slaten, B.L.: Objective assessment of fabric handle in fabrics treated with flame retardants. J. Test. Eval. **24**, 223 (1996)
9. Kim, J.O., Slaten, B.L.: Objective evaluation of fabric hand: part I: relationships of fabric hand by the extraction method and related physical and surface properties. Text. Res. J. **69**, 59–67 (1999)
10. Pan, N., Yen, K.C.: Physical interpretations of curves obtained through the fabric extraction process for handle measurement. Text. Res. J. **62**, 279–290 (1992)
11. Kawabata, S., Niwa, M.: Fabric performance in clothing and clothing manufacture. J. Text. Inst. **80**, 19–50 (1989)
12. De Boos, A., Tester, D.: SiroFast, Fabric Assurance by Simple Testing (1994)
13. Emtech Electronic GmbH. Tissue Softness Analyzer (2006)
14. Atlas, S.D.L.: Rycobel. Fabric Touch Tester, Instruction Manual (2014)
15. Vasile, S., Malengier, B., De Raeve, A., Louwagie, J., Vanderhoeven, M.: Assessment of Sensorial Comfort of Fabrics for Protective Clothing. In: 7th European Conference on Protective Clothing. ECPC, Izmir, Turkey (2016)
16. Musa, A.B.H., Malengier, B., Vasile, S., Van Langenhove, L., De Raeve, A.: Analysis and Comparison of Thickness and Bending Measurements from Fabric Touch Tester (FTT) and Standard Methods. Autex. Res. J. **18**, 51 (2017)
17. Vasile, S., Malengier, B., De Raeve, A., Musa, A.B.H.: FTT comfort indices of ring spun and air-jet knitted fabrics with post-treatments. In: Autex Conference Proceeding, Corfu, Greece (2017)
18. Liao, X., Li, Y., Hu, J., Wu, X., Li, Q.: A simultaneous measurement method to characterize touch properties of textile materials. Fibers Polym. **15**, 1548–1559 (2014)
19. Hu, J.Y., Hes, L., Li, Y., Yeung, K.W., Yao, B.G.: Fabric touch tester: integrated evaluation of thermal-mechanical sensory properties of polymeric materials. Polym. Test **25**, 1081–1090 (2006)
20. ASTM International. Standard Practice for Conditioning and Testing Textiles D 1776-04. vol. i (2004)

Development of a Smoothness Tester for Fabrics

M. S. Parmar, Nidhi Sisodia and Maheshwar Singh

Abstract An instrument has been designed and developed to measure the smoothness behavior of finished fabrics. The instrument is based on pendulum principle. The weight (hang on string) comprises a frictionless wheel movable along arc-shaped platform. The platform acts as a sample holder. When the weight is subjected to push, it swings back and forth in the platform. The amplitude of the swing reduces due to friction of the fabric. The amplitude is inversely proportional to the friction or roughness of the fabric. Various types of finished fabric samples are tested on the developed instrument. The results are compared with Kawabata system to verify the working of instrument. These results are also compared with the bending length and crease recovery behavior of the particular fabric sample. It is found that lesser the bending length more will be the smoothness. If the crease recovery angle is high, the fabric will be smoother.

Keywords Bending length · Crease recovery · Finishing · Kawabata system · Smoothness

1 Introduction

There are various characteristics of fabric that can influence fabric comfort. Fabric surface friction or smoothness is one of the important characteristics, which has direct relation with fabric comfort. It was observed that the customer buying decision is influenced by fabric smoothness behavior. The behavior of fabric surface smoothness depends on various factors starting from type of yarn (fiber compositions, yarn twist, spinning technologies, etc.), conversion of yarn into fabric (whether it is woven, knitted or nonwoven) and its construction particular, and finally on finishing treatments. The influence of these factors on fabric surface smoothness may be positive or negative. In one of such studies [1], knitted fabric was manufactured using carded and combed yarn. The study revealed that the surface friction of knitted fabric made out

M. S. Parmar (✉) · N. Sisodia · M. Singh
Northern India Textile Research Association, Sec-23, Rajnagar, Ghaziabad, UP, India
e-mail: drmsparmar@nitratextile.org

A. Majumdar et al. (eds.), *Functional Textiles and Clothing*,
https://doi.org/10.1007/978-981-13-7721-1_24

of carded yarn was higher than the knitted fabric made out of combed yarn. Another study [2] indicated that if the fabric cover factor increases, the fabric surface friction decreases considerable reduces as the surface of fabric become more uniform. It was also observed that fibers content also influences frictional characteristics of fabric. The fabric structure also influences fabric surface friction. Fabric structure having floats will have high frictional coefficients than plain structure fabric. Similarly, plain fabric has lower friction than twill fabric.

Determination of fabric handle using subjective assessment [3–5] is dependent on static and dynamic friction between the cloth surface and the thumb or finger. It is a well-known fact that the human fingers are very sensitive and identify a minor change in the fabric surface friction. However, the perception of surface roughness or smoothness varies from person to person. Therefore, objective measurement is always considered to be better option.

Objective measurement [6–9] of fabric surface friction behavior resolves any misunderstanding among the buyer and manufacturer about the surface smoothness property of particular fabric. It is well known that there are always disputes between buyer and manufacturer regarding feel of fabric as there is no quantitative method is available which can spell out the feel of the particular fabric. It also helps in process optimization to get desired smoothness property in the fabric.

Though fabric comfort property is influenced by its surface friction behavior, still there is no such type of instrument available in the textile industry which can measure fabric surface friction easily. Kawabata developed [10] the KES-FB4 for the measurement of surface friction and the surface roughness of the fabrics. This instrument is not reachable to the textile industry as it is very expensive. Researchers also tried to use Instron tensile tester with some modifications to measure the inter-fabric or fabric-to-metal friction [11], which again become costlier and complicated and thus make it unsuitable for the textile industry.

Hence, an indigenous cost-effective instrument is required to be developed to address the above problems, i.e., to determine the smoothness characteristics of fabric and can give indication on change in surface characteristics after the various pre-treatment and finishing processes. The data generated by this instrument shall help the finishers to take appropriate decision to alter the recipe or process to meet the required smoothness characteristics of the fabric.

2 Materials and Methods

The study was carried out in two parts. In the first part, the instrument was developed, and in the second part, its performance was evaluated by analyzing different types of fabric samples for their smoothness behavior.

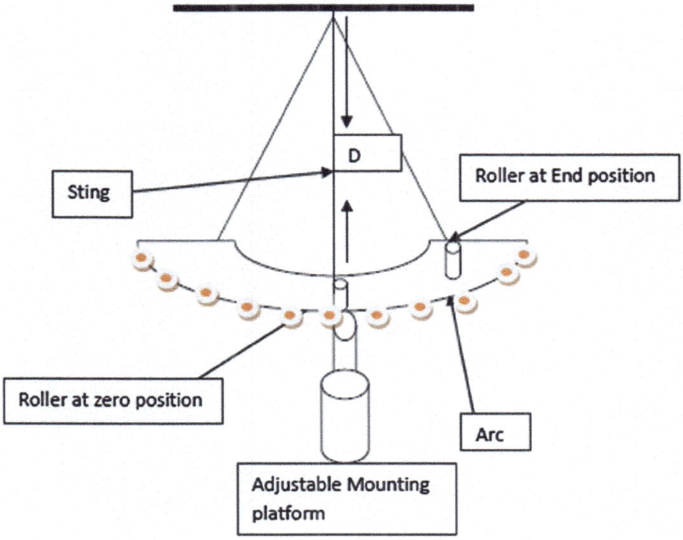

Fig. 1 Schematic diagram of working principal

2.1 Development of Instrument

The instrument is based on pendulum principle. The weight (hang on string) comprised of a frictionless wheel movable along arc-shaped platform. This acts as a sample holder. When the weight is subjected to push, it swings back and forth in the platform. The amplitude of the swing reduces due to the friction of the fabric. The amplitude is inversely proportional to the friction of the fabric. The schematic diagram of working principle is shown in Fig. 1. Line diagram and picture of instrument are given in Fig. 2. Whole assembly is in chamber in which air flow is constant.

The apparatus includes three chambers, namely top, middle, and bottom chamber. The description of the instrument chamber wise is discussed below:

Top chamber (T) accommodates display unit (1), on/off switch (2), geared motor with electromagnetic clutch, press button to actuate pendulum, rotor encoder in order to measure angle/amplitude and programmable logic controller (PLC) to control various parameters such as humidity and temperature. The display unit reflects information pertaining to humidity, amplitude, time of completion of cycle, air velocity, etc. The on/off switch is provided to switch on or off said apparatus. The geared motor with electromagnetic clutch controls oscillation of the roller hanging from the roof of the middle chamber with a rod.

The roller hanging from the roof by means of rod causes whole assembly to oscillate about the equilibrium position by swinging back and forth. This oscillation takes place with the help of geared motor. The electromagnetic clutch plays role to shift the roller assembly at the maximum angle on one side. When this roller assembly attains the maximum angle, it is released by means of a release button. Upon release

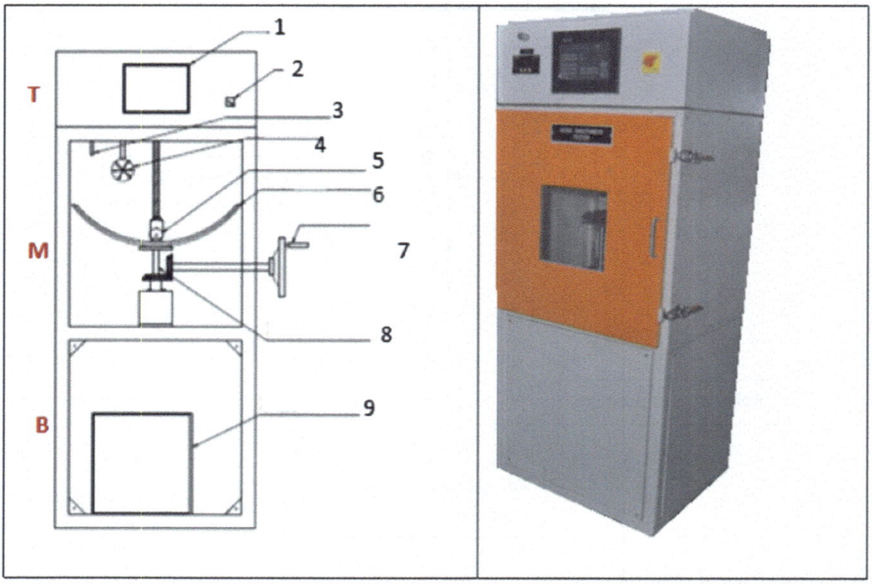

Fig. 2 Line diagram and picture of smoothness tester (Patent application no. 2053/DEL/2015 dated 07.07.2015)

of the assembly, it starts oscillating about the equilibrium position swinging back and forth. Said rotary encoder is provided to measure angle/amplitude of the roller assembly.

The middle chamber (M) embodies temperature and humidity sensor (3), anemometer (4), revolving roller assembly (5), arc-type sample holder (6), and screw arrangement and height adjustment (7 and 8). The above will be described herein below:

Bottom chamber (B): This chamber houses steam generator (9) to generate steam for changing humidity. Besides the above three chambers, an air-conditioning unit is also employed with the apparatus to maintain required temperature in the course of the testing.

2.2 *Evaluation Smoothness Property of Fabric Samples*

Preparations of samples: For the study, 100% cotton fabric was sourced from M/s Surya Processors Pvt Ltd, Ghaziabad. It was given pre-treatments (desizing, singeing, scouring, bleaching, and mercerizing) in the mill itself using standard recipe. The mercerized fabric sample was given various finishing treatment in the NITRA pilot plant so the effect of these finishing treatments can be assessed using NITRA Fabric smoothness tester. Following five types of finishing, chemicals at different

Table 1 Fabric samples with code number

Sample	Code	Sample	Code
Gray	G1	JinguardEco PCD (25 g/L)	T5
Singed	S1	JinguardEco PCD (45 g/L)	T6
Desized	D1	DPT (20 g/L)	T7
Scoured	SC1	DPT (60 g/L)	T8
Bleaching	B1	DPT (100 g/L)	T9
Mercerized	M1	ULTRA (30 g/L)	T10
Product 6000 (40 g/L)	T1	ULTRA (50 g/L)	T11
Product 6000 (50 g/L)	T2	ULTRA (70 g/L)	T12
Product 6000 (60 g/L)	T3	JinsofEco MAS Conc (40 g/L)	T13
JinguardEco PCD (5 g/L)	T4	JinsofEco MAS Conc (60 g/L)	T14

concentrations were used to finish mercerized fabric sample using standard recipe recommended by the supplier following pad-dry-cure method.

- Product-6000—hydrophilic nano-silicone softener (40, 50, and 60 g/L)
- JinguardEco PCD (water-repellent finish)—fluoro-alkyl-based emulsion (5, 25, and 45 g/L)
- DPT095—Resil—Modified polysiloxane micro-emulsion (20, 60, and 100 g/L)
- Ultra 196—Resil—Organo-modified polysiloxane, micro-emulsion (30, 50, and 70 g/L)
- JinsofEco MAS Conc—Concentrated silicone macro-emulsion (40 and 60 g/L).

The 20 samples so prepared are given in Table 1 with code numbers.

These fabric samples were analyzed for mass, thread density (EPI X PPI), tear and tensile strength, crease recovery angle and bending length of the fabric samples were tested as per IS 1964, IS 1963, ISO 13937-1, ISO 13934-1, IS 4681, and IS 6490 test methods, respectively (Table 2).

Evaluation of smoothness behavior of samples: For analysis of smoothness property, 20 specimens (15 cm × 15 cm each) per sample (10 specimens warpwise and 10 specimens weftwise) were cut and well ironed to remove wrinkles. These specimens were conditioned for 2 h in 65 ± 3% RH at 27 ± 2 °C. After conditioning, these were mounted on the sample holder fitted on the instrument, one by one. A constant load of 0.4 kg was applied on the specimen with the help of load cell during the test. After adjusting load, test was started by pushing start button. This initiated the movement of pendulum arrangement. The test is completed once the pendulum movement was stopped completely. Once the pendulum stops, the time taken to stop the pendulum in millisecond is displayed on the screen of instrument. Finally grading

Table 2 Fabric samples properties

Sample code	Mass (g/m^2)	End (inch)	Picks (inch)	Tensile strength (N)		Tear strength (g)		Crease recovery angle (warp + weft) (Degree)	Bending length (cm)	
				Warp	Weft	Warp	Weft		Warp	Weft
G1	128	120	72	576	259	1138	569	106	2.45	1.75
S1	124	122	72	560	270	1112	536	110	2.40	2.80
D1	118	126	78	520	230	1150	720	140	1.58	1.46
SC1	124	128	80	580	240	1064	676	144	1.55	1.42
B1	126	134	82	600	248	897	640	146	1.56	1.40
M1	127	136	80	681	258	977	670	156	1.40	1.38
T1	122	135	70	396	111	1342	823	173	1.26	1.16
T2	124	136	72	363	148	1380	790	178	1.22	1.12
T3	123	135	72	371	133	1340	773	180	1.18	1.10
T4	129	136	71	457	216	918	615	160	1.48	1.36
T5	121	136	72	452	185	919	652	162	1.44	1.30
T6	122	135	72	482	195	1001	685	164	1.40	1.28
T7	123	132	70	431	197	1380	892	180	1.30	1.24
T8	123	132	72	449	164	1496	940	182	1.22	1.20
T9	125	132	72	456	156	1516	947	186	1.18	1.16
T10	123	136	71	412	176	1338	913	188	1.18	1.14
T11	125	135	72	384	160	1340	849	182	1.16	1.12
T12	126	136	72	428	167	1404	821	184	1.12	1.10
T13	126	136	72	433	176	1428	959	186	1.20	1.16
T14	127	135	72	425	186	1426	949	190	1.14	1.12

Table 3 Grading system

Time (ms) to stop pendulum	Smoothness grade
Up to 400	Grade 1 (very poor)
401–500	Grade 2 (poor)
501–600	Grade 3 (good)
601–700	Grade 4 (very good)
>701	Grade 5 (excellent)

of specimen is also displayed in 1–5 Grades. Grade 1 means, sample is very rough surface, and Grade 5 means, sample is very smooth surface. Grading system is given in Table 3.

3 Results and Discussion

3.1 *Effect of Treatments on Bending and Crease Recovery*

The bending length is very important factor which determines the flexibility of the fabric. The bending length in both the warp and weft direction of the fabric is important in determining the flexibility of the fabric. The values of the bending length of untreated and treated cotton fabrics are given in Table 2. Untreated fabric, i.e., gray, shows the maximum bending length (warp wise: 2.45 cm and weft wise: 1.75 cm), and finished samples after treatment with softeners (T1 to T3 and T7 to T14) show lower bending length than other samples. Samples T4, T5, and T6 are finished with water-repellent finish in different concentration. These samples are found to be stiffer than sample finished with various softeners. The variation of bending length in both directions of singed (S1), desized (D1), scoured (SC1), and bleached (B1) samples is very less as shown in Table 2. The mercerized fabric sample (M1) is having lower bending length in both the directions than S1, D1, SC1, and B1. From the study, it is clear that the gray cotton fabric is stiffer than other samples in both warp and weft direction. This is due to the fact that the gray cotton fabric is having sizing chemicals as well as natural impurities. The greater bending length along the warp direction of all the samples (Table 2) reveals that the fabric is stiffer in the warp direction than in the weft direction. This can be due to higher density of fabric in the warp direction (ends/inch) than the weft direction (picks/inch). Greater the stiffness of the fabric along the warp direction reveals that the fabric has less bending elasticity along warp direction than the weft direction.

The value of crease recovery angle of untreated and treated cotton fabrics is given in Table 2. It is evident from Table 2 that the crease recovery angle is increasing from gray to finished sample. Untreated fabric shows minimum crease recovery angle which is periodically increased after the treatment such as desizing, scouring, bleaching, mercerization, and finishing with softener. It is clear from Table 2 that the crease recovery (dry state) of different treated samples is higher than untreated or gray fabric sample. This may be due to swelling of the fiber in the fabric. It appears that the treatment has developed the ability of the fabrics to recover from deformation. The materials, which have good crease recovery properties, exhibit excellent soft handle, draping, and appearance as well as a lack of flabbiness as washing proceeds. The gray fabric materials have less crease recovery angle tend to more limp and flabby on washing.

Both of these studies show that the application of softener reduces the bending length and improves the crease recovery angle of the samples. Silicone emulsion acts as a lubricating agent between the fibers in the yarns and between the yarns of the fabric, imparting softness to the material. This softness causes a reduction in bending length of fabric. The bending length and rigidity of the textile material are directly related to each other. The drop in bending length is thus indicative of reduced rigidity or improved softness of the fabric samples. Silicon softening capability comes from siloxane backbone's flexibility and its freedom of rotation along the Si–O bond. This freedom of rotation leads to unique flexibility of siloxane molecules [12]. The improvement in softness due to silicone softener application is also reflected by enhancement of crease recovery angles.

3.2 *Smoothness Property*

All the 20 samples were analyzed for smoothness behavior using newly developed smoothness tester. This study also revealed the changes occurred on the surface characteristics of the fabric after various processing treatment. Results of these samples are given in Tables 4a, 4b, 4c, and 4d. A comparison between smoothness grading and coefficient of friction (COF) obtained using Kawabata system is shown in Figs. 3 and 4.

From Table 4a, it is revealed that there is no significant change gray (G1) and singed (S1) sample in smoothness behavior in both warp and weft directions. There is little improvement after desizing (D1). This may be due to shrinkage in the fabric due to wet treatment. This shrinkage increases fabric density (ends/inch and picks/inch) as shown in Table 1. Increase in fabric density increased the fabric balance and fabric cover but decreased surface roughness [13], and thus, fabric surface becomes smooth. After scouring (SC1) and bleaching (B1), sample becomes harsher than desized sample (D1) as shown in the results (Table 4a). After mercerization (M1), time required to stop pendulum increases (warpwise: 550 ms and weftwise: 554 ms). It indicated that after mercerization, sample become smoother. It is a well-known fact that mercerizing process improves surface smoothness of cotton fabric [14]. The results obtained from NITRA smoothness tester were also compared with the results of coefficient of friction (COF) obtained using Kawabata system. It was found that there is similar trend of surface smoothness results from both the instruments. It is also clear that COF of mercerized fabric (M1) is lower than other fabric samples. It indicates that mercerized fabric is smoother than others. These samples are also

Table 4 Testing smoothness behavior of cotton fabric at various stages of wet processing

(a)

S. No.	Time required to stop pendulum (ms)											
	G1		S1		D1		SC1		B1		M1	
	Wp	Wt	Wp	Wt	Wp	Wt	Wp	Wt	Wp	Wt	Wp	Wt
1	470	420	419	411	519	449	448	439	432	439	599	559
2	486	434	429	419	519	419	439	440	418	438	539	569
3	488	428	439	418	518	459	439	438	428	420	599	559
4	447	430	418	429	528	479	440	448	438	428	589	569
5	445	440	399	420	529	418	438	439	418	438	560	611
6	466	430	399	419	519	478	436	449	448	439	539	559
7	454	434	418	419	519	479	432	448	438	429	500	558
8	442	442	419	429	522	479	441	438	438	428	579	559
9	439	438	420	428	519	479	439	449	429	439	500	499
10	447	444	429	418	528	468	438	448	419	439	499	499
Average	**458**	**434**	**419**	**421**	**522**	**460**	**439**	**444**	**430**	**434**	**550**	**554**
Grade	**2**	**2**	**2**	**2**	**3**	**2**	**2**	**2**	**2**	**2**	**3**	**3**
COF*	1.50	1.52	1.51	1.50	1.37	1.48	1.53	1.57	1.53	1.52	1.29	1.26

(b)

S. No	Time required to stop pendulum (ms)									
	T1		T2		T3		T4		T5	
	Wp	Wt	Wp	Wt	Wp	Wt	Wp	Wt	Wp	Wt
1	520	516	610	590	625	610	520	540	570	590
2	518	510	600	565	610	600	510	530	568	580
3	522	508	612	578	620	605	500	524	570	592
4	518	512	600	580	604	604	512	530	560	588
5	514	510	598	588	610	610	514	520	570	578
6	516	508	604	568	630	612	522	528	580	580
7	510	512	605	572	628	598	512	530	583	590
8	512	514	598	570	600	596	514	532	582	579
9	514	514	594	568	620	600	510	534	560	590
10	510	516	600	566	628	604	520	540	563	580
Average	**515**	**512**	**602**	**574**	**618**	**604**	**513**	**530**	**570**	**584**
Grade	**3**	**3**	**4**	**3**	**4**	**4**	**3**	**3**	**3**	**3**
COF*	1.32	1.34	1.17	1.20	1.12	1.14	1.34	1.28	1.20	1.18

(c)

S. No.	Time required to stop pendulum, millisecond									
	T6		T7		T8		T9		T10	
	Wp	Wt	Wp	Wt	Wp	Wt	Wp	Wt	Wp	Wt
1	590	536	510	500	528	490	566	556	604	610

(continued)

Table 4 (continued)

(c)

S. No.	Time required to stop pendulum, millisecond									
	T6		T7		T8		T9		T10	
	Wp	Wt	Wp	Wt	Wp	Wt	Wp	Wt	Wp	Wt
2	595	528	515	505	526	494	589	594	601	698
3	578	532	500	498	520	498	590	590	690	690
4	580	528	598	490	528	510	580	588	610	692
5	594	526	596	499	520	502	588	558	607	694
6	586	530	508	506	518	504	502	560	612	689
7	588	532	506	510	524	506	510	550	580	688
8	578	540	502	502	522	508	590	548	678	684
9	580	525	498	596	518	498	592	560	690	601
10	584	528	512	590	520	502	580	564	678	690
Avg.	**585**	**530**	**505**	**502**	**522**	**501**	**589**	**567**	**695**	**693**
Grade	3	3	3	3	3	3	3	3	4	4
COF*	1.24	1.26	1.35	1.35	1.34	1.38	1.16	1.22	1.04	1.04

(d)

S. No.	Time required to stop pendulum (ms)							
	T11		T12		T13		T14	
	Wp	Wt	Wp	Wt	Wp	Wt	Wp	Wt
1	610	602	650	646	710	701	740	680
2	620	621	652	642	1720	706	734	688
3	622	608	640	638	1725	690	729	690
4	625	610	642	635	702	698	720	680
5	605	615	647	632	704	692	724	682
6	610	616	646	630	710	696	725	698
7	609	618	647	640	722	697	722	690
8	606	610	652	645	710	694	718	675
9	620	611	654	633	715	699	732	680
10	623	612	640	638	716	695	730	682
Avg.	**615**	**612**	**647**	**638**	**713**	**697**	**727**	**685**
Grade	4	4	4	4	5	4	5	4
COF*	1.12	1.14	1.08	1.10	1.02	1.04	1.01	1.06

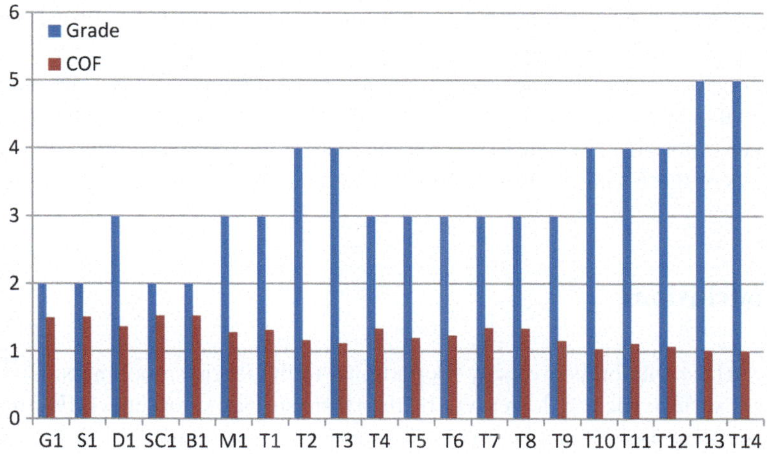

Fig. 3 Smoothnes grade versus COF of warpwise fabric samples

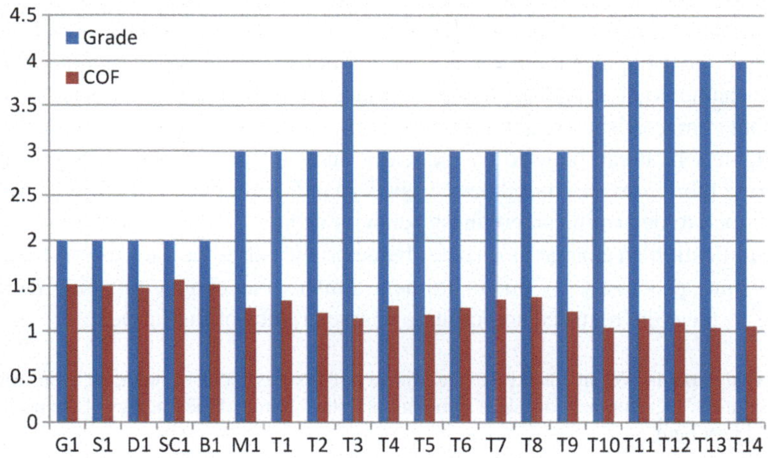

Fig. 4 Smoothness grade versus COF of weftwise fabric samples

graded as per Table 3. Mercerized fabric sample (M1) is having Grade 3 in both the direction. However, all the other samples are graded as 2 except warp direction of desized sample (D1), its grading is 3.

Samples coded as T1, T2, and T3 are treated with hydrophilic nano-silicone softener at different concentrations such as 40, 50, and 60 g/L, respectively. After treatment with this softener, smoothness property of fabric improved further in both warpwise and weftwise directions. It is also revealed from Table 4b that with the increase in softener concentration from 40 to 60 g/L, fabric smoothness also increases. Similarly other samples (T7 to T14) also treated with different softeners and show higher smoothness. It is a well-known fact that softener treatment improves fabric surface smoothness [15]. Samples coded as T4, T5, and T6 are obtained after treatment with water-repellent finishing agent. A water-repellent fabric is one in which the fibers are usually coated with a "hydrophobic" type of compound, and the pores are not filled in the course of the treatment. The latter types of fabrics are quite permeable to air and water vapor [16, 17]. Due to this reason, there is no improvement in smoothness properties of these samples from the mercerized fabric sample (M1).

4 Conclusions

Fabric surface friction is inversely proportional to the fabric smoothness. Higher is the fabric surface friction, lower will be the fabric surface smoothness. Friction is a resistance to motion. It can be observed whenever a textile material such as fabric is rubbed with itself or between the thumb and finger. The subjective assessment of fabric surface friction or smoothness mainly depends on the static and dynamic friction between the cloth surface and the thumb or finger. There is no doubt that the application of human fingers for subjective analysis of fabric surface smoothness is one of the most popular methods as they can detect even a small change in fabric surface characteristics. Although fabric surface friction or smoothness is one of the important characteristics which influence fabric comfort, there is no suitable device is available which can be used by the textile industry in their day-to-day work for measuring fabric surface smoothness. Hence, an indigenous cost-effective instrument is developed to determine smoothness behavior of fabric. The instrument is capable to give indication on change in surface characteristics after the various pre-treatment and finishing processes. By this instrument, finisher can change finishing recipe or process to meet the required smoothness characteristics of the fabric.

Acknowledgements We are thankful to the Ministry of Textiles, Govt. of India, for sponsoring the project entitled "Development of fabric smoothness tester." Some of the findings of the project are presented over here.

References

1. Okur, A.: Textile Asia **33**(8), 32–34 (2002)
2. Moorthy, R.R.: Surface friction characteristics of woven fabrics with nonconventional fibres and their blends. J. Text. Apparel Technol. Manag **9**, 3 (2015)
3. Das, A., Kothari, V.K., Vandana, N.: A study on frictional characteristics of woven fabrics. AUTEX Res. J. **5**(3), 133 (2005)
4. Ajay, J.O.: Fabric smoothness, friction and handle. Text. Res. J **62**, 52 (1992)
5. Mooneghi, S.A., Saharkhiz, S., Varkiani, S.M.: Surface roughness evaluation of textile fabrics: a literature review. J. Eng. Fibers Fabr. **9**(2) (2014)
6. Fujimoto, T.: Measurement of surface property using a special sensor developed for pile material. Indian J. Fibre Text. Res. **33**, 253–257 (2008)
7. Militky, J., Bajzik, V.: Characterization of protective clothing surface roughness. In: 2nd International Congress on Technical textiles (2004)

8. Militky, J., Mazal, M.: Image analysis method of surface roughness evaluation. Int. J. Clothing Sci. Technol. **19**(3–4), 186–193 (2007)
9. Kim, D., Shinjung, Y., Kim, E.: Objective measures for perceived touch of worsted fabrics. Int. J. Ind. Ergon. **35**, 1159–1169 (2005)
10. Kawabata, S.: The standardization and analysis of hand evaluation, 2nd edn (The textile machinery society of Japan), pp 31–35 and 63–70 (1980)
11. Bhuvana, G., Raghunathan, S.: Studied on frictional behavior of chitosan-coated fabrics. AUTEX Res. J. **6**, 216 (2006)
12. Chattopadhyay, D.P., Vyas, D.D.: Effect of silicone nano-emulsion softener on physical properties of cotton fabric. Indian J. Fibre Text. Res. **35**, 68–71 (2010)
13. Akgun, M.: The effect of fabric balance and fabric cover on surface roughness of polyester fabrics. Fibers Polym. **14**(8), 1372–1377 (2013)
14. https://textilelearner.blogspot.in/2013/06/mercerization-process-of-cotton-fabric.html]
15. Pattanaik, M.N., Ray, S.C.: Effect of silicone treatment on hand value of cotton handloom fabrics. Indian J. Fibre Text. Res. **41**, 298–305 (2016)
16. Rowen, J.W., Gagliardi, D.: Properties of water-repellent fabrics. J. Res. Nat. Bureau Stand **38**, 103 (1947)
17. Gulrajani, M.L.: Nano finishes. Indian J. Fibre Text. Res. **31**, 187–201 (2006)

Thickness Loss of Handmade Carpets After Dynamic Loading

Shravan Kumar Gupta, Kamal Kanti Goswami and Abhijit Majumdar

Abstract The effect of knot density, pile height, number of plies in pile yarn and pile yarn twist on thickness loss of Persian handmade wool carpets after dynamic loading has been studied. The relation between the process variables has been analysed with response surface methodology based on the Box–Behnken design of the experiment. Knot density is statistically significant for thickness loss after dynamic loading of 50, 100 and 200 impacts only. However, the effect of knot density on thickness loss becomes statistically insignificant when the numbers of impacts are very high. A number of plies in pile yarn and interaction between a number of plies and pile yarn twist are statistically significant for thickness losses of handmade carpets after dynamic loading of all impact levels. In general, all the samples show high thickness loss when the number of impacts is increased from 50 to 100. The rate of thickness loss with the increase in a number of impacts reduces between 100 and 200 impacts. After 200 impacts, the rate of thickness loss stabilizes to a small value.

Keywords Box–Behnken design · Dynamic loading · Persian handmade wool carpet · Response surface methodology · Thickness loss

1 Introduction

Carpet durability means the ability of a carpet to maintain its original properties after repetitive utilization [1]. Various forces like axial compression, bending, flattening and extension are operating on piles during carpet wear. Generally, these forces are created by means of dynamic loading for example walking or static loading via furniture reduce the thickness of carpet. Resistance to thickness loss caused by dynamic and static loads is one of the most important carpet durability parameters. Minimum thickness loss indicates better resilience and durability. It can be described

S. K. Gupta (✉) · K. K. Goswami
Indian Institute of Carpet Technology, Chauri Road, Bhadohi 221401, Uttar Pradesh, India
e-mail: shravangupta79@rediffmail.com

A. Majumdar
Department of Textile Technology, Indian Institute of Technology Delhi, New Delhi, India

A. Majumdar et al. (eds.), *Functional Textiles and Clothing*,
https://doi.org/10.1007/978-981-13-7721-1_25

with energy theory where the piles recover after bend if the energy stored greater than the resisting energy [2, 3].

Fibre, yarn and carpet manufacturing parameters are influencing to the thickness loss of carpets. When average fibre diameter and medullated fibre content in blends increase, then carpet resiliency significantly increases and thickness loss after dynamic loading reduces [4]. There are increment in compression and matting and also reduction in elastic recovery of pile yarn with an increase in percentage of slipe wool [5–7]. Celik and Koc examined the thickness loss of Wilton-type carpets by varying pile yarns (wool, acrylic and polypropylene) after dynamic loading. They concluded that when the number of impacts increases then thickness loss increases. They also reported that acrylic carpets have better ability to recover thickness loss after dynamic loading than wool and polypropylene carpets [8]. It has been concluded that pile density and resiliency increase, whereas compressibility decreases with the increase in the number of folds in yarn [9].

WIRA dynamic loading machine has been used in many researches to estimate carpet thickness loss after 50, 100, 150, 200, 500 and 1000 impacts and observations were associated with that of a carpet lay in a corridor. A sudden drop in carpet thickness was observed up to 200 impacts, when it was estimated with dynamic loading machine. Then a stable and lower rate of carpet thickness loss was observed during the remainder of impacts. The best association between corridor wearing trials and the test results of dynamic loading machine was established through 24 weeks with around 1,500 persons per week and 100 impacts [10, 11].

Some researchers have been discussed the effect of carpet manufacturing parameters on carpet thickness loss behaviours. They have been reported that knot density and pile height are the important factors to influence carpet thickness loss behaviours [5–7, 12–18]. In this research, an effort has been made to study the thickness loss behaviour of Persian handmade wool carpets after dynamic loading by developing a model using Box–Behnken response surface methodology. The per cent contribution of four variables (knot density, pile height, number of plies in pile yarn and pile yarn twist), their interactions as well as quadratic terms on thickness loss behaviour after dynamic loading has also been studied.

2 Materials and Methods

2.1 Materials

Pile, warp, thick weft and thin weft yarns were used for the developing of Persian handmade carpet samples as shown in Fig. 1. The pile yarns having nominal specifications 3.90 metric counts (Nm) and three dissimilar twist levels, i.e. 3.5, 4.0 and 4.5 twists per inch (tpi) were used. Pile yarns were manufactured from 100% wool fibres by woollen spinning system. In this method, double cylinder willow machine was used for opening of wool fibres. Consistency in mixing was attained by using

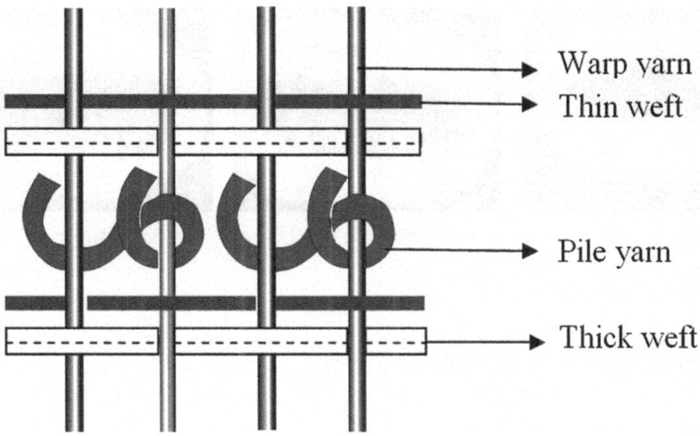

Fig. 1 Yarns used in Persian handmade carpets

Table 1 Properties of wool fibres

Wool fibre properties	Values
Mean fibre diameter (μm)	37.65 (24.4)
Mean fibre length (mm)	75.03 (28.5)

Values in the parenthesis indicates CV%

Table 2 Pile yarn characteristics

Sample No.	Stated metric count (km/kg)	Stated twist (inch^{-1})	Experimental metric count (km/kg)	Experimental twist (inch^{-1})	Twist direction
1	3.90	3.5	3.64 (7.98)	3.63 (9.51)	"S"
2		4.0	3.86 (8.29)	4.02 (9.61)	
3		4.5	3.96 (2.67)	4.62 (6.73)	

Values in the parenthesis indicates coefficient of variation (CV %)

the first stack mixing then two toppling processes of wool fibres. The conditioning of wool fibres was done at this step for 24 h. Then, THIBEU woollen (condenser) card was used for carding of the opened wool fibres and manufacturing of slubbings cake (bobbin). GSLTI woollen ring frame was used for processing of slubbings at a nominal draft of 1.1 and prescribed metric twist multiplier (α_m). Heat setting of ring frame cops was done in autoclave at 110 °C for 15 min to get perfect twist set yarns. The properties of wool fibres are presented in Table 1. The characteristics and images of pile yarns are presented in Table 2 and Fig. 2, respectively.

The characteristics and images of warp, thin weft and thick weft yarns are presented in Table 3 and Fig. 3, respectively.

(a) Sample no. 1 (b) Sample no. 2 (c) Sample no. 3

Fig. 2 Samples of pile yarns

Table 3 Warp, thin weft and thick weft characteristics

S. No.	Yarn	No. of plies	Resultant yarn count (Ne)	Plying twist ($inch^{-1}$)	Single yarn twist ($inch^{-1}$) (CV %)	Fibre content
1	Warp	6	0.9 (0.3)	5.5 (2.6)	11.5 (4.5)	100% cotton
2	Thin weft	2	1.6 (1.4)	5.1 (2.1)	8.5 (6.0)	60% cotton and 40% polyester
3	Thick weft	2	0.9 (0.9)	5.1 (1.6)	3.4 (14.6)	

Values in the parenthesis indicates CV%

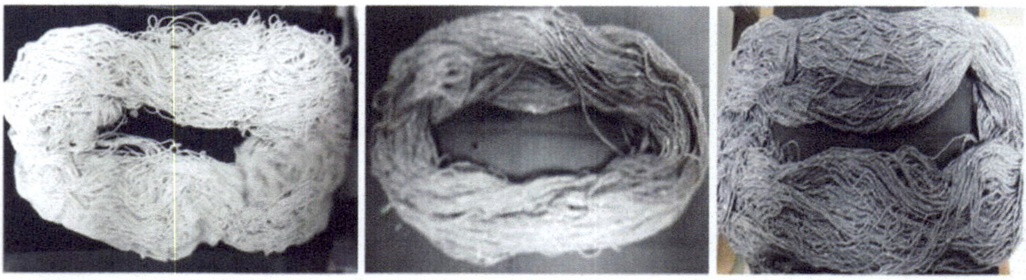

(a) Warp yarns (b) Thin weft (c) Thick weft

Fig. 3 Images of warp, thin weft and thick weft yarns

2.2 Manufacturing of Handmade Carpet Samples

Handmade carpet samples were produced by forming asymmetrical single knots known as Persian knots. This knot is also called Sehna knot. The Persian knots are created by covering the tuft around one warp yarn at an angle of 2π radians and then around nearby different warp thread at an angle of π radians as shown in Fig. 1. Two nearby warp threads are wrapped by a tiny section of pile yarn by hand. Thick and thin weft threads are picked throughout the entire warp area in alternate shedding, after completion of every row of knots. Then the knots are protected in position by

Fig. 4 Manufacturing of Persian handmade carpet sample

beating the weft threads with a comb. The procedure of knotting starts at the base of the loom. Continuously, the handmade carpet sample moves upside until it is completed as the knots and weft yarns are added as shown in Fig. 4.

2.3 *Design of Experimentation (DOE)*

Box–Behnken experimental plan was used for this investigation by using four factors with three levels. This has reduced the number of carpet samples to be produced by a great extent. With four factors and three levels, the number of samples would be 3^4, i.e. 81 if full factorial design was adopted. However, 27 samples were produced with Box and Behnken experimental design. The factors with their levels are depicted in Table 4. The values of variables at 0 levels are based on the general practice of the handmade carpet industry. Other values at -1 and -1 levels are one step higher and one step lower, respectively, than values at 0 levels.

Table 4 Coded and actual levels of Persian handmade carpet factors

Factor code	Factors	Levels		
		−1	0	+1
x_1	Knot density ($inch^{-1}$)	5	6	7
x_2	Pile height (mm)	10	13	16
x_3	Number of plies (pile yarn)	2	3	4
x_4	Pile yarn twist ($inch^{-1}$)	3.5	4	4.5

Table 5 Box–Behnken experimental sample plan

Sample No.	Knot density ($inch^{-1}$)	Pile height (mm)	Number of plies (pile yarn)	Pile yarn twist ($inch^{-1}$)
1	−1 (5)	−1 (10)	0 (3)	0 (4.0)
2	−1 (5)	+1 (16)	0 (3)	0 (4.0)
3	+1 (7)	−1 (10)	0 (3)	0 (4.0)
4	+1 (7)	+1 (16)	0 (3)	0 (4.0)
5	−1 (5)	0 (13)	−1 (2)	0 (4.0)
6	−1 (5)	0 (13)	+1 (4)	0 (4.0)
7	+1 (7)	0 (13)	−1 (2)	0 (4.0)
8	+1 (7)	0 (13)	+1 (4)	0 (4.0)
9	−1 (5)	0 (13)	0 (3)	−1 (3.5)
10	−1 (5)	0 (13)	0 (3)	+1 (4.5)
11	+1 (7)	0 (13)	0 (3)	−1 (3.5)
12	+1 (7)	0 (13)	0 (3)	+1 (4.5)
13	0 (6)	−1 (10)	−1 (2)	0 (4.0)
14	0 (6)	−1 (10)	+1 (4)	0 (4.0)
15	0 (6)	+1 (16)	−1 (2)	0 (4.0)
16	0 (6)	+1 (16)	+1 (4)	0 (4.0)
17	0 (6)	−1 (10)	0 (3)	−1 (3.5)
18	0 (6)	−1 (10)	0 (3)	+1 (4.5)
19	0 (6)	+1 (16)	0 (3)	−1 (3.5)
20	0 (6)	+1 (16)	0 (3)	+1 (4.5)
21	0 (6)	0 (13)	−1 (2)	−1 (3.5)
22	0 (6)	0 (13)	−1 (2)	+1 (4.5)
23	0 (6)	0 (13)	+1 (4)	−1 (3.5)
24	0 (6)	0 (13)	+1 (4)	+1 (4.5)
25	0 (6)	0 (13)	0 (3)	0 (4.0)
26	0 (6)	0 (13)	0 (3)	0 (4.0)
27	0 (6)	0 (13)	0 (3)	0 (4.0)

Twenty-seven (27) samples were manufactured with three levels of knot density, pile height, number of plies in pile yarn and pile yarn twist. The sample plan with coded levels and actual values of factors is shown in Table 5.

2.4 Testing Methods

2.4.1 Knot Density

A rule (Fig. 5) with least count of one millimetre was used for measurement of knot density in carpets. Standard IS: 7877 (Part III)—1976 (Reaffirmed 1997) was followed for this parameter. This characteristic was checked at the reverse side of the carpet in length and width directions. The average of 20 taken measurements was calculated for each sample.

2.4.2 Pile Height

Flat metal gauges of known height (Fig. 6) were used for measurement of pile height in carpets. Standard IS: 7877 (Part IV)—1976 (Reaffirmed 1997) was followed for this parameter. The average of 10 taken measurements was calculated for each sample.

2.4.3 Thickness Loss After Dynamic Loading

The length and width of the carpet samples used for this parameter were 125 and 125 mm. SDL digital carpet thickness gauge was used for measuring initial carpet thickness under 2 kPa pressures. Then the first dynamic loads of 50 impacts were applied on dynamic loading machine (Fig. 7). Standard BS/ISO-2094: 1999 was followed for this parameter. Weight piece and base plate are two main components of this machine. The total mass of weight piece is 1,279 ± 13 g. There are two rectangular cross section feet (length × width × depth = 51 mm × 6.3 mm × 9.5 mm) on its lower surface divided by an inner space of 38.1 mm. It is falling

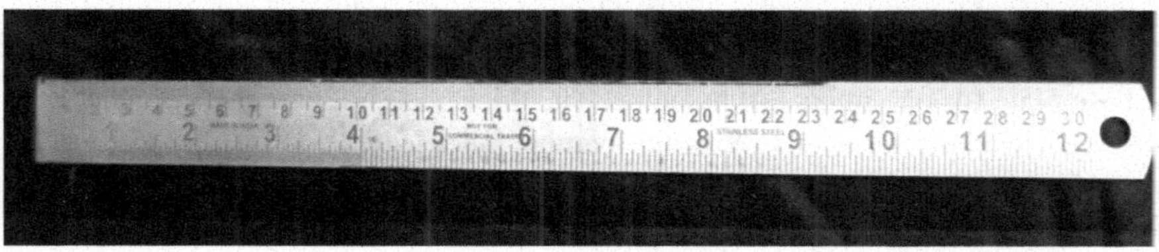

Fig. 5 Rule

Fig. 6 Metal gauges

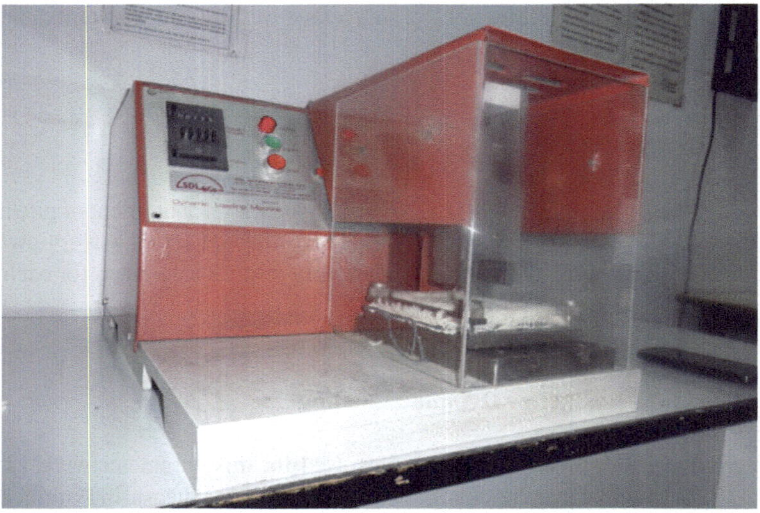

Fig. 7 Dynamic loading machine

freely on to the carpet sample from a height of 63.5 mm under gravity every 4.3 s. One impact means each fall of the weight piece. The dimension of steel base plate is 150 ± 0.5 mm $\times$ 125 ± 0.5 mm. There are two wide steel bars having dimension 150 $\pm$ 0.5 mm $\times$ 20 ± 0.5 mm at the sides of base plate. The carpet specimen is clamped at the sides of base plate with two steel bars, screwed at the ends of the base plate. There is 3.2 ± 0.1 mm forward traverse and 1.6 ± 0.1 mm backward movement of base plate for one impact of machine. The thicknesses of carpet specimens were observed instantly after this dynamic loading. Then carpet specimens were again tested for 100, 200, 500 and 1000 impacts under dynamic loading. The thickness loss was measured by taking difference between the original thickness and the thickness observed following the number of affirmed impacts under dynamic loading.

The average thickness loss was measured by the following equation:

$$Y = y - y_1$$

where

- y Initial carpet thickness at a pressure of 2 kPa before the application of dynamic load (mm).
- y_1 Carpet thickness after affirmed number of impacts under dynamic loading (mm).
- Y Difference between the initial thickness and remaining thickness after dynamic loading.

3 Results and Discussion

3.1 *Response Surface Model (RSM) for Thickness Loss After Dynamic Loading*

Handmade carpet characteristics (knot density, pile height, number of plies in pile yarn and pile yarn twist) and experimental values of thickness losses of Persian handmade carpets after dynamic loading are shown in Table 6. Thickness loss of carpets ranges from 0.14 to 3.32, 0.87 to 3.95, 1.25 to 4.56, 1.66 to 5.29 and 2.12 to 5.75 mm after dynamic loading of 50, 100, 200, 500 and 1000 impacts, respectively. This implies that thickness loss after dynamic loading can vary very widely depending on yarn and carpet construction parameters. Five response surface equations (Eqs. 1, 2, 3, 4 and 5) were developed for relating thickness losses of carpets after dynamic loading of 50 (y_1), 100 (y_2), 200 (y_3), 500 (y_4) and 1000 (y_5) impacts, respectively. The coefficient of determination (R^2) of Eq. 1 relating thickness loss of carpets after 50 impacts of dynamic loading is 0.813. This indicates that the model can elucidate around 81% variability present in the measured data. The coefficient of determination (R^2) of Eqs. 2, 3, 4 and 5 relating thickness loss of carpets after 100, 200, 500 and 1000 impacts of dynamic loading is 0.819, 0.799, 0.777 and 0.746, respectively. The mean absolute percentage error (*MAPE*) for these five equations is 29.53, 16.43, 13.17, 12.46 and 11.28, respectively. In general, the coefficient of determination of the predictive equation reduces as the number of impact increases. This implies that the thickness loss after dynamic condition becomes more complex as the duration of test increases.

$$\begin{aligned} y_1 &= 2.513 - 1.260x_1 + 1.016x_2 - 3.671x_3 + 1.970x_4 \\ &-0.093x_1x_2 - 0.268x_1x_3 + 0.300x_1x_4 - 0.133x_2x_3 + 0.067x_2x_4 \left[R^2 = 0.813\right] \\ &+1.640x_3x_4 + 0.143x_1^2 - 0.013x_2^2 - 0.052x_3^2 - 1.172x_4^2 \end{aligned} \tag{1}$$

Table 6 Thickness loss of carpets after dynamic loading

Sample No.	Thickness loss (mm)				
	After 50 impacts	After 100 impacts	After 200 impacts	After 500 impacts	After 1000 impacts
1	2.07	2.78	2.98	3.12	4.32
2	2.35	3.12	3.60	3.87	4.59
3	1.99	2.34	2.79	3.23	3.82
4	1.15	1.56	2.09	2.69	3.29
5	3.32	3.88	4.56	5.07	5.68
6	2.15	2.58	3.08	3.47	3.93
7	2.69	3.29	3.76	4.55	5.19
8	0.45	1.07	1.60	2.45	2.94
9	1.79	2.38	2.83	3.46	3.88
10	1.25	1.94	2.24	2.53	3.12
11	1.16	1.43	2.14	2.83	3.14
12	1.22	1.89	2.42	2.92	3.39
13	1.59	2.22	2.61	3.10	3.49
14	0.62	0.87	1.25	1.66	2.12
15	2.85	3.95	4.47	5.29	5.75
16	0.28	0.98	1.31	1.76	2.45
17	1.63	2.20	2.41	2.92	3.31
18	2.18	2.69	3.22	3.83	4.27
19	0.92	1.38	2.83	2.85	3.33
20	1.87	2.76	3.31	3.83	4.38
21	2.86	3.72	4.29	5.17	5.60
22	1.26	2.39	2.70	2.99	3.45
23	0.14	1.01	1.59	2.14	2.78
24	1.82	2.65	3.23	3.66	4.14
25	1.54	2.56	3.31	4.02	4.96
26	2.08	2.78	3.16	3.46	4.14
27	1.81	2.67	3.24	3.74	4.55

$$\begin{aligned} y_2 = {} & -4.195 + 0.848x_1 + 1.181x_2 - 3.340x_3 + 1.683x_4 \\ & -0.093x_1x_2 - 0.230x_1x_3 + 0.450x_1x_4 - 0.135x_2x_3 + 0.148x_2x_4 \; \left[R^2 = 0.819\right] \\ & +1.485x_3x_4 - 0.098x_1^2 - 0.030x_2^2 - 0.054x_3^2 - 1.300x_4^2 \end{aligned} \tag{2}$$

$$y_3 = -14.157 + 1.766x_1 + 2.303x_2 - 3.651x_3 + 2.125x_4 \\ -0.110x_1x_2 - 0.170x_1x_3 + 0.435x_1x_4 - 0.150x_2x_3 - 0.055x_2x_4 \left[R^2 = 0.799\right] \\ +1.615x_3x_4 - 0.162x_1^2 - 0.035x_2^2 - 0.117x_3^2 - 1.087x_4^2 \quad (3)$$

$$y_4 = -7.303 + 1.345x_1 + 2.370x_2 - 5.220x_3 + 0.477x_4 \\ -0.108x_1x_2 - 0.125x_1x_3 + 0.510x_1x_4 - 0.174x_2x_3 + 0.012x_2x_4 \left[R^2 = 0.777\right] \\ +1.850x_3x_4 - 0.154x_1^2 - 0.045x_2^2 - 0.014x_3^2 - 1.147x_4^2 \quad (4)$$

$$y_5 = -24.713 + 1.454x_1 + 2.347x_2 - 3.979x_3 + 8.610x_4 \\ -0.067x_1x_2 - 0.125x_1x_3 + 0.505x_1x_4 - 0.161x_2x_3 + 1.755x_3x_4 \left[R^2 = 0.746\right] \\ -0.212x_1^2 - 0.054x_2^2 - 0.183x_3^2 - 2.098x_4^2 \quad (5)$$

The ANOVA outcomes of the regression model are depicted in Table 7. The significance of the whole model and every parameter was checked with F-test and its related p-value. The model p-value of 0.0140 for thickness loss after 50 impacts

Table 7 Analysis of variance for thickness loss after dynamic loading

Parameters	p-value				
	After 50 impacts	After 100 impacts	After 200 impacts	After 500 impacts	After 1000 impacts
Model	0.0140	0.0116	0.0198	0.0322	0.0311
x_1-knot density	0.0320*	0.0186*	0.0449*	0.2305	0.1387
x_2-pile height	0.7143	0.7349	0.2642	0.3026	0.3197
x_3-number of ply	0.0002*	0.0001*	0.0002*	0.0004*	0.0006*
x_4-pile yarn twist	0.5439	0.2635	0.6170	0.8656	0.7699
x_1x_2	0.2922	0.3213	0.2767	0.3415	0.5700
x_1x_3	0.3133	0.4121	0.5680	0.7078	0.7215
x_1x_4	0.5660	0.4220	0.4670	0.4487	0.4749
x_2x_3	0.1415	0.1604	0.1461	0.1345	0.1832
x_2x_4	0.7009	0.4271	0.7806	0.9580	–
x_3x_4	0.0073*	0.0178*	0.0164*	0.0149*	0.0239*
x_1^2	0.5272	0.6848	0.5312	0.5946	0.4881
x_2^2	0.6019	0.2655	0.2341	0.1724	0.1279
x_3^2	0.8184	0.8225	0.6501	0.9608	0.5480
x_4^2	0.2080	0.1908	0.2999	0.3294	0.1010

*Indicates significant terms

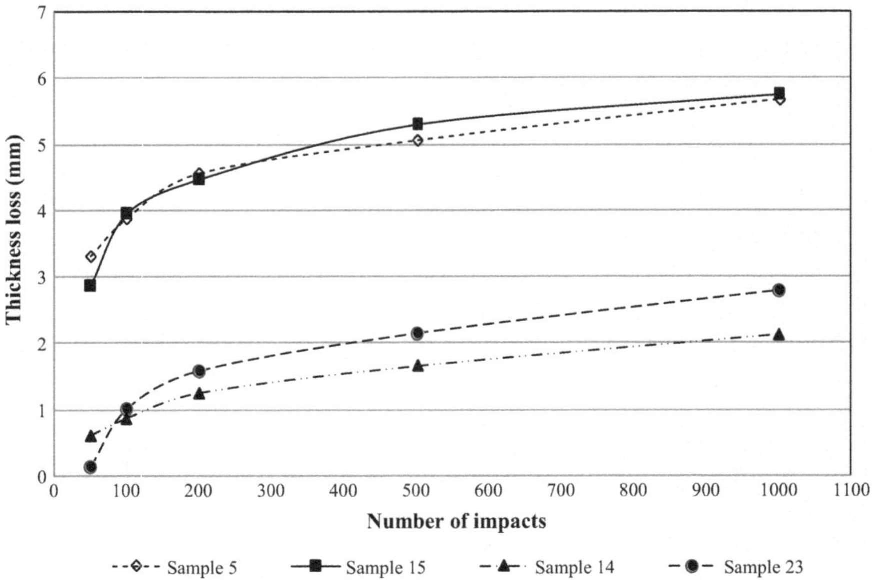

Fig. 8 Thickness loss behaviour of carpets after dynamic loading

indicates that the model is significant and there is only 1.4% chance that a p-value as great as this could happen because of chance. The model p-values for thickness losses after 100, 200, 500 and 1000 impacts are 0.0116, 0.0198, 0.0322 and 0.0311, respectively, which indicate the significance of respective equations.

From Table 7, it can be observed that x_1 (knot density) is statistically significant for thickness loss after dynamic loading of 50, 100 and 200 impacts only. However, effect of knot density on thickness loss becomes statistically insignificant when the numbers of impacts are very high. Pile height (x_2), which is a determining factor of carpet thickness, has an insignificant role on carpet thickness loss after dynamic loading. It is important to note x_3 (number of plies) and interaction term of x_3x_4 (number of plies × pile yarn twist) are statistically significant for thickness losses of handmade carpets after dynamic loading of all impact levels.

3.2 *Thickness Loss Behaviour After Dynamic Loading*

Figure 8 depicts the thickness loss behaviour of handmade carpets with increase in the number of impacts during dynamic loading. Four samples have been chosen for this purpose. Samples 5 and 15 have been chosen as they demonstrated the maximum thickness loss (5.68 and 5.75 mm, respectively) as shown in Table 6. On the other hand, samples 14 and 23 were chosen as they produced the minimum thickness loss (2.12 and 2.78 mm, respectively). In general, all the samples show high thickness loss when the number of impacts is increased from 50 to 100. The rate of thickness loss with the increase in number of impacts reduces between 100 and 200 impacts. After 200 impacts, the rate of thickness loss stabilizes to a small value. It is interesting to note that the slope of all four samples between 200 and 1000 impacts is similar. This implies that the change in thickness is influenced by the yarn and carpet constructional parameters at initial cycles of impacts. However, irrespective of the yarn and carpet construction parameters, the incremental loss of carpet thickness is practically the same after 200 impacts.

It should be noted here that samples 5 and 15, which show high thickness loss, have 2 plies in pile yarn. On the other hand, samples 14 and 23, which show low thickness loss, have 4 plies in pile yarn. This again bolsters the perception about the role of number of plies in pile yarn in determining thickness loss after dynamic loading.

3.3 Contribution of Yarn and Carpet Parameters

Table 8 also presents the per cent contribution of various terms towards the thickness loss of handmade carpets after dynamic loading. The most important parameter

Table 8 Per cent contribution of yarn and carpet parameters on thickness loss after dynamic loading

Parameters	Per cent contribution				
	After 50 impacts	After 100 impacts	After 200 impacts	After 500 impacts	After 1000 impacts
x_1-knot density	11.44	13.36	10.21	3.77	6.14
x_2-pile height	0.27	0.25	2.80	2.72	2.62
x_3-number of plies	52.07	54.31	54.04	56.21	50.97
x_4-pile yarn twist	0.75	2.46	0.55	0.06	0.21
x_1x_2	2.33	1.91	2.67	2.33	0.84
x_1x_3	2.18	1.29	0.73	0.33	0.31
x_1x_4	0.68	1.23	1.16	1.44	1.36
x_2x_3	4.82	4.06	4.92	6.04	4.88
x_2x_4	0.30	1.23	0.18	0.01	–
x_3x_4	20.24	13.61	15.87	18.96	16.15
x_1^2	0.83	0.31	0.85	0.72	1.26
x_2^2	0.53	2.46	3.22	4.93	6.55
x_3^2	0.08	0.12	0.43	0.01	0.94
x_4^2	3.46	3.45	2.37	2.44	7.70

that influences the thickness loss after dynamic loading of 50, 100, 200, 500 and 1000 impacts is x_3 (number of plies), accounting for 52.07, 54.31, 54.04, 56.21 and 50.97% of total contribution, respectively. Therefore, this finding implies that number of plies in pile yarn has a decisive role as far as thickness loss is concerned. The interaction term of x_3x_4 (number of plies × pile yarn twist) also has a significant influence on thickness loss and it accounted for 20.24, 13.61, 15.87, 18.96 and 16.15 of total contribution after dynamic loading of 50, 100, 200, 500 and 1000 impacts, respectively. As seen earlier in Table 8, the contribution of x_1 (knot density) seems to be diminishing at higher number of impacts. The role of pile height is miniscule in influencing the thickness loss after dynamic loading.

4 Conclusions

The thickness loss behaviour of Persian handmade wool carpets after dynamic loading has been investigated after 50, 100, 200, 500 and 1000 impacts. The numbers of plies in pile yarn as well as interaction term of number of plies × pile yarn twist are found to be statistically significant at all impact levels. These two parameters are also the major contributors, with contributions of ~50 and ~13–20%, towards the thickness loss of handmade carpets after dynamic loading. The role of knot density on carpet thickness loss diminishes as the number of impacts increases. The thickness loss of carpet is very high up to 100 impacts. However, after 200 impacts, the rate of change in carpet thickness with the increase in a number of impacts stabilizes to a small value. Therefore, the incremental loss of carpet thickness after dynamic loading should be similar irrespective of yarn and carpet construction parameters. The concluding remark of this research to carpet producers is lowest carpet thickness loss, i.e. better carpet wear performance can be achieved by increasing number of plies in pile yarn at initial dynamic impact levels.

Acknowledgements This research work has been supported by O/o Development Commissioner (Handicrafts), Ministry of Textile, Government of India through project titled "promotion of handmade carpet through improved and predictable wear performance of handmade woollen carpets" under research and development scheme.

References

1. Hamburger, W.J.: Mechanics of elastic performance of textile materials: development of an elastic performance coefficient in tension. Text. Res. J. **18**, 102–113 (1948)
2. Dayiary, M., Shaikhzadeh, N.S., Shamsi, M.: A new theoretical approach to the cut-pile carpet compression based on elastic stored bending energy. J. Text. Inst. **100**, 688–694 (2009)
3. Dayiary, M., Shaikhzadeh, N.S., Shamsi, M.: An experimental verification of cut pile carpet compression behaviour. J. Text. Inst. **101**(6), 488–494 (2010)
4. Gupta, N.P., Shakyawar, D.B., Sinha, R.D.: Influence of fibre diameter and medullation on woollen spun yarns and their products. Indian J. Fibre Text. Res. **23**(1), 32–37 (1998)

5. Moghassem, A.R., Gharehaghaji, A.A.: Evaluating pile yarn characteristics in hand woven carpet using stress-strain behavior in compression. IJE Trans. B Appl. **21**(3), 303–312 (2008)
6. Mirjalili, S.A., Sharzehee, M.: A study and investigation on the influence of static and dynamic loading on the properties of handmade Persian carpet (I)—the effect of static loading. Fibers Polym. **6**(2), 139–145 (2005)
7. Mirzalili, S.A., Sharzehee, M.: An investigation on the effect of static and dynamic loading on the physical characteristics of handmade Persian carpets: Part I, the effect of static loading. J. Text. Inst. **96**, 287–293 (2005)
8. Celik, N., Koc, E.: Study on thickness loss of wilton-type carpets under dynamic loading. Fibres Text. Eastern Europe **18**(1)(78) (2010)
9. Arora, R.K., Patni, P.C., Dhillon, R.S., Bapna, D.L.: Influence of tuft constitution on performance properties of hand-woven carpets. Indian J. Fibre Text. Res. **24**(2), 111–114 (1999)
10. Clegg, D.G.: A test for the assessment of compression during wear. J. Text. Inst. **53**, T347 (1962)
11. Clegg, D.G.: Correlation between floor trials on carpets and test on WIRA dynamic loading machine. J. Text. Inst. T646 (1965)
12. Cusick, G.E., Dawber, S.R.K.: Loss of thickness of carpet in floor trials. J. Text. Inst. **55**, 531–536 (1964)
13. Dunlop, J.I., Jie, S.: The dynamic mechanical response of carpets. J. Text. Inst. **80**, 569–578 (1989)
14. Dubinskaite, K., Lagenhove, L.V., Milasius, R.: Influence of pile height and density on the end-use properties of carpets. Fibres Text. Eastern Europe **16**(3)(68), 47–50 (2008)
15. Ishtiaque, S.M., Sen, K., Kumar, A.: New approaches to engineer the yarn structure; Part A: for better carpet performance. J. Ind. Text. https://doi.org/10.1177/1528083713505632 (2013)
16. Ishtiaque, S.M., Sen, K., Kumar, A.: Study of engineered carpet yarns structure: part B—by cross-sectional microtomy. J. Ind. Text. **44**, 605–624 (2015)
17. Onions, W.J.: An assessment of methods of test of carpets for flattening, change of appearance and long term wear. J. Text. Inst. **58**, 487 (1967)
18. Vangheluwe, L., Kiekens, P.: Resilience properties of polypropylene carpets. Text. Res. J. **67**(9), 671–676 (1997)

Part VII
Supply Chain Management and Sustainability

A New Collaborative Model for Demand-Driven Supply Chains: A Case Study on Textile Industry

Ke Ma, Sébastien Thomassey and Xianyi Zeng

Abstract As the increasing trend of customization in production, demand-driven supply chains became more and more important nowadays. However, it was still less addressed in previous supply chain research. Besides, traditional demand-driven supply chain model still had many defects. Therefore, a new collaborative model for demand-driven supply chains was proposed to solve current defects in this paper. A simulation model was built for the new collaborative model based on a case in garment supply chain. Simulation experiments show that various supply chain performance indicators were improved in the new model compared to the traditional model.

Keywords Demand-driven supply chain · Resource sharing · Discrete-event simulation · Garment industry · Supply chain collaboration

1 Introduction

In recent years, consumers always desire customization and personalization of products while making purchase decision. The trend of small series production is also increasing in manufacturing industry. Therefore, quick response and small series production play a significant role in today's supply chain. To meet this rising trend, it is not reasonable to follow traditional forecast-based supply chain model. Demand-driven supply chain (DDSC) model (see Fig. 1) is developed and employed more and more nowadays. Based on Verdouw et al. [1], companies in DDSC only produce and provide services based on customers' demands; thus, companies in the supply chain

K. Ma (✉)
ENSAIT, GEMTEX, 2 Allée Louise et Victor Champier, 59100 Roubaix, France
e-mail: ke.ma@ensait.fr

S. Thomassey
College of Textile and Clothing Engineering, Soochow University, Ganjiangdong Road 178, Suzhou, China

X. Zeng
Department of Business Administration and Textile Management, University of Borås, Allégatan 1, 50332 Borås, Sweden

A. Majumdar et al. (eds.), *Functional Textiles and Clothing*,
https://doi.org/10.1007/978-981-13-7721-1_26

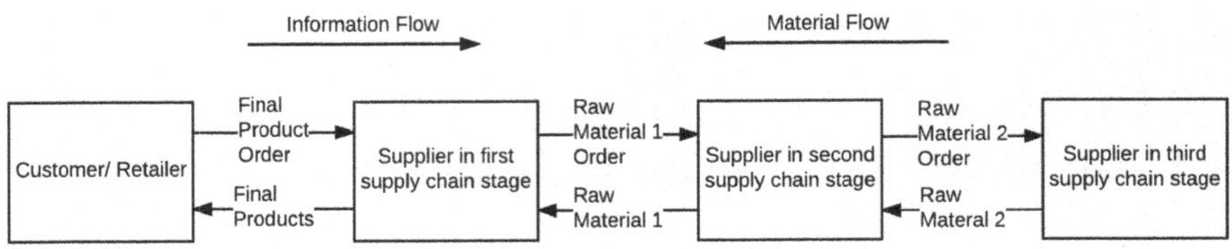

Fig. 1 Information flow and material flow of traditional DDSC model

must collaborate with each other [2]. However, this strategy is somehow neglected in previous supply chain management research. Companies can obtain many benefits by taking advantage of this strategy, e.g. decreased return from customers, reducing the risk of production based on forecasts and increased variety of products. Although the concept of DDSC seems to be easy to apply, in fact, the implementation is difficult [3]. This is because DDSC is totally different from traditional supply chain in many aspects. Moreover, there are still many defects in current DDSC model, e.g. too long lead time, unsmooth production flow, high cost and low efficiency. As it is relatively a new model compared to forecast-based supply chain model, DDSC is still less studied in previous research on supply chain management. DDSC has great potential in research and practice, and this study aims to fill this gap.

Inter-organizational supply chain collaboration (SCC) becomes important in today's supply chain management. It is almost impossible for a company to survive in a supply chain without collaboration with other companies nowadays. More and more researches regarding SCC were conducted as it could bring many benefits to companies, including reduced costs, enhanced profit margins and increased agility. [4]. In general, SCC is classified into two categories: horizontal collaboration and vertical collaboration. Vertical collaboration is "the collaboration when two or more organizations from different level or stages in supply chain share their responsibilities, resources and performance information to serve relatively similar end customers" [5]. Horizontal collaboration is "a business agreement between two or more companies at the same level in the supply chain (SC) or network in order to allow greater ease of work and cooperation towards achieving a common objective" [6]. A majority of SCC researches focused on vertical collaboration, e.g. information sharing [7, 8], coordinating contract [9, 10] etc. SCC also plays a key role in DDSC, the core process cannot be implemented without strong collaboration among suppliers in a demand-driven supply network [2]. Due to the vertical collaboration nature in DDSC, most DDSC studies discussed vertical collaborations, e.g. information sharing [11] and joint decisionmaking [2] within companies from different stages of DDSC. However, much less researches addressed horizontal collaboration in DDSC. Therefore, we filled the gap by integrating vertical collaboration and horizontal collaboration in this study. We developed a new model with a central order processing system for DDSC to address current defects in traditional DDSC model and to improve overall supply chain performance in DDSC. Discrete-event simulation technology was employed on a case in garment supply chains to demonstrate the feasibility and advantage of new proposed collaborative models. Several scenarios were also designed for simulation experiments to examine and compare the performance of collaborative model under different workload conditions, as in most cases different workload could lead to a huge difference in supply chain performance.

2 New Collaborative Model

As aforementioned, a new collaborative model integrating horizontal collaboration and vertical collaboration was proposed with the help of a central order processing system (COPS). The information flow of our new collaborative model is illustrated in Fig. 2. In the beginning, customers/retailers placed an order to COPS instead of direct supplier. COPS would calculate raw materials needed for this order based on internal database and created a series of sub-orders. Then, based on predefined criteria (idle capacity and priority of supplier), COPS would select a supplier in third supply chain stage for the production of raw material 2 which is the raw material of raw material 1. Once the production was done, this supplier informed COPS. COPS would select another supplier in second supply chain stage for the production of raw material 1. Supplier holding raw material 2 would deliver raw material 1 to the selected supplier in next supply chain stage. The same principle applies to the following supply chain stages until final products were produced and delivered to customers. Collaborative relationships are dynamic among companies in the new collaborative model, which is different from the fixed relationship between companies in traditional DDSC. All suppliers within COPS have potential to become partners. COPS plays a third-party role, acting as the "brain" in the new model, mainly realizing three functions: resource sharing, information sharing and joint decisionmaking.

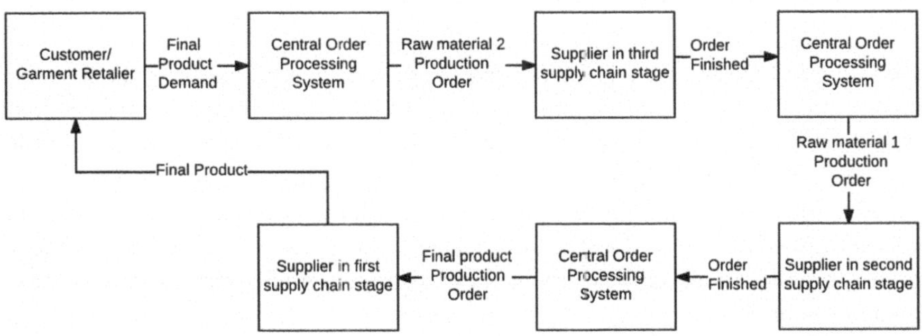

Fig. 2 Information flow of new collaborative model for DDSC

2.1 Resource Sharing

Considering the increasing sharing economy, resource sharing (RS) becomes a hot issue in real industry. In SCC research, RS is defined as that resources, e.g. raw materials, machines and operators, could be shared and leveraged among supply chain partners [12]. However, it was less discussed in previous SCC studies (e.g. transportation sharing [13] or inventory pooling [14], order sharing [15]), compared to other common SCC methods, such as information sharing, coordinating contract or joint decisionmaking. Therefore, considering the benefits of RS, e.g. activating idle resources for usage, we integrated RS into our proposed collaborative model as main implementation of horizontal collaboration. In the new model, resource could be shared among companies in the same supply chain echelon. It means that orders were shared among so-called competitors within the same supply chain echelon.

2.2 Information Sharing

Information sharing is a very common topic discussed in SCC research. It is regarded as one of the key elements for successful supply chain coordination [8]. In the traditional DDSC model, information was shared vertically to some extent. In our new collaborative model, information of each supplier, e.g. production capacity of each supplier for every type of products and status of each type of order, was shared to COPS vertically and horizontally. All information was updated in the system so that to help system distribute orders rationally.

2.3 Joint Decision-making

According to Pibernik et al. [16], SCC can be defined as a process of joint decision-making for aligning plans of individual supply chain members. Therefore, joint decision-making plays a significant role in SCC and it is another common topic discussed in SCC research. In our collaborative model, the order distribution process is a reflection of joint decision-making. COPS distributed each order based on defined criteria from an overall perspective (idle capacity and priority of supplier). If there is one or more order waiting in the queue for processing, initially, the capacity status of candidate suppliers for the specific order was checked. All candidate suppliers with idle capacities were selected if there is any. Then, the priority of those suppliers to corresponding order was checked. The supplier with highest priority was selected finally, and corresponding order was sent to this supplier. COPS would also inform current supplier which is holding raw material to send them to the new selected supplier, and collaborative relationships were built between them as well.

3 Simulation Study on a Case in Demand-Driven Garment Supply Chain

3.1 Simulation Modelling

It is difficult to experiment our new collaborative model in real-world industry. Therefore, it can only be experimented theoretically. Considering the complexity and stochastic nature of this model, traditional mathematical modelling method is not feasible either. Consequently, discrete-event simulation technology was employed in this study to examine the effect of new model. Garment industry is a typical industry which has more and more demand for small series and mass customization in production. DDSC model is a suitable solution for garment industry. Therefore, a case in demand-driven garment supply chains is representative, so that we built a simulation model according to a case in garment supply chain to examine new collaborative model. We collected relevant data based on historical data provided by a French garment company, including order size and order types. We also interviewed professionals working in garment industry and read relevant literature [17, 18] to calculate other necessary parameters for building the simulation model. SIMIO (SImulation Modelling framework based on Intelligent Objects) was utilized as the simulation engine in this study. A four-echelon garment DDSC (fabric manufacturers, dyeing workshop, garment manufacturers and garment retailer/customer) was simulated. In each echelon, three suppliers were available, while three types of garments were produced in the simulation model. A screenshot of the simulation model is shown below in Fig. 3.

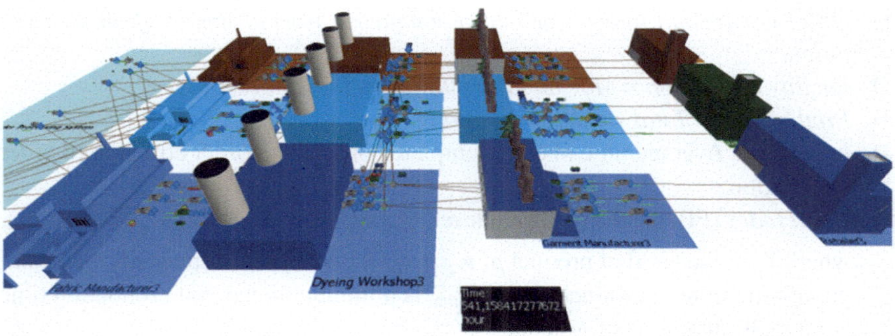

Fig. 3 Simulation model screenshot

3.2 Experiments Design

As introduced in the introduction section, we compared the supply chain performance of new collaborative model and traditional model for DDSC. Also, we were interested in the performance of new collaborative model under different workload conditions. Therefore, twelve scenarios were designed according to two aforementioned purposes, including T (traditional DDSC model) 2, N (new collaborative model) 2, T3, N3, T4, N4, T5, N5, T6, N6, T7, N7. The number in the name of each scenario means the order frequency of each type of product, which represents different workloads. For example, N3 means a new collaborative model under the condition where each retailer placed three different types of orders every three days, which is relatively a high workload. That is to say, on average, in a total of nine garment production orders were delivered to COPS every three days. After simulation experiments, the output results of each scenario of new collaborative model were compared to the corresponding scenario of traditional model, so that to evaluate the performance of new collaborative model and also its performance under different workloads.

4 Results

Each scenario was run for a duration of 20 weeks with 50 replications, respectively. We checked five key performance indicators (KPIs) for each scenario, including order completion rate, facility utilization, productivity, lead time and profit index, which are all common and essential KPIs in supply chain research. The definitions of selected KPIs in this study were introduced as follows:

(1) *Order completion rate = Number of completed orders/Number of all received orders*
(2) *Facility Utilization = Machine Running Time/Total Working Time*
(3) *Productivity = Total yield/Total working days*
(4) *Lead Time*: Average time of each completed order from order arrival until order completion
(5) *Profit Index* (PI): the overall profit level in a period of time. $PI = \sum_{p}^{n}(Y_p * K_p)$, where Y_p is the yield of product p; $K_p = \frac{PT_p}{PT_{min}}$, where PT_p represents mean unit processing time of product p and PT_{min} is minimum mean unit production time among all same type of products.

After experimenting each scenario for 50 replications, we got the average value of each checked KPIs, as shown below in Fig. 4. In general, improvements were obtained to a different extent in each scenario of new collaborative compared to corresponding scenario of traditional model. The improvements of all checked KPIs were significant under high workload condition (N2 vs. T2, N3 vs. T3). Under normal and low workload conditions, the increase is not apparent, but there are still some significant improvements in several checked KPIs, for example, order completion

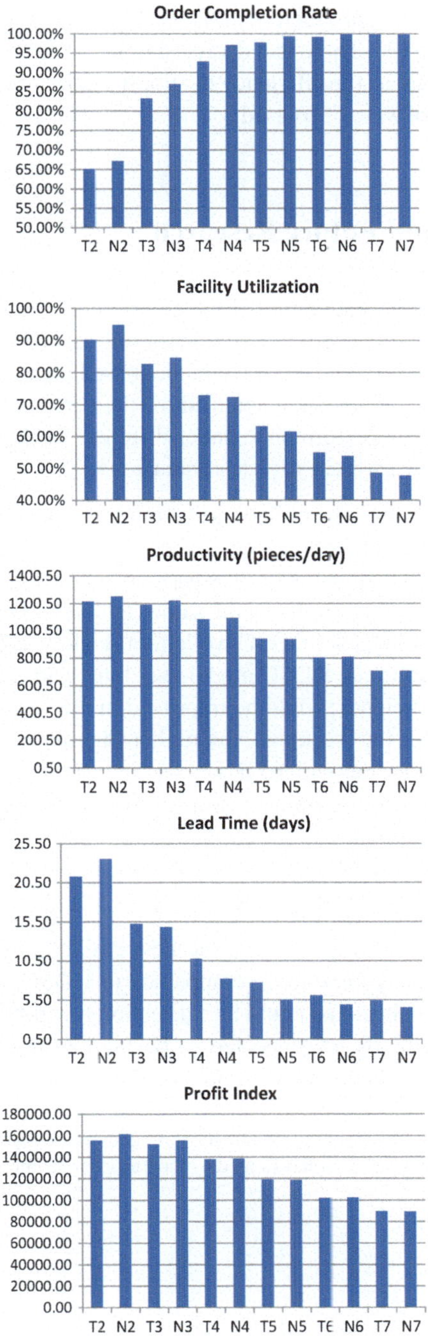

Fig. 4 Information flow and material flow of traditional DDSC model

rate (N4 vs. T4, N5 vs. T5), productivity (N4 vs. T4) and lead time (N4 vs. T4, N5 vs. T5, N6 vs. T6, N7 vs. T7). There is a decrease in terms of facility utilization under low workload condition, as the efficiency of supply chain was increased so that total machine running hours were decreased.

5 Conclusions and Future Research

In this study, a new collaborative model integrating horizontal and vertical SSC for DDSC was proposed to solve current issues in traditional DDSC model. A simulation model was built for the new collaborative model based on a case in garment DDSC. Simulation experiments were conducted to show the performance of new collaborative structure and also its performance under different workload conditions. Based on the simulation results, we got several conclusions:

- In general, the proposed new collaborative model obtained significant improvements in terms of multiple KPIs.
- The new model could help collaborative partners gain more profits and reduce the cost per product, as the cost remains the same (labour and machine cost) between traditional and new model for DDSC while total yield was increased.
- Current issues in traditional DDSC model, e.g. long lead time, unsmooth production flow and low efficiency, could be improved in the new collaborative model.
- In general, the higher the workload, the higher the improvements were obtained in new collaborative model with COPS.
- Only the performance of lead time was improved in new collaborative model under low workload condition.
- Facility utilization was decreased under low workload condition as the increase in efficiency leads to short working time.

According to the experiment results, the new collaborative model proposed in this study can bring many benefits to companies in DDSC, especially under high workload conditions. This model could be a direction for further research on DDSC. More optimization methods are expected to be merged into this model in future research so that to obtain better supply chain performances.

Acknowledgements This work is supported by the joint doctorate programme "Sustainable Management and Design for Textiles" which is funded by the European Commission's Erasmus Mundus programme .

References

1. Verdouw, C.N., Beulens, A.J.M., Trienekens, J.H., Van Der Vorst, J.G.A.J.: A framework for modelling business processes in demand-driven supply chains. Prod. Plan. Control **22**(4), 365–388 (2011)
2. Hadaya, P., Cassivi, L.: The role of joint collaboration planning actions in a demand-driven supply chain. Ind. Manag. Data Syst. **107**(7), 954–978 (2007)
3. Selen, W., Soliman, F.: Operations in today's demand chain management framework. J. Oper. Manag. **20**(6), 667–673 (2002)
4. Chen, L.T.: Optimal dynamic policies for integrated production and marketing planning in business-to-business marketplaces. Int. J. Prod. Econ. **153**, 46–53 (2014)
5. Chan, F.T.S., Prakash, A.: Inventory management in a lateral collaborative manufacturing supply chain: a simulation study. Int. J. Prod. Res. **50**(16), 4670–4685 (2012)
6. Bahinipati, B.K., Kanda, A., Deshmukh, S.G.: Horizontal collaboration in semiconductor manufacturing industry supply chain: an evaluation of collaboration intensity index. Comput. Ind. Eng. **57**(3), 880–895 (2009)
7. Inderfurth, K., Sadrieh, A., Voigt, G.: The impact of information sharing on supply chain performance under asymmetric information. Prod. Oper. Manag. **22**(2), 410–425 (2013)
8. Ganesh, M., Raghunathan, S., Rajendran, C.: The value of information sharing in a multi-product, multi-level supply chain: Impact of product substitution, demand correlation, and partial information sharing. Decis. Support Syst. **58**, 79–94 (2014)
9. Krishnan, H., Winter, R.A.: On the role of revenue-sharing contracts in supply chains. Oper. Res. Lett. **39**(1), 28–31 (2011)
10. Govindan, K., Popiuc, M.N.: Reverse supply chain coordination by revenue sharing contract: a case for the personal computers industry. Eur. J. Oper. Res. **233**(2), 326–336 (2014)
11. Cao, B.-B., Xiao, Z.-D., Sun, J.-N.: A study of the bullwhip effect in supply- and demand-driven supply chain. J. Ind. Prod. Eng. **34**(2), 124–134 (2017)
12. Cao, M., Vonderembse, M.A., Zhang, Q.Y., Ragu-Nathan, T.S.: Supply chain collaboration: conceptualisation and instrument development. Int. J. Prod. Res. **48**(22), 6613–6635 (2010)
13. Vilkelis, A., Jakovlev, S.: Outbound supply chain collaboration modelling based on the automotive industry. Transport **29**(2), 223–230 (2014)
14. Kurata, H.: How does inventory pooling work when product availability influences customers' purchasing decisions? Int. J. Prod. Res. **52**(22), 6739–6759 (2014)
15. Ma, K., Thomassey, S., Zeng, X., et al.: Int. J. Adv. Manuf. Technol. (2018)
16. Pibernik, R., Zhang, Y.Y., Kerschbaum, F., Schropfer, A.: Secure collaborative supply chain planning and inverse optimization—the JELS model. Eur. J. Oper. Res. **208**(1), 75–85 (2011)
17. Ormerod, A., Sondhelm, W.S.: Weaving: Technology and Operations. CRC (1995)
18. Broadbent, A.D.: Basic Principles of Textile Coloration, vol. 132. Society of Dyers and Colorists West Yorkshire (2001)

Customer Analytics in Fashion Retail Industry

Chandadevi Giri, Sebastien Thomassey and Xianyi Zeng

Abstract This paper aims to give an overview of customer analytics in fashion retail industry in the era of big data. Fashion retail industry has been facing significant challenges since last few years due to rapidly varying customer demands. Nowadays, customers are much more informed and connected because of social media and other channels on the Internet. They demand more personalized services, and perception is not sufficient to understand our customers. Therefore, we need data to understand our customers and meet their expectation. We will discuss how customer analytics can create value in fashion retail industry, strategies and methodology to examine the consumer data. Employing and investing in these methods and technologies, industry will benefit from improved revenues, improve in sales, higher customer retention rates and thereby it will sustain in the uncertain markets. Segments are created using recency value of the customers, and their future behavior is predicted using transition matrix.

Keywords Customer analytics · Big data · Segmentation · Consumer behavior · Fashion retail industry

1 Introduction

The present study is part of a Ph.D. project in sustainable design and management of textiles, focusing on e-commerce and consumer analytics to understand consumer behavior, and it is necessary to understand data apart from the perception of the consumer. The contemporary fashion industry is challenging, and integrating customer analytics with its business models will enable it to achieve business goals and high

C. Giri (✉) · S. Thomassey · X. Zeng
Gemtex, Ensait, 2 Allée Louise et Victor Champier, 59056 Roubaix, France
e-mail: chandadevi.giri@ensait.fr

C. Giri
University of Boras, 501 90 Boras, Sweden

Soochow, College of Textile and Clothing Engineering, Suzhou 21506, China

A. Majumdar et al. (eds.), *Functional Textiles and Clothing*,
https://doi.org/10.1007/978-981-13-7721-1_27

Fig. 1 Features of new customers

revenue growth. To illustrate how customer analytics can be effective, we have created customer segments on fashion apparel retail data to study customer behavior and the revenue generated by them. This will be discussed in detail in Sect. 5, customer data analysis.

Understanding consumer preferences has been a major challenge for fashion retailers. As per the market research of Statista [1], the revenue generated in the year 2017 from fashion globally is accounted for US$406,476 million globally and it is estimated to increase by 11.6% by 2022. In the fashion industry, the largest market segment is "Clothing" that contributes a market volume of US$272,599 million in 2017. Revenue generated in China itself contributed around US$164,219 million in 2017. Also, it is estimated that large no. of users will be buying fashion products online by 2022, and the clothing industry will have the maximum market share. Currently, China, the USA, and UK are the major players in the fashion industry, and they are expected to grow in the future. It is important for the retailers to invest in big data analytics to understand customer preferences in real time. Big data provides an opportunity to understand customers in more precise way, and this leads to the emergence of new analytics area which is termed as "customer analytics." "Customer analytics refers to the collection, management, analysis, and strategic leverage of a firm's granular data about the behavior(s) of its customers" [2]. A new type of consumers has evolved. The evolution of customer behavior can be seen in Fig. 1. Therefore, it is essential to study the customer insights from the generated data to understand the fashion market trends.

2 Research Problem and Goal

The premise of the research problem emerges from the fact that fashion retailers are continuously facing many challenges in terms of predicting customer behavior in real time and adopting new strategies to fulfill customer demands. In line with this premise, the goal of this research was to study the following:

1. How to classify customers based on their purchasing behavior evolving over time?
2. How to predict future customer movement and revenue generated by them?

The goal of the customer analytics is to address the above problems. It is important to understand whether consumer is creating value to the business, or whether they are satisfied with the services provided by the retail company, and their preferences in order to take appropriate actions to improve the services and products. Customer analytics can help retail companies to retain their customer base, increase revenue growth, and to predict consumer behavior, and eventually, the objective of value creation by each consumer could be achieved. Therefore, it is imperative to segment different group of the consumers as per their preferences and the value they create.

3 Customer Analytics

This section presents an overview of customer analytics strategy, scope, and methodology as shown in Fig. 2.

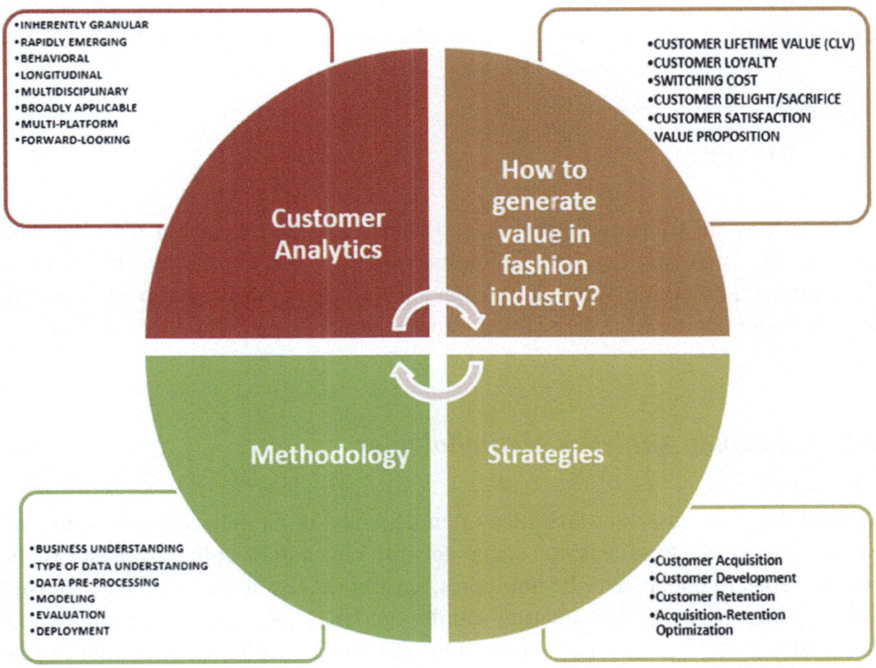

Fig. 2 Customer analytics, strategy, scope, and methodology

3.1 Feature of Customer Analytics

Company's business goals can define how customer analytics can be incorporated into their business intelligence. It could be focused on the prediction of customer behavior at an individual level without considering any other information, or it could be used by merging information from different systems to analyze, or it could be based on behavioral or longitudinal social network analysis. In other words, it depends on the business goals of the companies and which problem they are trying to address.

3.2 Value Generation in Fashion Retail Industry

Fashion retail industry is complicated, and it is quite difficult to understand the consumer choices toward the product. We can generate a high business value by identifying consumer lifetime value which can in turn help the fashion industry to achieve their profit goals. Customer Life Time Value (CLV) is the predicted value that businesses will derive from their entire relationship with their customers [3]. Well-known machine learning (ML) and probabilistic models such as Bayesian Inferences, Moving Averages, Regressions, and Pareto/NBD (Negative Binomial Distribution) can be used to predict CLV [4]. Customer segmentation can be done considering the demographic, geographic, behavioral factors using K-means clustering [5]. Association analysis can also be used for building recommendation system [6].

3.3 Strategy

All industries are consumer oriented, and consumers are crucial for their business success. The main strategy for fashion industry is to expand their customer base. As mentioned in the introduction, the dynamic nature of customers' buying pattern drives retailers to improve their strategy for customer acquisition and retention.

3.4 Customer Analytics Methodologies

Methodology for customer analytics often depends on the business problems that we are trying to address. Fashion social network data can be collected from tweets, boards, blogs to understand the hot topics, current trends, brands, events, criticism, etc. in fashion industry [7]. Internal sources are ERP, CRM, fashion e-commerce, etc., and external sources are cookies, plug-ins Adobe flash, etc. [8]. Data can be pre-processed to get into structured form; then, analytical method such as descriptive and predictive can be applied, and models can be evaluated. According to PwC and

Table 1 Customer analytics with ML methods and statistics

Analysis	ML methods and statistics
Future profitability	Neural networks
CLV	Statistics
Potential CLV	Multi-regression
CLV profiling	Supervised clustering
Churn	Decision trees/Classification
RFM profiling	Decision trees
Acquisition modeling	Neural networks
Response analysis and modeling	Neural networks
Response index	Statistics
ROI	Structured procedures
Campaigns	Regression and structured procedures

SAP retailer survey [9], 39% of retailers ranked "Ability to turn customer data into intelligent and actionable insight" one of their greatest challenges. There is a huge gap between the big data and fashion retail industry. Retailers are more concerned with the data collection. After collecting data, they do not know what to do with such a huge and highly complex customer data. There is a lack of systems for tracking their minute-wise inventory. Therefore, understanding business, data, and customers is very important but retailers have to invest on analytics for creating valuable insights from the chunks of data that could benefit the industry to improve their products and services. Fashion industry is in a greater need to use advanced business intelligence tools, data analytics platforms, big data tools for capturing and processing data. Table 1 lists different machine learning method which can be used for customer analytics to predict profitability, life cycle of the consumers, loyalty, and campaigns.

4 Customer Data Analysis

For this study, we used dataset from the apparel industry, which spans from 2015-10-01 to 2016-12-01. To create segments, we used "Recency" value of each customer as the main indicator. Segments are created based on the recency value for six months, and each segment is further classified as "Inactive," "Less Active," "Active," "Highly Active," and "New Customers." Given the 14-month time span of our dataset, we created two segments: one is 01-06-2016 to 01-12-2016, and the other is from 01-12-2015 to 01-06-2016. Using these two segments, we computed the transition matrix which is used for predicting the behavior and movement of customers within the five categories of each segment in the next one year. Based on the no. of the customers in each segment, revenue and cumulative revenue are calculated and predicted for next one year. No discount rate is considered for this study, but full price only.

4.1 Methodology

Transition matrix is used to compute the likelihood of the future occurrence depending on the current state. Let us assume that fashion retailers have X_n customers where X represents the total number of customers in a given segment at current state n, $X \in S$, where $S = \{$Inactive, Less Active, Highly Active, Active, New Customers], then the probability P_{nm} of the customer in next state m will be given as

$$P_{nm} = P(X_{t+1} = m | X_t = n) \tag{1}$$

Thus, the transition matrix is conditioned on the present state, and the previous and the upcoming conditions are independent. In our study, we predicted the number of customers in each segment in an organization and their transition to the next state. Suppose the states are 1, 2, 3, … r, where r represents the row, then the transition matrix for the different segments can be represented in matrix form as shown in Eq. 2. Thus, the probability of the customers in a segment in state m conditioned on state n can be represented by Eq. 3, and sum of all probabilities in a row will be equal to 1.

$$P(S) = \begin{bmatrix} S_{11} & S_{12} & S_{13} & \dots & S_{1r} \\ S_{21} & S_{22} & S_{23} & \dots & S_{2r} \\ \dots & \dots & \dots & \dots & \dots \\ \dots & \dots & \dots & \dots & \dots \\ S_{r1} & S_{r2} & S_{r3} & \dots & S_{rr} \end{bmatrix} \tag{2}$$

$$\sum_{m=1}^{r} P(S_{nm}) = \sum_{m=1}^{r} P(X_{t+1} = m | X_t = n) = 1 \tag{3}$$

4.2 Descriptive Analysis of the Data

Dataset is comprised of 5,770,844 customer transaction data with 1,020,923 distinct customers. For the customer analysis, we considered three variables: "Customer ID," "Purchase Amount," and "Date of Purchase", see Table 2. The statistics show that average spending by each customer is 38.60 units, and maximum spending for each transaction is 199.90 units. The maximum time lapse with last transactions is 426.77 days (approx. 14 months).

We have calculated three variables for customer data, i.e. recency, frequency, and monetary Value (RFM). Recency is the no. of the days lapsed between customers' recent transaction date and the last transaction date. The bigger the recency value, the less active the customers are. Frequency is defined as the no. of the purchases made by the customers in a given period of time, and monetary value is the amount of money customers spent in each transaction. RFM is calculated for each customer. We can see from Table 3 that the average recency of a customer is 180 days, and

Table 2 Summary statistics of customer data

Purchase_amount	Date_of_purchase	Days_since
Min.: −4.08	Min: 2015-10-01	Min.: −0.2292
1st Qu.: 19.95	1st Qu.: 2016-01-10	1st Qu.: 120.7708
Med Median: 34.95	Median: 2016-04-22	Median: 222.7708
Mean: 38.60	Mean: 2016-04-25	Mean: 219.4961
3rd Qu.: 53.15	3rd Qu.: 2016-08-02	3rd Qu.: 325.7708
Max.: 199.90	Max.: 2016-12-01	Max.: 426.7708

Table 3 Summary statistics of recency, frequency, and monetary (amount) for customer

Recency	Frequency	Amount
Min.: −0.2292	Min.: 1.00	Min.: 0.00
1st Qu.: 80.7708	1st Qu.: 2.00	1st Qu.: 25.82
Median: 155.7708	Median: 3.00	Median: 37.84
Mean: 180.4872	Mean: 5.65	Mean: 40.93
3rd Qu.: 288.7708	3rd Qu.: 6.00	3rd Qu.: 51.35
Max.: 426.7708	Max.: 1555.00	Max.: 199.90

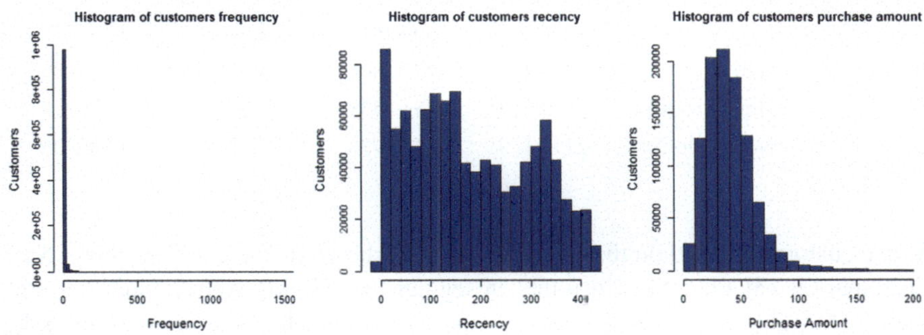

Fig. 3 Histogram of recency, frequency, and purchase amount

maximum recency is 426 days. The average spent amount per transaction is 40.93 units while the maximum spent amount for each transaction is 199.90 units. According to the frequency, customers had made at least one purchase and on an average five purchases. Histogram of RFM values is depicted in Fig. 3.

4.3 Segmentation Based on Recency

Two segments were created for the six-month interval based on recency values. In each segment, classes were assigned as Inactive, Less Active, Active, and Highly

Table 4 Segmentation class based on recency value

Recency (days)	Customer class
>180	Inactive
≤180 and >90	Less active
≤90 and >50	Active
≤50	Highly active

Table 5 Average recency, frequency, and purchase amount for each class within Segment 1

	Segment 1	Recency	First_purchase	Frequency	Amount
1	Inactive	296.47006	317.42518	3.286801	40.69390
2	Less active	131.86588	217.59175	5.252555	34.16796
3	Active	69.98107	195.52441	6.867103	44.75028
4	Highly active	21.90170	301.30038	15.947734	43.73544
5	New customers	23.62203	25.01742	3.271787	62.07015

Table 6 Average recency, frequency, and purchase amount for each class within Segment 2

	Segment 2	Recency	First_purchase	Frequency	Amount
1	Inactive	212.44278	214.87399	2.388128	52.89656
2	Less active	135.97081	150.84615	3.346653	33.82017
3	Active	68.74415	117.36808	4.974818	43.37889
4	Highly active	26.68699	175.74538	12.261362	39.70714
5	New customers	24.02350	25.32776	2.695572	43.46406

Active customers based on the recency values as shown in Table 4. The class "New Customer" is calculated as Customer in segment = "Highly Active" and first purchase $\leq$ 50. This will help us to identify the customers who were not present in the Segment 2 but are newly added in the Segment 1. Highly Active customers are those whose recency is less than or equal to 50 days while inactive customers are those whose recency is more than 180 days. Recency, frequency, and average purchase amount for Segments 1 and 2 can be seen in Tables 5 and 6.

In Fig. 4, we have depicted how new customers who were absent in Segment 2 are now evolved in the Segment 1. As the logic behind our segmentation is to identify the similarity between the customers in different segments and understanding their behaviors after 6 months, we grouped them together according to their recency criteria for each segment. From the Segment 1, it is evident that new customers have been acquired that were not present in the Segment 2. Those who were new in Segment 2 transferred into another category "Inactive" and "Less Active" in Segment 1. Same can be observed for the highly active consumer, no. of highly active customers reduced in the segmentation 1, which are current customers.

The concept of transition matrix is employed to identify the probabilities of customers changing their segments. In other words, it is important to measure the likeli-

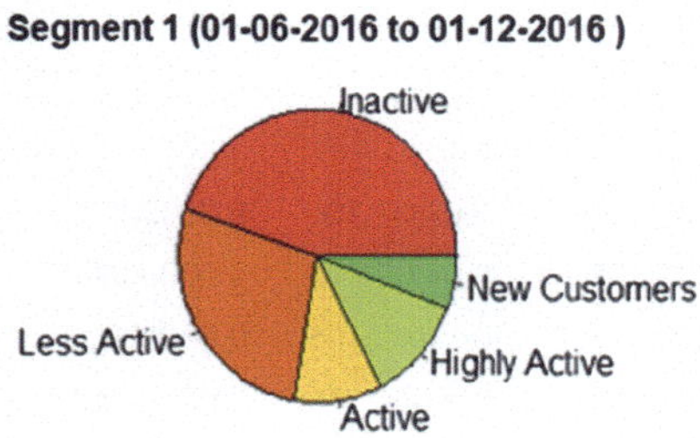

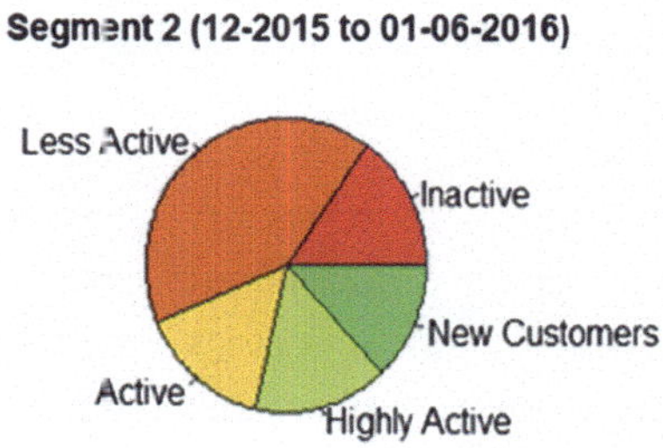

Fig. 4 Number of customers of each class in Segment 1 and Segment 2

Table 7 Transition matrix as per class of customers in Segment 1 and Segment 2

	Inactive less	Active	Active	Highly active	New customers
Inactive	83,293	12,661	4869	8986	0
Less active	208,564	48,854	15,118	24,620	0
Active	64,694	19,288	8322	16,226	0
Highly active	37,687	25,518	13,197	36,721	0
New customers	64,861	14,182	5430	9628	0

hood of inactive customers from the Segment 2 to moving to the Less Active, Active and Highly Active classes of the Segment 1. Therefore, transition matrix enables us to find the number of customers in each class of the Segment 1. The probabilities from transition matrix are crucial in directing the future marketing campaigns and in targeting the potential customers. Transition matrix showing the probabilities of customers evolving in the next segment (Segment 1) is depicted in Fig. 4.

We have created transition matrix for the two segments to understand how the customers have changed their status from Segment 2 to Segment 1 in Table 7. The classes shown horizontally are Segment 1, and the one vertically is Segment 2. Based on vertical classes, we will see the movement of the customer in the horizontal classes For example, in the first row, the customers who were inactive in segment 2, 83,293 customers remain inactive, 12,661 becomes less active, 8986 becomes active, and 4869 becomes highly active in Segment 1. So, we can say that very less inactive customers joined the groups: “Active” and “Highly Active.” Now, if we see what happened to the new customers in Segment 2, the fifth row of the transition table, 64,861 customers became inactive and 14,182 customers became less active, while only few remained in the groups: “Active” and “Highly Active.”

After we divided the rows in Table 7 by the sum of the customers for a given class, we get the transition probabilities as shown in Table 8. From Table 8, we can interpret that 75% customers who were inactive in Segment 2 will remain inactive in Segment 1, while about 8% inactive customers in Segment 2 will become highly active in Segment 1. For new customers in Segment 2, about 10% of them will remain

Table 8 Transition matrix by probability

	Inactive	Less active	Active	Highly active	New customers
Inactive	0.75852617	0.11530020	0.04434063	0.08183300	0.00000000
Less active	0.70186703	0.16440523	0.05087563	0.08285210	0.00000000
Active	0.59609325	0.17772045	0.07667926	0.14950705	0.00000000
Highly active	0.33315064	0.22557747	0.11666063	0.32461126	0.00000000
New customers	0.68927004	0.15071041	0.05770396	0.10231560	0.00000000

highly active, 5% will remain active, and about 69% will remain inactive in Segment 1. Likewise, we can interpret the results for other classes too.

4.4 Prediction Based on Transition Matrix for the Next 6 Months and 12 Months

It is often necessary to predict the no. of new customers in the next time period as they can potentially contribute to the revenue growth. Based on the prediction, managers design their marketing campaigns targeting new customers and the customers that are less likely to evolve, or in other words who are not exhibiting any movement to the active segments can be removed from the target group. Based on the results of transition matrix, we predicted the no. of the customers for the next 6 and 12 months and the revenue generated by them as shown in Figs. 5 and 6. Table 9 shows the predicted value of total number of customers in each class of the segments after 6 and 12 months. We can perceive that the number of the inactive users will be increasing in next one year, which means that new customers, active customers, and highly active customers from Segment 1 (current customers) will move to either Inactive or Less Active group of the same segment. As we have seen that the probability of "Inactive" customers shifting to "Active" group is quite less. New customers will be evolving in the future for 6 and 12 months of prediction. However, Fig. 5 shows the probability of the customer's transition to other groups after 6 and 12 months. We can see from Table 9 that the no. of the customers in "Inactive" and "Less Active" class is increasing, while in "Active" group is decreasing. No. of "Highly Active" users remain approximately the same.

Intuitively, if the number of customers becoming active in the Segment 1 is higher, it means that the revenue generated by them will be higher. The future revenue or the revenue that will be generated in the next segment or in future segments can be predicted by looking at the classes of segments to which they belong. Table 10 shows revenue generated by each class in a segment currently and after 6 and 12 months. From Table 10, it is evident that the inactive customers cannot generate revenue as indicated by values "0," whereas new customers are significant to high revenue generation. However, we could not predict the number of new customers and revenue

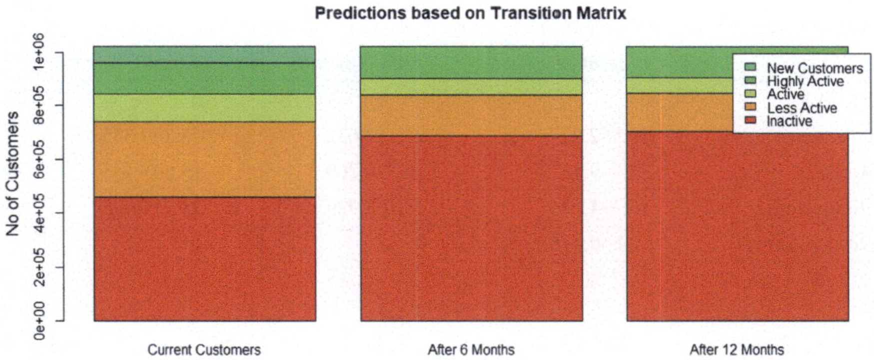

Fig. 5 Prediction using transition matrix for no. of customers for each class for 6 and 12 months

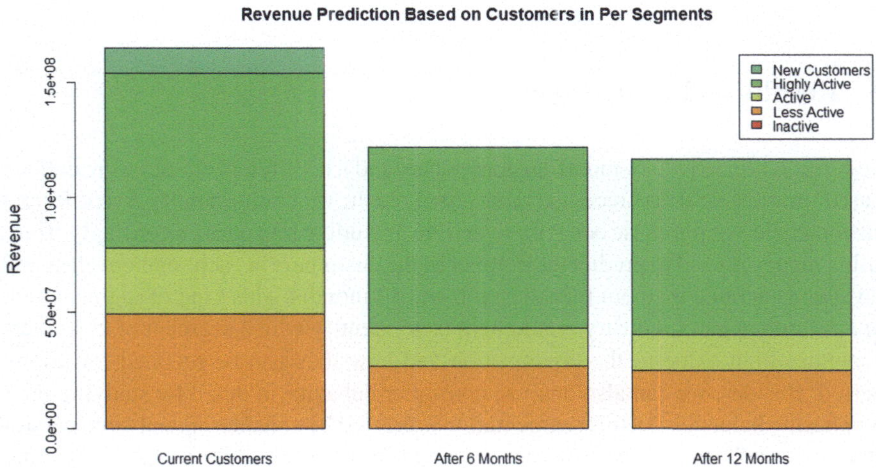

Fig. 6 Prediction of revenue generated by each class after 6 and 12 months

Table 9 Number of customers in each class after 6 and 12 based on Segment 1

	Current customers	After 6 months	After 12 months
Inactive	459,099	687,299.59	704,646.68
Less active	279,777	153,078.56	142,277.96
Active	103,584	59,794.53	56,935.07
Highly active	118,107	120,750.32	117,063.29
New customers	60,356	0.00	0.00

Table 10 Revenue generated by each class after 6 and 12 months based on Segment 1

	Current customers	After 6 months	After 12 months
Inactive	0	0	0
Less active	49,552,927	27,112,632	25,199,675
Active	26,369,141	16,376,269	15,593,133
Highly active	76,394,171	78,103,929	75,719,075
New customers	10,850,470	0	0

generated by them in the future time periods (6 and 12 months) because the Customer IDs of the new customers will be unique and different from the Customer IDs in the dataset we used for the study.

5 Results and Conclusion

Segment is created on 14-month customer data, and five different classes were defined based on the calculated recency value. As a result, we could identify five different customer classes purchase behavior over time, including revenues, accordingly. Transition matrix is used to predict the number of the customers in each segment class and revenue generated by them for the next 6 and 12 months. This kind of segmentation in a fashion apparel industry would help us to identify which segment of customers generates high value to the organization and how they can be retained for a long period. Besides, we can also analyze consumer behavior in detail by studying their purchasing behavior. As this segmentation is created by considering only one parameter "recency" value of the customer, it could be further improved by including other parameters: "Frequency" and "Monetary Value." We will consider machine learning methods for predicting the customer's behavior in our future work.

The fashion industry is dynamic and sensitive to quick changes in the customer behavior with the seasons and trends. Customer analytics will help the fashion industry to quickly respond to changing customer preferences. By applying customer analytics, we can analyze the buying behavior of the consumers and their preferences. Furthermore, it will also benefit supply chain and inventory management and it will be easy for the retailers to make decisions based on real-time tracking systems, reducing losses and helping the company to operate more environmentally friendly. Customer Analytics in fashion retail industry will help to customize the profiles of their consumer, enhance the personalized recommendation services, more loyalty programs, will give the opportunity to know their customers better than before and will help the business to create value from it. Thus, focusing on customer analytics in the era of big data, industries could be benefitted more than ever in the history.

Acknowledgements We are grateful to SMDTex—Sustainable Design and Management of Textiles, Erasmus Mundus Joint Doctoral commission for creating good research environment to take out this research and their consistent support. We would like to show our deep gratitude to Mr. Giuseppe Craparotta, Senior Data Scientist, Evo Pricing for helping us with the data. Without him, it would have been difficult for us to execute this research work. We would also like to thank you Evo Pricing and their innovative team for their contribution to the research.

References

1. Statista—The Statistics Portal for Market Data, Market Research and Market Studies. (n.d.). Retrieved from https://www.statista.com 15 Oct 2017
2. Home—Wharton CAI. (n.d.). Retrieved from http://wcai.wharton.upenn.edu/ 16 Oct 2017
3. Strickland, P.J.: Data Science and Analytics for Ordinary People. S.l.: Lulu com (2015)
4. Wilkinson, J.W., Trinh, G., Lee, R., Brown, N.: Can the negative binomial distribution predict industrial purchases? J. Bus. Ind. Market. **31**(4), 543–552 (2016). https://doi.org/10.1108/jbim-05-2014-0105
5. Li, Z.: Research on customer segmentation in retailing based on clustering model. In: 2011 International Conference on Computer Science and Service System (CSSS) (2011). https://doi.org/10.1109/csss.2011.5974496
6. Tan, P., Steinbach, M., Kumar, V.: Introduction to Data Mining. Pearson, Nueva Delhi (India) (2016)
7. Data is Fashionable [Web log post]. (n.d.) (2016)
8. Curto, J., Braulio, N.: Customer Analytics. Editorial UOC (2015)
9. Verhoef, P., Kooge, E., Walk, N.: Creating Value with Big Data Analytics. Routledge (2016)

Recent Developments in Recycling Silk Saris

S. Nivedita and Gargi

Abstract Silk is a natural protein fibre. It accounts for less than 1% of the total textile fibre produced worldwide and about 1.3% of total fibre production in India. About 75% of raw silk is used for making saris, the most preferred traditional attire amongst women in the country. Saris made of silk with intricate woven designs on handlooms are unique and exquisite. However, once a sari becomes old or damaged, its value is reduced to that of scrap. Little attempts are made to repurpose/recycle the garment. This paper discusses various options for maintaining the value of used silk saris through reusing/recycling the material. It can be concluded that this garment was traditionally designed with a zero-waste, cradle-to-cradle concept. By extending the life of a sari, maximum benefits can be extracted from it which benefits environment as well as economy and leads to sustainable development.

Keywords Pattern design · Sari ribbon · Compostable textiles · Cradle to cradle · Silk flock

1 Textile Recycling

Over 100 billion garments are produced worldwide annually and about the same quantity of used garments are discarded. Eighty-five per cent of this ends in landfills or is burnt, causing pollution and wastage of natural resources [1]. One of the best ways to manage textile waste is by recycling, that is, converting waste material into new and useful material. Both pre- and post-consumer waste materials can be recycled. In fact, today the textile recycling industry is rapidly growing. 'Textile waste management' and 'textile recycling technology' are aiming at zero-waste production. The new design principles go beyond recycling and take a 'whole system' approach to the vast flow of resources and waste. They include recycling wastewater from textile production, reclaiming and reuse of textile chemicals, recycling all wastes:

S. Nivedita (✉) · Gargi
Central Sericultural Germplasm Resources Centre, Hosur 635109, India
e-mail: nive_cstri@rediffmail.com

A. Majumdar et al. (eds.), *Functional Textiles and Clothing*,
https://doi.org/10.1007/978-981-13-7721-1_28

raw material, yarn and fabric wastes, designing ‘easy-to-recycle’ textile products and recycling used products.

Textile recycling has both economic and environmental benefits. For example, by recycling 1 kg of used clothes, 6000 l of water consumption, 0.3 kg of the use of fertilizers, 0.2 kg of the use of pesticides and 3.6 kg of CO_2 emissions can be reduced according to a study made at the University of Copenhagen [2, 3]. Also, the use of virgin material can be minimized. Therefore, recycling textiles has been identified as one of the key technologies for sustainability and resource efficiency along with greater use of natural and biomass-based fibres, process technologies which are less energy-intensive and less water-intensive, bio-chemistry-based textile processing and using substitutes for toxic/restricted textile auxiliaries [4].

2 Silk Sari Recycling

2.1 Lifespan of a Silk Sari

Silk, the queen of textiles, is a natural protein fibre produced by the silkworm. It accounts for less than 1% of the total textile fibre produced worldwide and about 1.3% of total fibre production in India [5]. India is the largest consumer of silk in the world. Consumers today have greater buying power but less time to take care of their silk saris. New saris are bought frequently and the older ones discarded. Traditionally, silk saris were often handed down by a mother to her daughter. A silk sari can last for two to three generations but tends to lose strength and weaken over a period of time. It may cut along the folds or develop holes and tears. Most silk saris eventually end up with the scrap dealer who sells them by truckloads for the price of rags. On an average, silk saris today last for 12–18 months.

By recycling, it is possible to extend the life of a silk sari and to extract maximum benefits from this scarce and valuable fibre.

2.2 Silk Sari Pattern Design

While the beauty, grandeur and elegance of a silk sari are well known, its ‘zero-waste’ pattern design and suitability to recycling are less known and appreciated. A sari is a rectangular seamless garment, and no portion of it is cut out as scrap during its manufacture. The garment comes in only one size that fits all without any alteration. A traditional silk sari is always in trend, and the need to discard it due to fashion change does not arise. It is designed for sustainability. However, due to accidental damages like tear, stain or colour change the sari may not be suitable for wearing. Such a sari can be recycled either as a whole or in well-defined portions such as the border, body and the pallu (free end of a sari). If the sari has pure zari work, then the precious metals, silver and gold, can be reclaimed by a melting process.

In short, the sari pattern is ideally designed on the principles of a circular economy for maximizing use and minimizing waste with plenty of recycling options. Some of the options tried out by the authors are presented in this paper.

2.3 Upcycling Used Silk Sari

Upcycling is to create a new product of higher quality or value than the original. Some of the upcycling options for used silk saris are making other garments, tapestry, curtains, table/bedcovers, runners, wall panels, file folders, accessories, etc. (Fig. 1).

2.4 Downcycling Used Silk Sari

Downcycling involves breaking an item down into smaller elements and creating usable products from them. Used silk saris can be cut down into small pieces, sari ribbons, sari yarn or into fibrous web (Fig. 2) which can then be used to make recycled products such as rugs, bags, accessories, wadding (Fig. 3).

2.5 Other Possibilities

(i) **Flock silk**: The fibre length is cut to 0.25–0.50 mm by milling. The fibres are electrically charged so that they stand up. They are then propelled and anchored in adhesive applied on any substrate such as metal, plastic, textile to impart a velvety look and feel. Used silk saris can be a source of these short fibres. Being old, there are weak and easy to cut.

(ii) **Silk powder/nanomaterials**: The particle size in silk powder is in the range of 1–100 μm. Over-degummed silk, cut into small pieces, is pulverized using rotary and planetary ball milling [6]. The average particle size of ultra-fine powder is around 200 nm. In nanomaterials, the particle size in the nanometre range. They have a range of applications as biomaterials in medical, cosmetic and biotechnology fields. Silk saris stripped off the colours may be suitable precursor for making silk powders.

(iii) **Compostable textiles** [7]: Silk is a biodegradable material, capable of being broken down rapidly by the action of micro-organisms in the environment into natural materials and returned to the earth as resources without causing harm or releasing heavy metals. In a commercial composting facility, silk will break down into humus.

Fig. 1 Upcycling options for used silk saris (# 1 to 8 were developed by the authors)

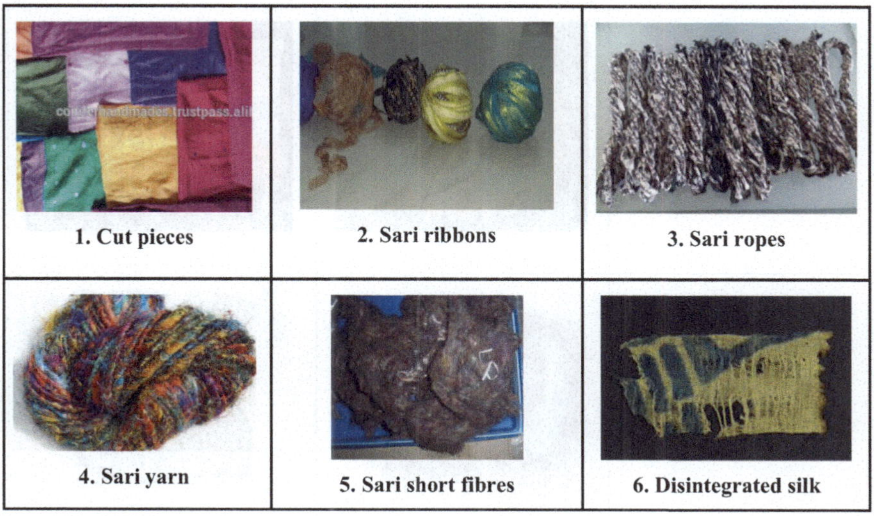

Fig. 2 Silk sari elements (# 1 to 3 were used by the authors)

3 Conclusion

Silk saris provide several options for reusing until complete degradation. This is typically ‘cradle-to-cradle’ concept which aims to reuse valuable fibres via closed loop manufacturing methods (see Fig. 4).

Recycling saris into beautiful products generates employment as well as market returns, as the material cost is drastically low compared to virgin silk material. A new silk sari costing around Rs. 5000/- is sold at only Rs. 100/- by the scrap dealer, which is perhaps less than the cost of other natural materials like cotton and jute. Extracting maximum benefits from silk by extending its life aids sustainable development. In conclusion, it can be said that the principles of a circular economy in the textile industry can be best demonstrated by the silk sari. The products developed are displayed in the museum at Central Sericultural Germplasm Resources Centre, Central Silk Board, P.B No. 44, Thally Road, Hosur-635 109, for spreading awareness [8].

1. Products made from Silk cut pieces	 Silk rug	 Door decoration
2. Products made from Sari ribbon / yarn	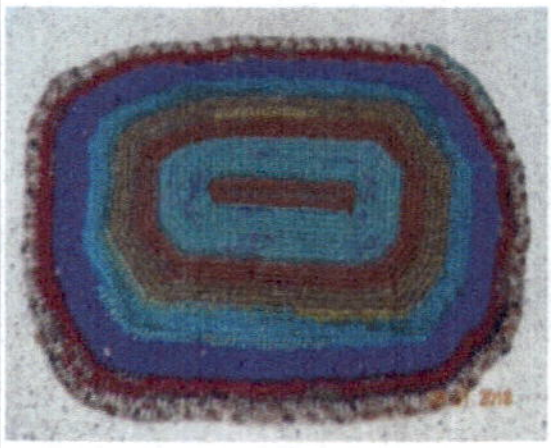 Silk floor covering	 Silk mats
3. Products made from Sari ropes		Macramé wall hanging
4. Products made from Sari yarn	 Poncho	 Accessory
5. Products made from Short fibres	 Wadding	 Felt

Fig. 3 Downcycling options for used silk saris (# 1 to 3 were developed by the authors)

Fig. 4 Cradle-to-cradle principle of silk saris

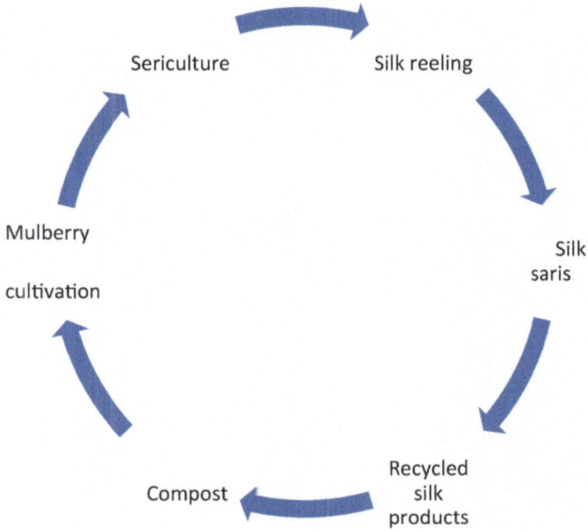

References

1. Rick LeBlanc: Textile recycling facts and figures. https://www.thebalance.com. Updated 31 Jan 2017
2. RockPool Recycling, Recycling Facts. http://www.rockpoolrecycling.com/textiles/php. Last accessed 2017
3. Eryuruk, S.H.: Life cycle assessment method for environmental impact evaluation and certification systems for textiles and clothing. Elsevier BV (2015)
4. Senthil Kumar, P., Gunasundari, E.: Sustainable wet processing—an alternative source for detoxifying supply chain in textiles. In: Muthu, S.S. (ed.) Detox Fashion, Textile Science and Clothing Technology, vol. VII, 137p, pp. 37–60. Springer Nature Singapore (2018)
5. Textile Industry, Overview, Ministry of textiles, GoI. http://texmin.nic.in/textile-data
6. Nivedita, S., Sivaprasad, V.: Biomedical applications of mulberry silk and its proteins: a review. J. Inst. Eng. India Ser. E **95**, 57 (2014). https://doi.org/10.1007/s40034-014-0034-0
7. ASTM D6400—Compostable Product Test—Composting
8. silkgermplasm.com

Awareness of Green Manufacturing in Apparel Industry

Ankur Saxena and Ajit Kumar Khare

Abstract Green manufacturing involves the control and reduction of hazardous substances in the design, manufacturing and application of products or processes. The alarming situation of global warming has, in recent years, caused researchers and practitioners to devote attention towards the impact of garment and textile industry on the environment. Significant research work has been done in studying the impact and reduction of carbon footprint in different manufacturing industries. Delhi and National Capital Region (NCR) is a significant and large emitter of greenhouse gases and most of it because of the standardised and acceptable processes in industrial production. Hence, it becomes imperative to investigate the emissions in the given region. This paper has discussed the present level of awareness about green manufacturing in apparel industry in Delhi and National Capital Region. A research instrument was prepared to collect the data from garment manufacturing factories in selected geographical area and having substantial size and volume. Responses were collected on different aspects of green manufacturing like challenges, benefits, understanding and threats in apparel industry of Delhi and NCR. Responses were statistically analysed and interpreted to present a broader picture of awareness level about green manufacturing.

Keywords Global warming · Green manufacturing · Apparel manufacturing process

1 Introduction

1.1 Global Warming and Its Effect on Climate

The effects of climate change, along with pollution and the depletion of non-renewable natural resources, have given rise to environmental awareness [1]. Since the early twentieth century, the average surface temperature of the earth has increased

A. Saxena (✉)
National Institute of Fashion Technology, 342017 Jodhpur, India
e-mail: ankur.saxena@nift.ac.in

A. K. Khare
National Institute of Fashion Technology, 410210 Mumbai, India

A. Majumdar et al. (eds.), *Functional Textiles and Clothing*,
https://doi.org/10.1007/978-981-13-7721-1_29

by about 0.8 °C (1.4 °F), with about two-thirds of the increase occurring since 1980. Researchers indicate that during the twenty-first century, the global surface temperature is likely to rise by a further 1.1–2.9 °C (2–5.2 °F) for the lowest emissions scenario and 2.4–6.4 °C (4.3–11.5 °F) for the highest emission scenario. Warming of the climate system is irreversible, and scientists are in the view that most of it are caused by increasing concentrations of greenhouse gases produced by activities such as deforestation and the burning of fossil fuels [2]. In 2014, greenhouse gas emission was 50 Giga tonnes (Gt.) globally which include 196 countries, around 5% of total countries are responsible for more than 85% of the total emission. Industrialisation, energy and agriculture are the major factors contributing more than 80% of the total global emission. In the global context, India is the fourth largest emitter of carbon dioxide at 1.65 Gt. per year, after China (6.9 Gt. per year), USA (5.2 Gt. per year) and European Union (2.5 Gt. Per year) [3]. As per the report of World Resources Institute, India is producing over 7% of global greenhouse gases. Industries, agriculture and power generation are the main factors and produce more than 90% of Indian emission [4]. According to Mehta et al., the textile industry affects the environment by releasing carbon dioxide into the atmosphere. The weaving and spinning sector contributes a lot to emission in carbon dioxide. However, the apparel sector contributes 3–4% of overall emission and increasing with a rapid growth [5]. To reduce the greenhouse gas emission and further to minimise the adverse effect on the environment, the importance and effect of green manufacturing have created a much wider debate in recent years.

1.2 Green Manufacturing and Its Advantage

Researchers have given multiple definitions of Green Manufacturing; one of the definitions given by Das et al. states that these are manufacturing methods that support and sustain a renewable way of producing products and services that do no harm to you or the environment. Green manufacturing is defined as the design, processing and commercial use of materials processes and products, which are economical and sustainable while minimising pollution and risk to human health and the environment. Green manufacturing is more of a philosophy rather than an adopted process or standard [6].

Green manufacturing is a need for sustainable development and a means of competitive advantage for the firms [7]. It is considered that firms which successfully implement green manufacturing are preferred by customers. These manufacturing organisations continuously strive hard to innovate strategies to adopt the green practices to reorganise the supply chain structures, so that benefit of strategic green manufacturing practices can be achieved in the product management [8–11]. The value of investing in green technology and green transformation is considered as a topic of argument between researchers and managers [12]. Hoffman states that environmental and green attempts in manufacturing should move from being an environmental management approach to an environmental strategy. This will create a win-win situation by which manufacturers can improve their environmental performance while achieving economic gains [13].

1.3 Green Manufacturing Practice in Apparel Industry

Although literature is available on the use of green manufacturing or reducing carbon footprint in different industries around the globe [14–16], researchers talk about green manufacturing in several industries like cement and automobiles [17] In the context of global apparel industry, Eryuruk et al. have calculated carbon emission during a lifecycle of an apparel product. He has divided the life cycle from design to reuse and calculated emission for each process [18]. Guo-Ciang et al have presented a model for Taiwan textile and apparel industry in which drivers for supply chain management were identified [19]. Several researchers have discussed evaluating present green manufacturing practices in different department of apparel and textiles globally [20, 21].

2 Need of the Research

It is pertinent to mention that global warming and greenhouse gas emission are an area of concern for the world and for India. Researchers have agreed that one of the main reasons for global warming is industrialisation, and green manufacturing can be considered an effective tool to control and minimise the GHG (Green House Gases) emission from Industries. In India, textile industry and garment industry are a significant contributor to Indian economy of industrial production [22]. A significant fraction of the apparel industry is in Delhi/NCR which is also the highest emitter of GHG because of industrialisation. As per a book written by T. V. Ramachandra et al. "Sector-Wise Assessment of Carbon Footprint across Major Cities in India", eight cities including all metros were studied and their sector-wise greenhouse gases emissions were calculated. Industrial GHG (Green House Gases) emission per capita was highest in Delhi NCR during 2014–15 [23].

As per Prof. Mukesh Sharma, the annual per capita tCO2e emission in the city of Delhi was estimated as 2.26 ton in 2016, which is 1.5 times of national average [24]. Also, as per the research by Dr. Deveraja under Indian Council of Social Science-Ministry of HRD, Government of India, Delhi region is considered a significant cluster of apparel manufacturing in India and contributes 8.7% in export and 10.9% in domestic market [25]. Statement of T. V. Ramachandra and MOT (Ministry of Textiles) report confirms that there is a problem of GHG emission in Delhi/NCR. Industrialisation in general and apparel industry of the region in particular is the major cause for the same.

3 Methodology and Research Design

3.1 Methodology

To investigate the level of awareness and attitude of garment manufacturers of Delhi/NCR towards green manufacturing, an itemised instrument was used. This instrument was prepared based on initial responses to a checklist and brainstorming sessions, held with selected decision makers of garment factories situated in Delhi and NCR. Figure 1 shows the steps used to achieve this objective.

3.2 Sample Size

According to the research published in "Journal of industrial statistics" in 2015 about efficiencies and policies of Indian apparel industry, a low scale of operations prevents the adoption of modern technology in Indian apparel industry, and size of the firm may significantly affect the efficiency and policy making of the firm in a positive manner. The researcher has also divided the garment factories into four quadrants according to their sizes and policies [26]. Learning from this, it was decided to include the factories which are in quadrant three and four, i.e. factories having more than 400 sewing machines. There are around 800 such factories in Delhi/NCR (Source: Apparel Resources yellow pages). It was suggested by the research committee to have minimum 10% (80 factories) responses in order to obtain valid opinion. To ensure minimum required responses, around 200 respondents of different apparel manufacturing units using purposive sampling were approached, out of which 81 responses were found complete and included for analysis using SPSS, a statistical tool. The questionnaire was responded by middle management of garment manu-

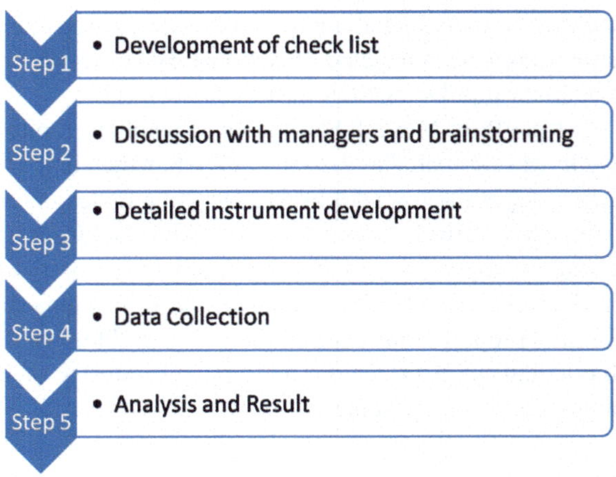

Fig. 1 Methodology process flow

facturing organisation; some of the profiles were production manager, sustainability officer, Operation Head, etc.

4 Data Collection and Analysis

India is a significant and large emitter of greenhouse gases and most of it because of the industrial production. Hence, there is a certain need to reduce these emissions, justifying thereby the focus is on investigating the awareness and attitude towards green manufacturing in the apparel industry of Delhi/NCR. A detailed research instrument, which covered all the possible aspects and concept of green manufacturing, was prepared to achieve this objective.

4.1 Checklist and Brain Storming Session for Research Instrument Development

To understand the intensity of final questionnaire a checklist was prepared and pilot tested on 30 respondents, a brainstorming session was also conducted with six decision makers of different garment factories in Delhi/NCR. Observations of those checklists and brainstorming session are as follows:

- Based on the responses, it emerges that the respondents have a basic understanding about the terms like global warming, carbon footprint and greenhouse gases.
- The respondents have very less information or understanding on the technical and monetary aspects of green manufacturing.
- The respondents have very little idea about possibility and advantages of green manufacturing in the garment Industry.

4.2 Data Analysis

Received responses were coded for analysis. Responses were analyzed through SPSS. Phase one of the instrument included general information like size, mode of business and environmental policy. As per the responses received, 34.6% respondents are working in domestic market; however, 65.4% are working in both domestic and exports. 100% respondents have more than 500 machines and having an environmental policy in their factory documents. All of them have agreed to make their responses public for research purpose.

Analysis Regarding Corporate Objective To investigate the attitude, respondents were asked to rank their corporate objective. Six options were given, based on the

Table 1 Descriptive statistics for corporate objective

Corporate objective	Mean	Rank
Profit	2.86	2
Growth	1.63	1
Survival	3.70	4
Maintain competitive advantage	3.20	3
Global concern	5.23	6
Social responsibility	4.37	5

Table 2 Cross-tabulation profit versus global concern

	Global Concern					
	2	3	4	5	6	Total
Profit	0	0	1	3	9	13
	0	0	0	5	22	27
	3	0	0	12	11	26
	0	0	0	0	3	3
	0	8	0	4	0	12
	3	8	1	24	45	81

responses received from respondents regarding the corporate objective growth is the most important corporate objective. However, global concern and social responsibility are having rank 5 and 6, respectively, and coming out as the least important corporate objective, which is an alarming situation. Descriptive statistics for corporate objectives are presented in Table 1.

Cross-Tabulation Analysis Within Corporate Objective From the cross tabulation shown in Table 2, it is observed that 70% respondents who have given growth as rank 1 have also given global concern as rank 5 or 6. Similar kinds of results were found between growth versus social responsibility, profit versus social responsibility and profit versus global concern. This indicates that respondents who have growth or profit as primary objective are not much concerned about global concern or social responsibility or one can easily assume that manufacturers consider global concern or social responsibility as a hindrance in making profit and growth. Table 2 is presenting cross-tabulation data between profit and global concern.

Priorities of Respondents Priorities are internal objective which help the organisation to achieve its overall objective. To further investigate about the working of

Table 3 Descriptive statistics for priorities

Priorities	Minimum	Maximum	Mean	Rank
Cost	1	9	3.81	2
Flexibility	2	10	6.38	8
Environmental consciousness	1	10	6.67	9
Quality and reliability	1	4	2.23	1
Innovation	1	10	5.47	5
Delivery	1	10	4.78	3
Morale	1	10	6.16	7
Customer	2	9	5.78	6
Productivity	1	9	5.19	4
Sustainability	5	10	8.53	10

garment factories, they were asked to rank their priorities. Total 10 priorities were given to each respondent and asked them to rank from 1 to 10. Responses are tabulated and presented in Table 3. It reflects that matters related to quality, cost, etc. are the priorities of garment manufacturers as maximum respondent has awarded them rank 1 or 2. Issues like sustainability and environmental consciousness are not reported as priorities. This reconfirms the results of previous question regarding corporate objective. This outcome further strengthens the point that manufacturers are of the view that green manufacturing is not a cost-effective phenomenon, and they will not able to compete in business with green manufacturing

Level of Concern on Green Manufacturing Issues To investigate the attitude of garment manufacturers of Delhi/NCR about green manufacturing, level of concern on nine issues were explored and were analyzed through cross-tabulation and descriptive analysis. Respondents had to rate any issue on a scale of 1–5 where 1 states not concerned and 5 is highly concerned. Concern level of respondents for each issue is presented in Table 4.

Table 4 Level of concern on different issues related to green manufacturing

S. No.	Issue related to green manufacturing	Mean value	Concern level
1	Level of concern on climate change	4.2	Concerned
2	Level of concern on global warming	4.1	
3	Level of concern on air pollution	4.3	
4	Level of concern on water pollution	3.6	
5	Level of concern on industrial pollution	3.7	
6	Level of concern on ozone hole	2.3	Partially concerned
7	Level of concern on soil pollution	2.9	
8	Level of concern on carbon footprint	2.8	
9	Level of concern on carbon credits	1.9	Not concerned

It is evident that many of the manufacturers are having their concerns about the issues though there was a mixed response towards ozone hole, soil pollution and carbon footprint because of limited awareness.

Attitude Towards General Statements To investigate the awareness and attitude of garment manufacturers of Delhi/NCR towards green manufacturing, they were asked to rate their agreement level on 12 general statements. Table 5 explains their agreement level on general statements.

As per the responses presented in Table 5, it is evident that respondents are aware of the phenomenon of green manufacturing but are not aware of India's role in it. They are also in disagreement with the statement that "garment industry is a significant contributor towards green manufacturing".

They have no idea about the possibilities of using green manufacturing in apparel industry and how they can be moving towards green with their existing setup. They have an opinion that it is not cost effective to invest in green manufacturing setup as it will not give them much monetary benefits. They also think that garment industry is very complicated to implement green manufacturing.

Table 5 Agreement level on general statement about green manufacturing

S. No.	Statement	Mean	Mode	Consolidated level of agreement
1	Global warming is a major environmental issue that needs a critical attention	3.8	4	Towards agreement
2	Greenhouse gas emission is a major reason for causing global warming	3.9	4	
3	It is not cost effective to make a green apparel product	3.9	4	
4	Garment industry is too complicated to think about green manufacturing	3.9	4	
5	Indian apparel industries are significant contributor of greenhouse gas emission	2.6	2	Towards disagreement
6	It is worth investing in green manufacturing in terms of profit	2.6	2	
7	Garments with lesser carbon footprint will give you an edge on your competitors	2.7	2	
8	India is a major contributor of greenhouse gas emission	3.1	3	No idea
9	In coming time, sustainability will be the requirement to survive in garment industry	3.1	3	
10	I do not think it is possible to convert existing setup into green setup	3.1	3	
11	It is very much possible to calculate energy input per garment	3.1	3	
12	Sustainable consumption or reuse of waste is possible in garment industry	3.1	3	

Green Manufacturing Advantages To investigate the awareness of respondents regarding green manufacturing benefits and to further investigate and reconfirm the attitude of manufacturers towards green manufacturing, respondents were asked to record their level of agreement on eight possible benefits of green manufacturing; responses are presented in Table 6.

As per the responses received, it can be inferred that green manufacturing may be an effective tool in brand building in the long term but it is not cost effective; they are of the view that one might be innovative and popular while using green manufacturing, but they are not sure of any other benefit about green manufacturing".

Table 6 Possible benefits of green manufacturing

S. No.	Benefit	Mean	Mode	Consolidated label of agreement
1	Improved company or brand image	4.14	4	Towards agreement
2	Competitive advantage	3.71	4	
3	Employee satisfaction, morale and retention	3.56	3	
4	Product, service and market innovation	4.2	4	
5	Cost saving	2.7	3	Towards disagreement
6	New source of revenue or cash flow	3.2	3	No idea
7	Effective risk management	3.1	3	
8	Enhanced stakeholder relations	3.25	3	

5 Findings

- Given that the Indian garment industry is run by entrepreneurs who operate on order-to-order basis, cost and profit turn out to be the primary objective, and an overwhelming 64% put this factor at the top of the list. This negates the thought of social responsibility and global concern. It therefore comes as no surprise that 70% of the respondents put these factors of global concern and social responsibility at the bottom of the preference list of green manufacturing.
- The findings also confirmed the hypothetical understanding that green manufacturing would mean more investment, and 67% of the respondents did not see green manufacturing as a value for money proposition and observed this to be a non-effective cost factor while 85% of the respondents saw the green manufacturing initiative as erosion to the market/financial competitiveness of the organisation. Along the same lines, as a re-affirmation, it was found that 71% of the respondents from the garment manufacturing companies in Delhi/NCR region viewed environmental consciousness and sustainability as the least among priorities and ranked them 9th and 10th in the list, respectively.
- While the economic factors of the organisation drove the responses, it was also the researcher's interest to understand the attitude of the respondents towards green manufacturing. It was conveniently assumed that a positive attitude could be the first step towards future change.
- Against 12 green manufacturing statements for the respondents to rate in agreement or disagreement, 68% of the respondents in Delhi and NCR region are in agreement that global warming is an environment issue and needs attention. Their knowledge on global warming, GHG emission and green manufacturing was no suspect as 60% of the respondents did agree to knowing its effects on environment. While knowledge on the subject matter is widespread, 86% of the respondents were at

a loss on their roles and responsibilities and any controlling mechanism to ensure green manufacturing.

6 Conclusion

Apparel manufacturers of Delhi/NCR are partially aware of global warming and green manufacturing. Environmental consciousness and sustainability are not in the priority of Apparel manufacturers in Delhi/NCR. Maximum respondents are not aware of their role as a manufacturer in global warming so do not feel the need of green manufacturing in apparel industry. They also think that green manufacturing is not cost effective, and they may lose their financial competitiveness because of green manufacturing.

References

1. Douglas, J.: Building adaptation. J. Clean. Prod., 99–110 (2006) (Butterworth-Heinemann)
2. International Panel of climate control (IPCC) Expert Meeting on the Sciences of Alternative Metrics, Oslo, Norway (2009)
3. Climate Change and Resource Sustainability: An Overview of Actuaries. Canadian Institute of Actuaries (2015)
4. Climate Change Report, Worls Research Institute (WRI), Zeneva (2015)
5. Mehta, M.V.: Emerging Issues in Apparel Trade. Apparel Export Promotion Council, Delhi (2014)
6. Das, M.: Performance measurement of green manufacturing criteria of Indian SME's. Int. J. Eng. Res. Technol., 2913–2920 (2013)
7. Venkatesh, V.G.: Factor influencing sucessfull implementation of green manfacturing. Res. Gate, 1–16 (2015)
8. Adner, R.: Evaluation of sustainable development in manufacturing industries. J. Clean. Prod. (2006)
9. Bordoloi, S.A.: Design for control: a new perspective. Int. J. Prod., 346–358 (2008)
10. Gerrard, J.A.: Is European end-of-life vehicle legislation living up to expectation? Assesing the impact of the ELV Directive on green innivation and vehicle recovery. J. Clean. Prod.,17–27 (2007)
11. Tan, X.C.: A decision-making framework model of cutting tool selection for green manufacturing and its application. J. Adv. Manuf. Syst., 257–260 (2008)
12. Deif, A.M.: A system model for green manufacturing. J. Clean. Prod. 1553–1559 (2011)
13. Hoffman, A.: Competitive Environmental Strategy, pp. 1564–1578. Island Press (2000)
14. R. Shirley, Jones, C., Kanmen, D.: A household carbon footprint calculator for islands: case study of the United States Virgin Islands. Ecol. Econ., **80**, 8–14 (2014)
15. Juan Cagiao, B.G.: Calculation of the Corporate Carbon Footprint of the Cement industry (2011)
16. Shailee, G., Acharya, D.J.: A review on evaluating green manufacturing for sustainable development in foundry industries. Int. J. Emerg. Technol. Adv. Eng. (2014)
17. Kannan, G.: Analyzing supplier development criteria for an automobile industry. Ind. Manage. Data Syst., 43–62 (2010)
18. Eryuruk, S.H.: Greening of the Textile and Clothing Industry. Fibres Text. East. Europe (2012)

19. Wu, G.-C., Ding, J.-H., Chen, P.-S.: The effects of GSCM drivers and institutional pressures on GSCM practices in Taiwan's textile and apparel industry. Int. J. Prod. Econ., 618–636 (2014)
20. Baskaran, V.: Indian textile suppliers'sustainability evaluation using the grey approach. Int. J. Prod. Econ., 647–658 (2014)
21. Caniato, F.: Environmental sustainability in fashion supply chains: an exploratory case based research. Int. J. Prod. Econ., 659–670 (2015)
22. Ministry of Textiles: Annual Report-2016. Ministry of Textiles, New Delhi (2016)
23. Ramachandra, T.V, Shwetmala, K.: Assessment of Carbon Footprint in Different Industrial Sectors. Springer (2015)
24. Sharma, D.M.: Comprehensive Study on Green House Gases (GHGs) in. IIT Kanpur, Kanpur (2016)
25. Devaraja, D.: Indian Textile and Garment Industry. Bangalore. Indian Council of Social Science- Ministry of HRD, Govt. of India (2012)
26. Chattopadhyay, S.: Efficiency of Indian garment manufacturing units in the post MFA period. J. Ind. Stat., 58–75 (2015)

Sustainable Production by Modifying Reduction Clearing in Polyester Dyeing

S. D. Kiruthika, R. Ugamoorthi, C. Venkatachalapathi and S. Ramarethinam

Abstract Conventional reduction clearing of dyed polyester fabric is carried out by using the sodium dithionate under strong alkali conditions. Technical issue is the sensitivity of sodium hydrosulfite to air oxidation in an alkaline medium at high temperature, so that an excess is used to compensate for the loss. In our present study, non-ionic ethoxylated surfactants, as separate and in combination, with different HLB values have been used for wash-off process. Fatty alcohol ethoxylated surfactant and castor oil ethoxylated surfactant with different HLB 9–17 as separate compound and in combination has been used for wash-off process at different pH range (4.5–12 pH), time (5–20 min) and temperature (50–100 °C). Our findings indicate that surfactant can be efficiently used in neutral pH to replace the sodium-dithionate-based reduction-clearing process in the polyester dyeing. Thereby, eliminates the pH adjustment after dyeing from acidic to strongly alkaline condition. Surfactant blends with all HLB value give better results than individual surfactants. Our study results confirmed better color fastness properties and delta E of the fabric <1 with surfactant wash-off process. Also, the effluent load in terms of TDS (96%) and BOD and COD (30–40%) has been reduced. And the use of caustic lye to adjust pH and sodium dithionate for reduction-clearing process is completely eliminated. Therefore, wash-off treatment of polyester by surfactants emphasizes the substantial energy and chemical savings as well as lower environmental impact than traditional process.

Keywords Reduction clearing · Ethoxylated surfactant · Polyester fabric

1 Introduction

Limited water solubility and tendency for particles in the dispersion to aggregate during dyeing, some residual dye commonly remains on the fiber surface after the dyeing phase. The surface deposits may have an adverse effect on the color and

S. D. Kiruthika · R. Ugamoorthi (✉) · C. Venkatachalapathi · S. Ramarethinam
Eastman Export Global Clothing Private Limited, Erode 638005, India
e-mail: ugamoorthi@eastmanexports.com

A. Majumdar et al. (eds.), *Functional Textiles and Clothing*,
https://doi.org/10.1007/978-981-13-7721-1_30

fastness properties of the dyed fabric, and thus an after-treatment process to remove them is commonly introduced. Normally, for reduction-clearing process, sodium hydrosulfite, sodium hydroxide, and a surfactant are employed at 80 °C for 30 min. In this method, the use of sodium hydrosulfite has certain environmental disadvantages. Textile wastewaters are characterized by extreme fluctuations in many parameters such as chemical oxygen demand (COD), biochemical oxygen demand (BOD), pH, color, and salinity. The composition of the wastewater will depend on the different organic-based compounds, chemicals, and dyes used in the dry- and wet-processing steps [2, 3]. Sodium hydrosulfite will also contribute to effluent load in terms of COD, BOD, and TDS. In addition, after treatment requires pH adjustment from the acidic condition during dyeing to the strongly alkaline-clearing condition, followed by final neutralization. Another technical issue is the sensitivity of sodium hydrosulfite to air oxidation in an alkaline medium at high temperature, so that an excess is used to compensate for the loss. Sulfur containing degradation products (sulfites and sulfates) from sodium hydrosulfite with potentially toxic effects are corrosive, which cause severe damage in waste lines [4]. Few by-products formed like Na_2S and NaHS, pollute air through the formation of H_2S [5].

Use of conventional reduction-clearing agent, there are environmental disadvantages inherent in the process [4]. As a consequence, there is an interest in alternatives to traditional reduction clearing. The results of Mausner et al. research and testing confirm that the primary and secondary alcohol ethoxylates, the alkyl alkanolamides, and the alkylamine oxides are all highly biodegradable [6]. By considering the advantages of surfactant, our study focused on replacing sodium hydrosulfite in reduction-clearing process with different surfactants.

2 Materials and Methods

2.1 Materials

Fabric: 100% polyester interlock fabric made up of 100 DNR polyester with 170 g/m^2 has been selected for our study. This fabric is selected based on the maximum production in our industry.

Dyes, Auxiliaries, and Surfactants: Table 1 indicates the list of dyes and auxiliaries used for dyeing and Table 2 indicates the list of surfactants considered for study.

In addition to the above, acetic acid, soda, and caustic lye are used for pH adjustment and sodium hydrosulfite is used for reduction clearing (control). In addition, wetting agent, pH buffer, and dispersing agent are used. These dyes and auxiliaries were selected by considering the usage in our industry.

Table 1 Dyes and auxiliaries used for dyeing

Product name	Usage
C.I.Disperse blue 79	Disperse dye
C.I.Disperse orange 44	Disperse dye
C.I.Disperse red 167	Disperse dye
C.I.Disperse red 311	Disperse dye
C.I.Disperse yellow 126	Disperse dye

Table 2 Non-ionic surfactants and its properties

Product name	Contents	CAS No.	Mol. formula	Physical form	HLB
C.O 8.8	Castor oil ethoxylate	61791-12-6	$C_{57}H_{10}O_9(CH_2CH_2O)_n$	Yellowish liquid	8.8
F.A 17.8	Fatty alcohol ethoxylate	68439-50-9	$C_{14}H_{30}O$	White waxy solid	17.8
F.A 11.5	Fatty alcohol ethoxylate	68439-50-9	$C_{14}H_{30}O$	Clear and transparent liquid	11.5
F.A 15.6	Fatty alcohol ethoxylate	68439-50-9	$C_{14}H_{30}O$	Clear and transparent liquid	15.6
LCSFG 8.5	Lauryl alcohol and Cetosteryl alcohol ethoxylate, Ethoxlyated fatty glycerides	9002-92-0, 9004-95-9, 61791-2-6		White to off white cream	8.5

2.2 *Methods*

Pre-treatment and Dyeing: Pre-treatment is carried out with wetting agent and soda at 90 °C for 30 min. Dyeing program is done as shown in Fig. 1. In this method, dyeing was performed by raising the dye bath temperature from 40 to 90 °C with gradient of 2.5 °C/min, then raising from 90 to 130 °C with the gradient of 1.5 °C/min, holding this temperature for 60 min, and cooling to 70 °C with the gradient of 4 °C/min.

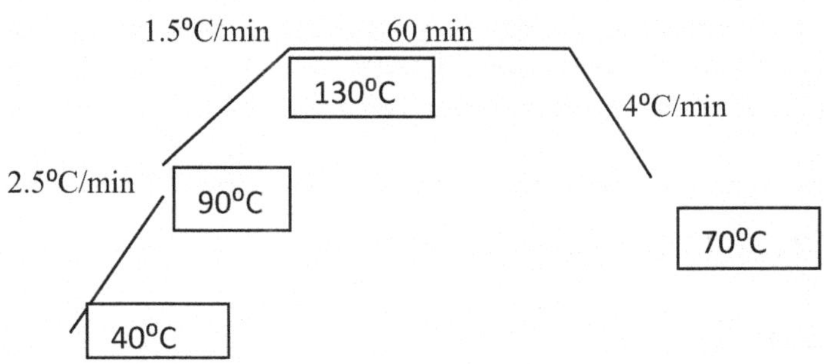

Fig. 1 Dyeing program

Table 3 Surfactant combination

S. No.	Surfactant A	A (%)	Surfactant B	B (%)	Resultant HLB	Solvent (10 ml)
1	F.A 15.6	17.64	C.O 8.8	82.36	10	H_2O
2	F.A 17.8	23.8	F.A 11.5	76.19	13	H_2O
3	F.A 17.8	79.36	F.A 11.5	20.63	16.5	H_2O
4	F.A 15.6	24.39	F.A 11.5	75.61	12.5	H_2O
5	F.A 15.6	47.1	C.O 8.8	52.9	12	H_2O
6	F.A 15.6	23.5	C.O 8.8	76.4	14	H_2O
7	F.A 17.8	55.5	F.A 11.5	44.4	15	H_2O
8	F.A 17.8	48	F.A 11.5	52	13.1	H_2O
9	F.A 17.8	6	C.O 8.8	94	9.3	H_2O

After treatment: For control 2 RC-Normal reduction clearing process with sodium hydrosulfite and caustic lye at 80 °C for 20 min @ 12 pH is done. With individual surfactants listed in Table 2, 2RC process has been done at 80 °C for 20 min @ 4 pH and 7 pH. Surfactant combinations listed in Table 3 is used to treat the dyed sample at 60, 70 and 80 °C with the time interval of 10, 15, 20, and 25 min.

Drying: The samples were then dried at 120 °C.

Testing: 0.1% methanol extraction of surfactant is analyzed in GC-MS using ISO 14369 method. Cloud point is determined for non-ionic surfactants as per ASTM D 2024-65 procedure. Color fastness to washing-ISO 105-C06:2010 (test C2S), color fastness to rubbing-ISO 105-X12:2002, color fastness to water-ISO 105 E01 and color fastness to acidic and alkaline perspiration-ISO 105-E04:2009. Color measurements are carried out using Datacolor spectrophotometer for illuminant D65 and the 10° observer DL*, Da*, Db*, DC*, DH* and DE with metamerism values were obtained. Absorbance for each water sample after reduction clearing is determined under visible range of spectrophotometer and maximum wavelength and absorbance were noted. Total dissolved solids were measured using conductivity methods and results are in ppm. Photometric measurement of chemical oxygen demand carried out using USEPA APPROVED method. USEPA-labeled 5210B standard test method is used for the determination of biological oxygen demand.

3 Results and Discussion

3.1 GC-MS Analysis

Five different surfactants were analyzed in GC-MS as per the required surfactant analysis techniques [7] (Table 4).

Table 4 GC-MS mass spectrum

C.O 8.8	
Mass spectrum	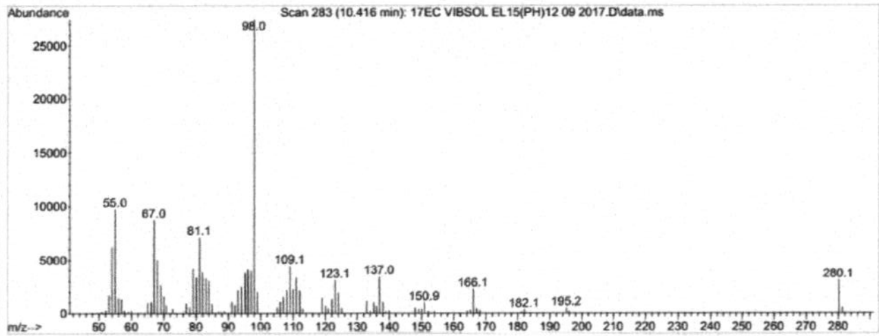
F.A 11.5	
Mass spectrum	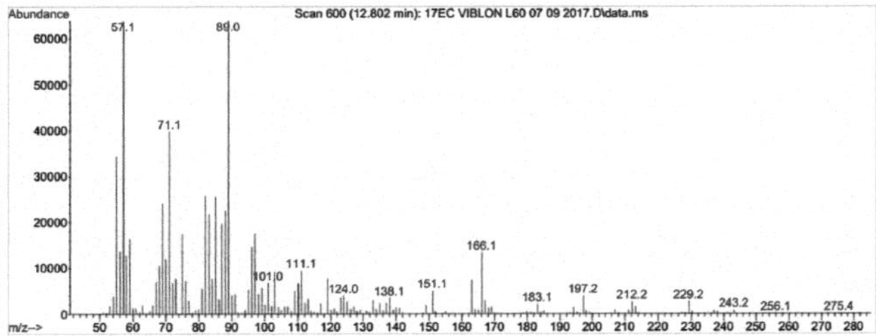
F.A 15.6	

(continued)

Table 4 (continued)

C.O 8.8	
Mass spectrum	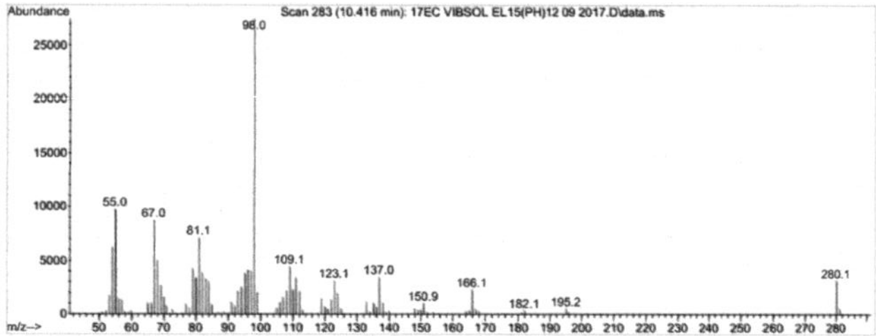
Mass spectrum	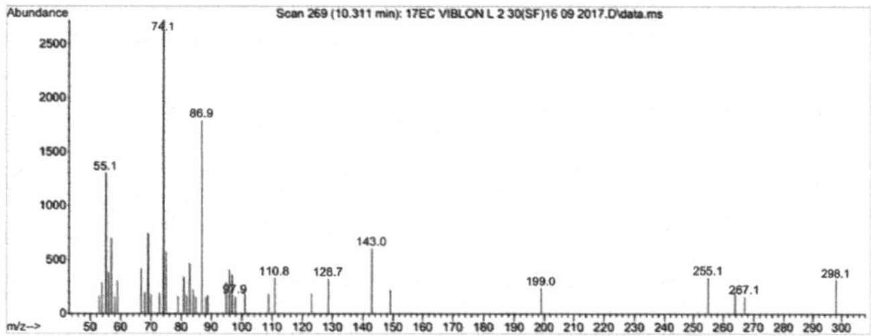
F.A 17.8	

(continued)

Table 4 (continued)

C.O 8.8	
Mass spectrum	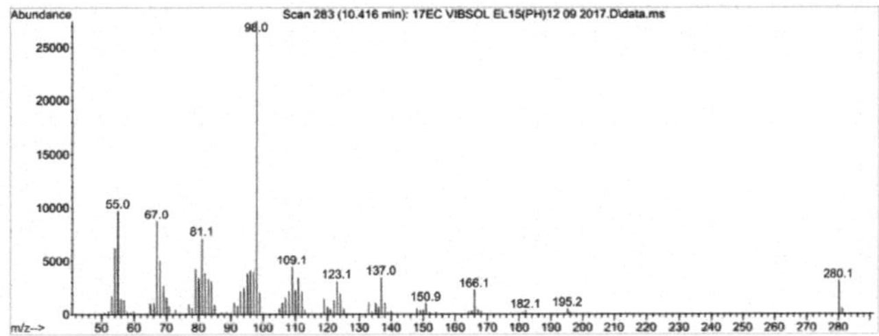
Mass spectrum	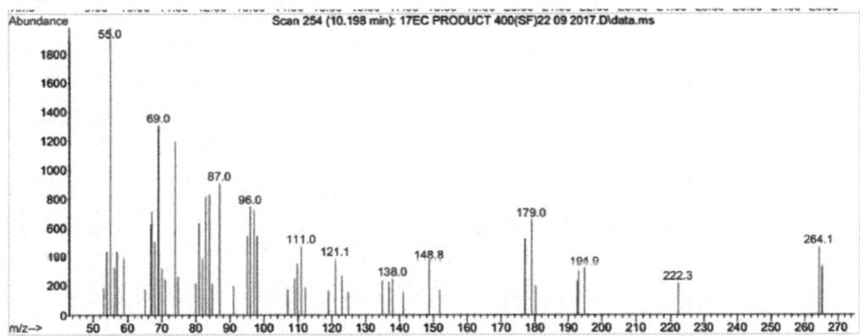
LCSFG 8.5	

(continued)

Table 4 (continued)

C.O 8.8	
Mass spectrum	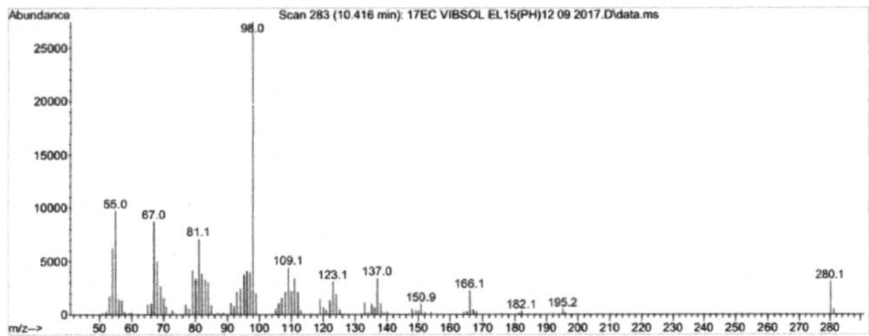
Mass spectrum	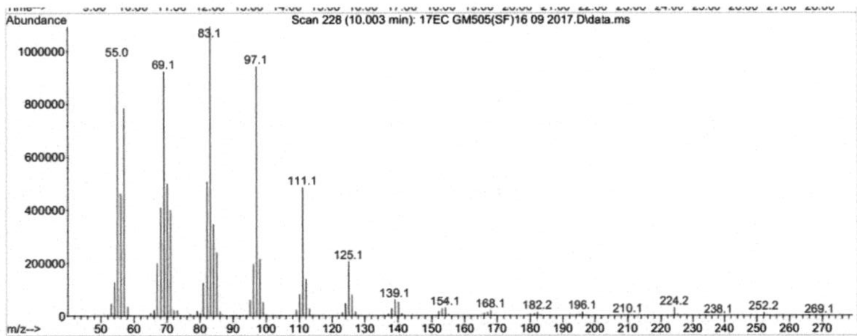

Table 5 GC-MS analysis of surfactants

Surfactant name	Molecular weight	CAS No.	Chemical name
C.O 8.8	280	127062-51-5	13-Hexyloxacyclotridec-10-en-2-one
F.A 11.5	314	2136-72-3	Ethanol, 2-(octadecyloxy)-
F.A 15.6	298	112-61-8	Octadecanoic acid, methyl ester
F.A 17.8	296	112-62-9	9-octadeconoic acid (2)-methyl ester
LCSFG 8.5	240	64437-47-4	Hexadecen-1-ol, trans-9-

Table 6 Cloud point of individual surfactant

S. No.	Auxiliary	pH	Turbidity NTU	Cloud point in temperature	
				Trial 1	Trial 2
1	C.O 8.8	7.06	246	72	72
2	F.A 11.5	6.57	1.8	72	72
3	F.A 15.6	3.92	2.2	140	140
4	F.A 17.8	6.7	0.5	140	140

From the obtained GC-MS results, the largest peak determines the following compounds in the surfactant considered for our study (Table 5).

Analysis again proves the chemical composition of the surfactant considered for our study. It confirms that F.A 11.5, F.A 15.6, and F.A 17.8 are fatty-alcohol-based ethoxylated surfactants; C.O 8.8 as castor-oil-based ethoxylated surfactant and LCSFG 8.5 as lauryl-alcohol- and cetosteryl-alcohol-based ethoxylated surfactants. The molecular weight of all the surfactants is in the range of 240–315.

3.2 *Cloud Point*

Changes in test solution monitored at each 10 °C temperature change up to 150 °C. Test results tabulated as Tables 6 and 7.

C.O 8.8 surfactant becomes a white turbid solution in preparation stage. So, its turbidity is in higher range. F.A 11.5 reaches the cloudy white form from clear solution at 70 °C and dark cloud at 72 °C. No changes in F.A 17.8 and F.A 15.6, even at 140 °C, it is a stable as clear liquid

Cloud point determines that these surfactants and its combinations can be treated at high temperature. But C.O 8.8 and F.A 11.5 as an individual surfactant can be treated only up to 70 °C.

Table 7 Cloud point of combination of surfactant

S. No.	Surfactant	pH	Turbidity NTU	Cloud point in temperature	
				Trial 1	Trial 2
1	C.O 8.8	6.35	249	128	128
	F.A 15.6				
2	F.A 11.5	6.20	9	105	105
	F.A 17.8				
3	F.A 11.5	6.02	9	150	150
	F.A 17.8				

3.3 Application Test

3.3.1 pH Optimization

All the five samples were treated at 80 °C for 20 min at pH 4.5 and pH 7. Tables 8 and 9 show the color fastness test results of the treated samples. Tables 10 and 11 show the absorbance results of the treated water at 4.5 pH and 7 pH. Spectral measurements of the sample are tabulated in Tables 12 and 13 pH and TDS of the treated water sample is tabulated in Tables 14 and 15.

From the observed CF results, it is understood that surfactant works better in 7pH than in 4.5 pH. So, all the surfactant works better in neutral pH, thereby it is not necessary to carry out pH adjustments during after-treatment process. Comparatively, C.O 8.8 and F.A 11.5 give better results than other surfactants.

From the observed absorbance value, it is understood that reduction is done while using sodium hydrosulfite, thereby reducing the color intensity in the wash-off water. Whereas, the color intensity of surfactant-based wash-off water shows that unfixed dye molecules are removed based on surfactant chemistry. Also, the results prove that unfixed dye removal using surfactant at 7 pH is higher than at 4.5 pH.

Spectral measurement, especially, delta E value shows better color strength and required color depth at the neutral pH. Surfactant treated at 4.5 pH affects the color depth and strength.

TDS of all the surfactant is less than 50, whereas the TDS of the traditional process is higher than 1000. Thereby, by utilizing—for the replacement of Alkali medium sodium dithionate from reduction clearing process by surfactant in after treatment process reduces the TDS upto 96%.

3.3.2 Time and Temperature Optimization

Based on the obtained results of individual surfactants, which show better result at 7 pH, the combination of surfactant trial was done at 7 pH and the obtained results were tabulated in Table 16.

Table 8 Color fastness results of fabric RC treated at 4.5 pH

S. No.	Chemical used	Conc. (gpl)		Wash	Water	Alkali	Acid
1.	C.O 8.8	2	Acetate	4	4	4	4
			Cotton	4	4	4	4
			Nylon	4	4	4	3.5
			Polyester	4	4	4	4
			Acrylic	4	4	4	4
			Wool	4	4	4	4
			Color change	4	4	4	4
2	F.A 17.8	2	Acetate	4	3.5	3.5	3.5
			Cotton	4	4	4	4
			Nylon	4	3.5	3.5	3.5
			Polyester	4	4	4	4
			Acrylic	4	4	4	4
			Wool	4	4	4	4
			Color change	4	4	4	4
3.	F.A 15.6	2	Acetate	4	3.5	3.5	3.5
			Cotton	4	4	4	4
			Nylon	3.5	3.5	3.5	3.5
			Polyester	4	4	4	4
			Acrylic	4	4	4	4
			Wool	4	4	4	4
			Color change	4	4	4	4
4.	F.A 11.5	2	Acetate	4	4	4	4
			Cotton	4	4	4	4
			Nylon	4	4	4	4
			Polyester	4	4	4	4
			Acrylic	4	4	4	4
			Wool	4	4	4	4
			Color change	4	4	4	4
5.	LCSFG 8.5	2	Acetate	4	4	4	4
			Cotton	4	4	4	4
			Nylon	4	4	4	4
			Polyester	4	4	4	4
			Acrylic	4	4	4	4
			Wool	4	4	4	4
			Color change	4	4	4	4

(continued)

Table 8 (continued)

S. No.	Chemical used	Conc. (gpl)		Wash	Water	Alkali	Acid
6.	Hydros + NaOH	2	Acetate	4	4	4	4
			Cotton	4	4	4	4
			Nylon	4	4	4	4
			Polyester	4	4	4	4
			Acrylic	4	4	4	4
			Wool	4	4	4	4
			Color change	4	4	4	4

From the observed color fastness results, it is understood that, compared to the individual surfactants, the combination of surfactants gives better results.

Further, the surfactants treated at different times (10, 15, 20, 25 min) and temperatures (60, 70 and 80 °C) are tabulated below. Since 12.5 and 13 resultant HLB is close by, for further temperature and time optimization trials, only 13 HLB is considered (Table 17).

All the observed results show similarity with the traditional reduction-clearing process. The change in spectral values is also in the acceptable range (Table 18).

Observed results were similar to the results observed with 80 °C. Thereby, further reduction in temperature can be done. Also, the results were better even with the treatment time of 10 min. The difference in delta E value and metamerism found to be similar to the traditional sodium hydrosulfite reduction-clearing process (Table 19).

From the observed results, the combination of surfactants works better even at 60 °C with the treatment time of 10 min.

3.3.3 Ecological Considerations

While considering the ecological parameters, all the surfactants are readily biodegradable >70%. EC50 (72 h) and LC50 (48 h) are greater than 0.8 mg/L in fatty-alcohol-based surfactants (F.A 11.5, 15.6, and 17.8). Whereas, castor-oil-based surfactant (C.O 8.8) and lauryl-alcohol- and cetosteryl-alcohol-based surfactants (LCSEG 8.5) does not have any environment-related toxicity. While considering the BOD and COD, it is reduced up to 30% and 55%, respectively, while using castor-oil-based surfactant (C.O 8.8) and lauryl-alcohol- and cetosteryl-alcohol-based surfactants (LCSEG 8.5).

Table 9 Color fastness results of fabric RC treated at 7 pH

S. No.	Chemical used	Conc. (gpl)		Wash	Water	Alkali	Acid
7	C.O 8.8	2	Acetate	4	4	4	4
			Cotton	4	4	4	4
			Nylon	4	4	4	4
			Polyester	4	4	4	4
			Acrylic	4	4	4	4
			Wool	4	4	4	4
			Color change	4	4	4	4
8	F.A 17.8	2	Acetate	4	4	4	4
			Cotton	4	4	4	4
			Nylon	4	4	4	4
			Polyester	4	4	4	4
			Acrylic	4	4	4	4
			Wool	4	4	4	4
			Color change	4	4	4	4
9	F.A 15.6	2	Acetate	4	4	4	4
			Cotton	4	4	4	4
			Nylon	4	4	4	4
			Polyester	4	4	4	4
			Acrylic	4	4	4	4
			Wool	4	4	4	4
			Color change	4	4	4	4
10	F.A 11.5	2	Acetate	4	4	4	4
			Cotton	4	4	4	4
			Nylon	4	4	4	4
			Polyester	4	4	4	4
			Acrylic	4	4	4	4
			Wool	4	4	4	4
			Color change	4	4	4	4
11	LCSFG 8.5	2	Acetate	4	4	4	4
			Cotton	4	4	4	4
			Nylon	4	4	4	4
			Polyester	4	4	4	4
			Acrylic	4	4	4	4
			Wool	4	4	4	4
			Color change	4	4	4	4

(continued)

Table 9 (continued)

S. No.	Chemical used	Conc. (gpl)		Wash	Water	Alkali	Acid
12	Hydros + NaOH	2	Acetate	4	4	4	4
			Cotton	4	4	4	4
			Nylon	4	4	4	4
			Polyester	4	4	4	4
			Acrylic	4	4	4	4
			Wool	4	4	4	4
			Color change	4	4	4	4

Table 10 Absorbance results of the RC-treated water at 4.5 pH

S. No.	Product name	RC 1 @ pH 4.5		RC 2 @ pH 4.5	
		Absorbance	Wavelength (nm)	Absorbance	Wavelength (nm)
1	C.O 8.8	1.159	440.2	0.467	381.7
2	F.A 17.8	1.818	382.8	0.138	480.4
3	F.A 15.6	1.007	381.6	0.182	480.3
4	F.A 11.5	1.734	426.8	0.552	480.3
5	LCSFG 8.5	0.971	628.5	0.557	622.4
6	Hydros + NaOH	0.063	628.6	−0.088	622.3

Table 11 Absorbance results of the RC-treated water at 7 pH

S. No.	Product name	Trial 1 @ pH 7		Trial 2 @ pH 7	
		Absorbance	Wavelength (nm)	Absorbance	Wavelength (nm)
7	C.O 8.8	1.649	453.7	0.604	553.9
8	F.A 17.8	1.970	454.2	0.201	464.2
9	F.A 15.6	1.510	467.9	0.177	469.4
10	F.A 11.5	2.366	479.6	0.477	469.6
11	LCSFG 8.5	1.376	628.5	0.265	622.4
12	Hydros + NaOH	0.022	640.6	0.020	658.5

Table 12 Spectral measurement of the fabric treated at 4.5 pH

Samples	DL*	Da*	Db*	DC*	DH*	ΔE	Strength (%)
C.O 8.8	0.79	0.06	0.61	−0.47	0.39	1.00	92.77
F.A 17.8	0.98	0.06	0.64	−0.50	0.41	1.17	90.53
F.A 15.6	0.69	0.10	0.66	−0.48	0.47	0.96	92.73
F.A 11.5	0.50	0.07	0.54	−0.42	0.36	0.74	94.81
LCSFG 8.5	0.05	0.02	0.07	−0.06	0.05	0.09	97.89
Hydros + NaOH	−0.03	−0.05	0.08	−0.09	−0.00	0.10	99.95

Table 13 Spectral measurement of the fabric treated at 7 pH

Samples	DL*	Da*	Db*	DC*	DH*	ΔE	Strength (%)
C.O 8.8	−0.20	0.15	−0.08	0.14	0.10	0.26	100.30
F.A 17.8	0.01	0.19	−0.06	0.15	0.14	0.20	98.8
F.A 15.6	−0.13	0.15	−0.09	0.15	0.09	0.22	99.95
F.A 11.5	−0.24	0.06	−0.14	0.15	−0.01	0.28	101.37
LCSFG 8.5	−0.04	0.16	−0.02	0.09	0.13	0.16	98.66

Table 14 pH and TDS after treatment—result analysis (at pH = 4.5)

Sample No.	Chemical name	Trial 1		Trial 2	
		pH	TDS	pH	TDS
1	C.O 8.8	5.8	594	5.82	52
2	F.A 17.8	7.21	198	4.89	40
3	F.A 15.6	7.00	188	5.04	57
4	F.A 11.5	7.14	210	6.30	51
5	LCSFG 8.5	8.72	182	7.16	40
6	Hydros + NaOH	12.6	9590	12.6	1004

Table 15 pH and TDS after treatment—result analysis (at pH = 7)

Sample No.	Chemical name	Trial 1		Trial 2	
		pH	TDS	pH	TDS
7	C.O 8.8	8.5	186	7.75	63
8	F.A 17.8	7.7	178	7.98	40
9	F.A 15.6	8.7	258	7.64	42
10	F.A 11.5	9.2	172	7.93	38
11	LCSFG 8.5	8.9	179	6.7	40
12	Hydros + NaOH	12.6	5530	12.6	1027

Table 16 Combination of surfactants—color fastness results

Trial No.	Chemical used	CONC. (GPL)	pH		Wash	Water	Alkali	Acid
65	C.O 8.8–82.36% F.A 15.6–17.64% HLB = 10	2	7	Acetate	4	4.5	4.5	4.5
				Cotton	4	4.5	4.5	4.5
				Nylon	4	4	4	4
				Polyester	4	4.5	4.5	4.5
				Acrylic	4	4.5	4.5	4.5
				Wool	4	4.5	4.5	4.5
				Color change	4	4	4	4
77	F.A 11.5–76.19% F.A 17.8–23.8% HLB-13	2	7	Acetate	4.5	4.5	4.5	4.5
				Cotton	4.5	4.5	4.5	4.5
				Nylon	4.5	4.5	4.5	4.5
				Polyester	4.5	4.5	4.5	4.5
				Acrylic	4.5	4.5	4.5	4.5
				Wool	4.5	4.5	4.5	4.5
				Color change	4	4	4	4
81	F.A 11.5–20.63% F.A 17.8–79.36% HLB-16.5	2	7	Acetate	4.5	4.5	4.5	4.5
				Cotton	4.5	4.5	4.5	4.5
				Nylon	4.5	4.5	4.5	4.5
				Polyester	4.5	4.5	4.5	4.5
				Acrylic	4.5	4.5	4.5	4.5
				Wool	4.5	4.5	4.5	4.5
				Color change	4	4	4	4
82	F.A 11.5–75.61% F.A 15.6–24.39% HLB-12.5	2	7	Acetate	4.5	4.5	4.5	4.5
				Cotton	4.5	4.5	4.5	4.5
				Nylon	4.5	4.5	4.5	4.5
				Polyester	4.5	4.5	4.5	4.5
				Acrylic	4.5	4.5	4.5	4.5
				Wool	4.5	4.5	4.5	4.5
				Color change	4	4	4	4

Table 17 CF and spectro results of samples treated at 80 °C and different time ranges

Auxiliaries	Time in min.	Color fastness						Spectrophotometer color quality			
		Wash	Water	Per. Acid	Per. Alkali	Wet rub	Dry rub	dC	dH	dE	Metamary
Sodium dithionate	10	4	4	4	4	4	4–5	0.34	−0.88	0.95	1.34
	15	4	4	4	4	4	4–5	0.42	−0.86	0.98	1.31
	20	4	4	4	4	4	4–5	0.49	−0.84	0.98	1.3
	25	4	4	4	4	4	4–5	0.45	−0.88	0.99	1.39
F.A 11.5–20.63% F.A 17.8–79.36% HLB-16.5	10	4	4	4	4	4	4–5	0.29	−0.73	0.82	1.33
	15	4	4	4	4	4	4–5	0.47	−0.73	0.87	1.33
	20	4	4	4	4	4	4–5	0.51	−0.71	0.89	1.34
	25	4	4	4	4	4	4–5	0.43	−0.64	0.8	1.32
F.A 11.5–76.19% F.A 17.8–23.8% HLB-13	10	4	4	4	4	4	4–5	0.19	−0.64	0.67	1.28
	15	4	4	4	4	4	4–5	0.29	−0.67	0.82	1.36
	20	4	4	4	4	4	4–5	0.35	−0.68	0.77	1.39
	25	4	4	4	4	4	4–5	0.35	−0.66	0.76	1.32
C.O 8.8-82.36% F.A 15.6–17.64% HLB = 10	10	4	4	4	4	4	4–5	0.24	−0.69	0.89	1.32
	15	4	4	4	4	4	4–5	0.38	−0.7	0.83	1.34
	20	4	4	4	4	4	4–5	0.43	−0.67	0.89	1.39
	25	4	4	4	4	4	4–5	0.39	−0.72	0.79	1.35

Table 18 CF and spectro results of samples treated at 70 °C and different time ranges

Auxiliaries	Time in min.	Color fastness						Spectrophotometer color quality			
		Wash	Water	Per. Acid	Per. Alkali	Wet rub	Dry rub	dC	dH	dE	Metamary
Sodium dithionate	10	4	4	4	4	4	4–5	0.35	−0.73	0.81	1.29
	15	4	4	4	4	4	4–5	0.39	−0.84	0.95	1.33
	20	4	4	4	4	4	4–5	0.44	−0.83	0.95	1.27
	25	4	4	4	4	4	4–5	0.48	−0.73	0.88	1.37
F.A 11.5–20.63% F.A 17.8–79.36% HLB-16.5	10	4	4	4	4	4	4–5	0.18	−0.71	0.84	1.3
	15	4	4	4	4	4	4–5	0.28	−0.7	0.77	1.29
	20	4	4	4	4	4	4–5	0.42	−0.66	0.79	1.31
	25	4	4	4	4	4	4–5	0.13	−0.83	0.84	1.32
F.A 11.5–76.19% F.A 17.8–23.8% HLB-13	10	4	4	4	4	4	4–5	0.35	−0.68	0.77	1.38
	15	4	4	4	4	4	4–5	0.37	−0.66	0.76	1.34
	20	4	4	4	4	4	4–5	0.23	−0.6	0.65	1.34
	25	4	4	4	4	4	4–5	0.24	−0.66	0.73	1.34
C.O 8.8–82.36% F.A 15.6–17.64% HLB = 10	10	4	4	4	4	4	4–5	0.28	−0.63	0.78	1.29
	15	4	4	4	4	4	4–5	0.28	−0.68	0.73	1.31
	20	4	4	4	4	4	4–5	0.31	−0.69	0.72	1.34
	25	4	4	4	4	4	4–5	0.34	−0.61	0.79	1.32

Table 19 CF and spectro results of samples treated at 60 °C and different time ranges

Auxiliaries	Time in minutes	Color fastness						Spectrophotometer color quality			
		Wash	Water	Per. Acid	Per. Alkali	Wet rub	Dry rub	dC	dH	dE	Metamary
Sodium dithionate	10	4	4	4	4	4	4–5	0.36	−0.82	0.9	1.35
	15	4	4	4	4	4	4–5	0.3	−0.81	0.9	1.32
	20	4	4	4	4	4	4–5	0.3	−0.89	0.98	1.3
	25	4	4	4	4	4	4–5	0.37	−0.74	0.84	1.29
F.A 11.5–20.63% F.A 17.8–79.36% HLB-16.5	10	4	4	4	4	4	4–5	0.13	−0.79	0.8	1.26
	15	4	4	4	4	4	4–5	0.26	−0.77	0.81	1.3
	20	4	4	4	4	4	4–5	0.15	−0.75	0.77	1.27
	25	4	4	4	4	4	4–5	0.15	−0.75	0.88	1.29
F.A 11.5–76.19% F.A 17.8–23.8% HLB-13	10	4	4	4	4	4	4–5	0.05	−0.78	0.78	1.35
	15	4	4	4	4	4	4–5	0.23	−0.66	0.7	1.34
	20	4	4	4	4	4	4–5	0.27	−0.65	0.7	1.33
	25	4	4	4	4	4	4–5	0.15	−0.73	0.74	1.38
C.O 8.8–82.36% F.A 15.6–17.64% HLB = 10	10	4	4	4	4	4	4–5	0.1	−0.76	0.79	1.35
	15	4	4	4	4	4	4–5	0.32	−0.72	0.73	1.34
	20	4	4	4	4	4	4–5	0.25	−0.72	0.81	1.34
	25	4	4	4	4	4	4–5	0.19	−0.78	0.76	1.36

4 Conclusion

To efficiently remove the deposited disperse dyes from the fabric surface, oligomer deposits from the polyester fabric; to reduce the COD, BOD, TDS in the effluent water; to reduce the environmental disadvantages and to avoid the generation of aromatic amines [1] caused by the use of dithionate-based wash-off process, our work focused on the usage of surfactants. Also, the castor-oil-based surfactant, lauryl- and cetosteryl-alcohol-based surfactants are compared against the fatty-alcohol-based surfactant and traditional reduction-clearing agent (sodium hydrosulfite). Initially, the surfactant under study has been analyzed in GC-MS and confirmed the presence of castor oil ethoxylated molecule, lauryl alcohol, and cetosteryl alcohol ethoxylates and fatty alcohol ethoxylates. The cloud point for each surfactant has been analyzed to determine the maximum effective temperature of surfactants and found that C.O 8.8 and F.A 11.5 form cloud at 72 °C, thereby limiting its temperature range up to 70 °C. Other surfactants found to be effective up to 140 °C. During application, the performance of castor oil ethoxylated surfactant is not affected at 80 °C, because it may be effectively utilized even before reaching 72 °C.

Individual surfactant application for after treatment at 4.5 and 7 pH was analyzed to freeze the effective pH range. Better color fastness results proved that individual surfactant works better at 7 pH. The absorbance study in the wash-off water proves that the unfixed dyes are removed using surfactant chemistry. There is no reduction or oxidation of dyes, thereby eliminating the formation of aromatic amines during wash-off process. Spectrophotometer analysis shows that, by using the surfactant, the color shade and color strength in the fabric is not affected. Same color with delta E value less than 1 is obtained which is similar to that of traditional reduction-clearing process. By using the surfactant, in general, TDS in the wash-off water is reduced by 96%, thereby reducing the effluent load.

Combination of surfactant at different times (10, 15, 20, 25 min) at different temperature ranges (60, 70 and 80 °C) is carried out at 7 pH. And, it is found that surfactants work better even at 60 °C temperature and in 10 min time. Further studied were on to determine the minimum effective operative time and temperature of the surfactants. Though the bio-degradability of all the surfactants is >70%, ecological toxicity is not applicable for castor oil ethoxylated surfactant.

From all the observed parameters, castor oil ethoxylated surfactant (C.O 8.8) individually and in combination; and lauryl alcohol and cetosteryl alcohol (LCSEG 8.5) combination surfactant is equivalent with the fatty-alcohol-based surfactants (F.A 11.5, 15.6, and 17.8). Finally by considering all the processing parameters, ecological parameters, reduction in effluent load, efficiency, etc., Castor oil ethoxylated surfactant is found to be the best surfactant than fatty alcohol based surfactant. Our study concludes that castor oil ethoxylated surfactant individually and in combination can be used effectively for polyester wash-off process instead of traditional sodium hydrosulfite.

References

1. Burkinshaw, S. M.: The reduction clearing of dyed polyester. Part 1: Colour strength. Dyes Pigm. **26**(3) (2008)
2. Dos Santos, A.B., Cervantes, F.J., van Lier, J.B.: Review paper on current technologies for decolourisation of textile wastewaters: Perspectives for anaerobic biotechnology. Bio-resour. Technol. **98**(12), 2369–2385 (2007)
3. Dos Santos, A.B.: Reductive decolourisation of dyes by thermophilic anaerobic granular sludge. Ph.D. thesis, Wageningen University, The Netherlands (2005)
4. ul Aleem, A., Christie, R. M.: The clearing of dyed polyester. Part 1. A comparison of traditional reduction clearing with treatments using organic reducing agents. Coloration Technology © 2016 Society of Dyers and Colourists, pp. 280–296. https://doi.org/10.1111/cote.12217
5. Chavan, R.B.: Environment friendly dyeing processes for cotton. Indian J. Fibre Text. Res. **26**, 93–100 (2001)
6. Mausner, M., Benedict, J.H., et al.: The status of biodegradability testing of nonionic surfactants. J. Am. Oil Chem. Soc. **46**(9), 432–440 (1969) https://link.springer.com/article/10.1007/BF02544362
7. Schmitt T. M.: Analysis of surfactant. In: Surfactant Science Series, vol. 96

Part VIII
Traditional Textile Arts and Crafts

Crafting Lives: Redefining Culture and Artisan Lives Through the Revival of Crafts in the State of Punjab, India

Simrita Singh and Anu H. Gupta

Abstract Craft is traditionally an important part of the culture connecting to the past and shaping the future. This is particularly true in case of Punjab which has cultural and traditional rich heritage where skill is given a significant place which produces symbols of Punjabi character and identity like *phulkari*, *durrie*, *jutti*, etc. With commoditization and industrialization coming to India after independence, there have been many cultural trends leading to the decline of the crafts of the state. The efforts through government, NGOs and designers intervention have given a new lease of life to the crafts. Hence, an investigation was done to see the organizations which are making a difference clinched alongside restoring of the crafts of Punjab and how artisans are affected with this revival. The main thrust of such efforts has been given to skill and quality up gradation, design development, trainings, raw material, prototype production, trendy colour palettes, etc. Many innovative and unique products are generated out of the existing craft techniques by the two way process of communication, design and product development between designers and artisans. Therefore, handcrafted contemporary products are revitalizing markets increasing the choice for the customers maintaining heritage status in global era. Though many crafts are revived and many are in process of experimentation, many need more efforts to be revived and the welfare and status of artisan needs more attention.

Keywords Craft · Revival · Artisans

1 Tracing the Tradition of Handicraft Skill

Human instinct for decorative art form is primordial to human civilization. History of the culture of art has recorded ritual and social significance of these material

S. Singh
NIIFT, Mohali, Punjab, India
e-mail: singhsimrita@hotmail.com

A. H. Gupta (✉)
UIFT & VD, Panjab University, Chandigarh, India
e-mail: guptaanupu@gmail.com

A. Majumdar et al. (eds.), *Functional Textiles and Clothing*,
https://doi.org/10.1007/978-981-13-7721-1_31

manifestations of human skills and creativity. He wanted to enhance his abilities, and for this, he opted to ornament his surroundings, his body and every object of personal and ritual significance. To celebrate this sense of control, he introduced in his articles the element of design or adornment … this was the arrival of the artisan!

The expressions of man's creativity since times immemorial have led to creation of objects with hands that are commonly addressed as handicrafts. Chattopadhyay [7] attributes the tradition of handicrafts to man's intense desire and craving to embellish a given material or object to make it more decorative. According to The Director, Publication Division, Ministry of Information and Broadcasting, New Delhi [31], "A handicraft is the work of a craftsman who uses manual skill and often traditional technique, his own tools, materials and designs, and depends only on his own labour and what assistance he can get from his family. His independence and individuality naturally find expression in his work". In Indian subcontinent, skill holds significant value, and hence, the craft has been practised for centuries to adorn various important possessions of man which varied from textiles for temples, houses to drapes for cattle to ornamentations on the walls and floors, etc. [9, 15, 24]. The worth and importance of exquisite handcrafted products within a society are measured by the value placed upon them in the community: they are treated with great respect and handed down from generation to generation and becomes a significant part of the culture [13].

The Punjab artistic traditions can be said to be an amalgamation of a number of influences that have embedded their stamp historically and geographically. Beginning from the times of Alexander the Great up to the times of the Great Maharaja Ranjit Singh and even much after his times, the Punjab cultural scene has seen major and subtle incorporations, adaptations and adoptions of peculiar cultural and artistic traits. The harmonious coexistence of artisans of various communities, cultures and religions made a long-lasting impact and influence on the craft of the region. Punjab has rich traditional colourful handicrafts ranging from heavily and richly embroidered hand-woven textiles to captivating woodcraft and inlay, *jutti*s, doll making, carpet and *durries*.

2 Livelihood and Emergence of Commercial Skill

Crafts have always occupied a functional and aesthetic value in Punjab. It was produced for the day-to-day usage: *Phulkari* as a wrap in winters, *durrie* as a bed or floor spread, *jutti* as a footwear, etc. These articles were crafted either by women at home or by the craft persons living in the community. These products occupied an important place a value in the pre-colonial times because these creations were interwoven in the cultural construct of the society, for instance, *durrie* being an important part of the trousseau led women to spin yarn at home and many a times they would further give it to a village weaver for making the rug.

The great and continuous impact of invasions and conquests from all directions, the impact of the ancient trade routes and migrations till recent times like the partition of India and Pakistan in the Indian history have all contributed and created a craft history of vibrant colours and forms coupled with a variety of creative skills and techniques. We know that commercialization has taken place, since the earliest trade routes. As an example, Crafting and embroidering for livelihood have changed the contours of traditional embroidered handicrafts that were done only for love by the family and for the family or community. Vatsyayan [33] in a profound statement reflects: "traditional embroideries have become the only instruments of livelihood and have been displaced from lifecycle, intra community dialogue and intra-regional communication". Abraham [2] established the relationship between the movement of the British into India and the commercialization of the craft. He argues that due to the development of factory system, large scale or mass production was the immediate, prompt and most consequential cause for the decline of Indian handicrafts. In the late 1860s and early 1870s, many craft traditions were in decline [20].

Frater [12] emphasizes that the profound change in the transformation of art into a commodity cannot be underestimated. With industrialization coming to India after independence and change of many cultural trends like veiling, availability of mill made bed sheets, rugs, toys, footwear at a price lesser than the handcrafted village products and their availability in a variety of colours and designs led to the decline of many crafts of Punjab. Even, the artisans are forsaking crafts in search of more lucrative vocations. Many of the craft forms were being confined to the pages of history, as there were no takers. Crafts are revived for two reasons as suggested by Frater and in our view these are interrelated. If crafts are resurrected for reviving the tradition, these also bring back the traditional craftsperson and create renewed opportunities for employment.

In order to revive the handcrafted items, many efforts were done by the government by setting up museums, handloom and handicraft boards. Many schemes were floated so that artisans stick to the crafts. Many craft bazaars were created where these articles can be sold to the consumers. But repeatedly same products, same colour combinations, same style with lesser design variations did not help popularize the craft. Many trainings were started where artisans were given training with a stipend and financial support by various state agencies like DC (Handicrafts), Ministry of Textiles, NABARD[1], KVIC[2], etc., and subsequently a place to sell the crafted products. Due to globalization and changes in social set-ups and lifestyles, enhanced economic independence, incremented cultural influence, quick progression of information and communication technology, there is a demand for innovative products. There is an increase in international trade and trade in between the domestic boundaries of cultural products and services such as movies, music, food and decorative items. Statistics of the handicrafts exports implies that amid the present day of globalization, the local handicraft products of our country have adequate open doors in the local/domestic market as well as at the worldwide front. The exposure to new culture and cultural goods frequently brings about changes in local cultures, traditions and the products produced.

---

[1] National Bank For Agriculture and Rural Development.

[2] Khadi and Village Industries Commission.

Hence, a study was undertaken to see the agencies which are helping in reviving or sustaining or flourishing the craft and how artisans are placed in this scenario and finally how are they affected with the revival. The geographical locale of the crafts was initially identified, and these locales were visited as indicated in Map 1. Hence, data was generated from the artisans involved in different crafts, NGOs who are involved in training artisans or getting work done from these artisans, designers, shopkeepers/vendors who sell the craft. Case studies and in-depth interviews of the stakeholders were taken to see the present status of different crafts of Punjab and how it has impacted the creator's life and the culture. Not only this, various markets or the selling points of these crafts like craft fairs, museums which display and promote craft were also visited.

3 The Popular Crafts of Punjab

3.1 Phulkari *and* Bagh

These embroidered textiles (Fig. 1) are the most vibrant and well-known-age-old craft of Punjab. The Punjabi women are known for embroidery with superb imagination.

Map 1 Field sites in Punjab

This form of textile craft is also synonymously used as a bridal textile forming part of the brides' trousseau. It has its important place in the Punjabi wedding ceremonies.

3.2 *Durrie*

Durrie weaving is also a great handicraft, primarily based in Nakodar, Punjab. It is a flat woven rug used not only as a floor mat or as a carpet but also as a heavy spread on beds or cots. It was a tradition to use a very rough yarn often called as '*baan*' to weave bed or charpoys in Punjab. So covering the rough surface with durries was a common feature. Weaving of durries has been practised in rural Punjab for preparing products for regular usage or for gifts. Bridal durries hold a special place prepared for daughter's trousseau and embellished with motifs inspired from flora, fauna as well as day-to-day items.

3.3 *Jutti*

This is leather footwear (Fig. 2) commonly worn by men and women of Punjab and neighbouring regions which is an internationally renowned craft of Punjab. The exquisite *tilla* work (golden embroidery), beadwork, thread embroidery in geometric and floral patterns enhance the look of the Punjabi *jutti* even further. Globalization and awareness of the ongoing trends have given a boost to the craft.

3.4 *Paranda*

It is a traditional hairdressing accessory (Fig. 3) that belongs to the makeup (*shringaar*) of a Punjabi lady. Most of the Punjabi women has long hair which

Fig. 1 *Phulkari* products

Fig. 2 *Jutti* with different types of ornamentation

Fig. 3 *Paranda*—a traditional hair accessory: woman wearing paranda in her braid (left) and tassels of paranda hanging on a rod

are plaited and in the end braided with *paranda* in vibrant colours and silken tassels hanging.

3.5 Wood Inlay

Wood inlay is the most elaborate and exquisite of the handicrafts belonging to the state of Punjab. The uniqueness of the craft lies in the way white material is inserted into

Fig. 4 Details of inlay work (left) and products of inlay work (right)

designed depressions to create objects of utility and decoration (Fig. 4). Primarily, this craft has its base in Hoshiarpur.

4 The Punjabi Swag

Crafts are an essential part of the character and culture of different societies, regardless of whether regional, national or ethnic and these are associated with a lifestyle [11]. Contemporary Indian designers have endeavoured to incorporate such elements in their creations giving a new look to the traditional craft items and hence made a paramount contribution in linking traditional craft with modern markets, for engendering an incipient vigilance and a new desirability [19] and in generating employment opportunities for craftspersons and remuneration for their skills [18]. Reubens [27] advocated such relationship between the designer and the craftsperson as a collaborative innovation, and Kumar and Dutta [18] addressed it as symbiotic because a designer draws inspiration from the craftsperson and the later innovative inputs and better productivity.

Designer's creation and innovation can give a new direction to the fashion industry by contributing to new trends [6]. Such trendy products with a touch of tradition and local provenance enjoy wider authenticity and become increasingly sought after. Articles with an authentic local association are progressively valued for their irregularity and rarity [21]. Customers would like to possess such craft pieces which have interesting stories about the creators and personal connections behind objects. Designers connect their creations with the hard work of the artisans sitting in the local pockets and hence claim to empower the craftspersons and take the credit of sustaining the crafts.

The hi-fashion garments churned by the designers reflecting the handcrafted techniques not only boost the textiles of the region but also throw opportunities for the local artisan to have more orders through the popularity gained on the ramp. This is what happened when Manish Malhotra, Indian fashion designer, experimented with *Phulkari* and presented it in a contemporary way on saris, anarkalis, kurtas and structured clothing at Wills India Fashion Week, 2013. Not only this but later on in 2017, an exhibition at Philadelphia Museum of Art displayed traditional *phulkari*s and *baghs* from the Jill and Sheldon Bonovitz collection, museum's collection and high fashion ensembles by the same designer's 2013 collection. The background was traditional vintage heritage *phulkari*s with a spectacular display of designer ensembles as the foreground. The message conveyed was that "this age old folk tradition has entered the realm of high fashion through designers" [26]. Many designers based in Punjab are also experimenting with the embroidered craft, giving a contemporary and high-end look with a traditional touch to the wraps being produced.

5 Empowerment and Upliftment of Artisans

With expanded globalization, items are ending up increasingly commoditized, and artisans discover their items rivalling merchandize from everywhere throughout the world. Design intervention is a process that includes designing of new products; redesigning subsisting products, with vicissitudes in shape, size, colour, surface manipulation, function and utility; exploring new markets and reviving lapsed markets; applying conventional and traditional skills to address new opportunities and challenges; and the introduction of new materials, new procedures, new instruments and technologies.

On a commercial front brands like 1469 and Urban Theka are exposing the traditional craft/heritage/art and its milieu to the generation today, which till some decades ago were stacked away in storerooms of most Punjabi households. So the old-fashioned culture of Punjab has made an entry once again in the modern households of Punjabis but with a twist. Such renewed interest in craft with contemporary designs and their making, as revealed by a London-based designer Sebastian Cox, not only helps in retaining skills but also offers designers different ways of working with materials [35].

Harinder Singh of 1469 in an interview with Deepak [10] elaborates his concept of reviving or redefining the culture of Punjab through crafts:

> "I realised that Punjabis were losing pride in their mother tongue, so we decided to make the language fashionable by manufacturing good-quality T-shirts with catchy Punjabi slogans. They were lapped up not just by people from the state but also savvy customers, residing in India as also tourists, looking for alternative clothing. The brand is showcasing its products - traditional and cultural finery with handcrafted *phulkari*, quirky T-shirts, Gurumukhi calligraphy, handmade *chai* cups, *phulkari* based accessories and souvenirs."

Another entrepreneur of Urban Theka elaborates their reason for establishing their brand "idea is to merge Punjab's age-old rural culture with its modernity" [34]. Such initiatives to promote Punjabi culture not only help artisans to gain livelihood but production of such products connect people or the buyers to socio-cultural traditions,

Fig. 5 Crafts fair—a platform for artisan and customer interaction

and cultivates preservation of cultural diversity and identity. As business sectors open up in urban metros inside India and abroad, craft producers can keep selling their products and hence generating livelihoods [36].

6 Steps and Medium for Revival of Crafts in Redefining the Culture of Punjab

The national and international craft fairs and *melas* held by various private and government organisations have been very effective in the exchange of cultural and traditional artefacts.[3] The artisans get a ready platform to showcase their talent and handicraft products to masses at all levels. This exposure surely builds up and boosts the confidence of the artisans and popularizes their craft. Such fairs are further popularized through electronic and social media playing an important role in promoting crafts. Ajanta Ellora Festival, India Art Fair, Kala Ghoda Arts Festival, Surajkund Crafts Mela, Taj Mahotsav, Goa Carnival, Rajasthan International Folk Festival, Special Handicrafts Thematic Exhibitions at Established Malls/Metro Cities and International Craft Exchange Programmes are some of the famous fairs which attract people across the globe. Gandhi Shilp Bazaars, crafts bazaars, exhibitions are being organized in metropolitan cities/state capitals/places of tourist/historic places or commercial interest/other places. This provides a direct marketing platform to the handicrafts artisans from various parts of the country (Fig. 5). According to the annual domestic marketing plan, a total of 64 Gandhi Shilp Bazaars, 53 crafts bazaars and 63 exhibitions are scheduled for 2017–18 [4].

[3]All India handicrafts week is celebrated every year in all states from 8th to 14th December [3].

These crafts fairs are held at national level, regional level and at local level usually in collaborations with organizations like EPCH,[4] CEPC,[5] COHANDS,[6] NCDPD[7] and State Handicrafts Corporations, etc.

The National Handicrafts and Handlooms Museum, popularly known as the Crafts Museum, organizes workshops by master craftsmen/craftswomen where these are invited to demonstrate their skills to the visitors and are able to sell their creations. Special events such as folk craft festival of India organized to advance painstaking work items created by the qualified ace craftsperson and furthermore to make a brand image for Indian handiworks are also a point of exchange between craftspersons and the consumers. Such occasions additionally encourage the artisans to have diverse cross-cultural interaction between the local and global [16] and hence know the demand of the globalized market.

6.1 Research and Development through Academia

According to a report by Crafts Council [8]: "Education and training in crafts is of far-reaching significance: it produces creators of the future, prepares those with crafts skills for the wider creative economy and beyond, and builds up the haptic and imaginative abilities so imperative for all youngsters and their learning".

The fashion and design colleges churn out a substantial number of ready fashion and lifestyle designers. Most fashion school curricula have a craft, traditional arts as a major part of the syllabi which encourage the students to take projects, internships and placements in the handicraft sector. Integrating the curriculum with the practices of the artisan can help regenerate the community and village culture while improving the worth of products [17]. Many of such researchers and design pass-outs are selected as empanelled designers with Ministry of Textiles or they work in crafts sector with NGOs and hence further work with various artisans to develop new designs and products for the market.

Design institutes work in collaboration with the Ministry of Textiles as an outreach programme and many undertake projects related to handicrafts. A leading newspaper highlighting travel study of students at the craft locale mentions:

> "Craft research is an important part of the design syllabus for which the students travel and learn about the craft from the artisans themselves …" [25]

In an interview with a director of NGO who is working with the embroidery of Punjab narrated:

> "Every year we get two to three students from Design institutes as interns. We usually ask them to initially study the traditional crafts, designs and then innovate or revive the old

[4]Export Promotion Council for Handicrafts.

[5]Carpet Export Promotion Council.

[6]Council of Handicrafts Development Corporations.

[7]National Centre for Design And Product Development.

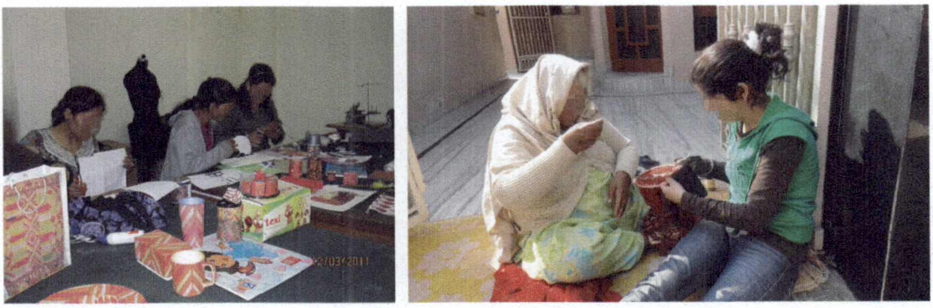

Fig. 6 Explorations in the design institute and in the field with artisan

Fig. 7 Design college students with one of the major players of the craft

> designs with new colour combination or a fresh look. This serves two purposes: one is that the new generation gets in touch with the roots preserving the rich heritage of the state and second we are able to revive or redo the old motifs which are lost. These students not only work on motifs but also on product diversification, hence help us to give something new to the market."

Research and education in the field of crafts (Figs. 6 and 7) also help the budding designers to develop an appreciation of the craft and craft maker and hence can help them in contributing in the culturally rich industry which can benefit the artisan and the society [23].

7 Handicraft–Tourism Integration

Tourism is another factor responsible for crafts revival and popularity. In this globalized world, Subrahmanian [30] points out a positive relationship between the demand

Fig. 8 Virasat-e-Khalsa

for handicrafts and the growth of world tourism. Hargrove [14] in his book on Cultural Heritage mentions about 2009 Mandala Research report which says that cultural heritage traveller during a leisure trip prefers to attend arts and crafts fair or festival and likes to visit art museums and galleries. Intercultural contact amongst tourists and host nationals can be viewed as a novel type of diverse cooperation and a unique form of cross-cultural interaction.

As indicated by the Union Ministry of Tourism, the tourist advents in the state of Punjab have dramatically increased with the quantity of household voyagers going up from 1.05 crores in 2010 to 2.57 crores in 2015 and the quantity of outside sightseers shooting up from 1.37 to 2.42 lakhs in the corresponding time period [28]. Liebel and Roy [19] as well as UNESCO World Report [32] claim that intervention in promoting tourism has been prodigiously prosperous in reclaiming sustainability for traditional craft skills. Crafts and global tourism are socially, culturally and financially essential: the former, by giving creative and artistic shape to decorative or domestic objects, and the latter, by giving access to the diversity of cultures in their natural settings.

Museums and historical monuments and places are space for intercultural learning. Virasat-e-Khalsa (Fig. 8) is a multi-gallery, state-of-the-art museum built on Sikh history. The galleries of the museum showcase the history beginning from the lives of the Gurus, the traditional occupations, lifestyles, rituals, kingdoms, folk tales, religious sites, rich cultural history of the region rendered in 3D and depicted along with and through the crafts and art traditions of Punjab which have been languishing since long. This has truly given a boost to the traditional arts and crafts of Punjab. It is now one of the most visited museums in India. Nearly 10 million persons have visited it since its inauguration in November 2011 [5].

With the coming up of heritage destinations like *Sadda Pind* in Amritsar and *Havelis*[8] all-over Punjab and neighbouring region, the long-standing demand for a cultural hub showcasing the life of a typical Punjab village is fulfilled. *Sadda Pind* is a Punjabi village living museum spread across 12 acres of land, the beautifully landscaped area which has within its boundaries a virtual village with entwining narrow lanes popularly known as *gallis*. The village has it all to one's surprise: a

[8]A Punjabi restaurant.

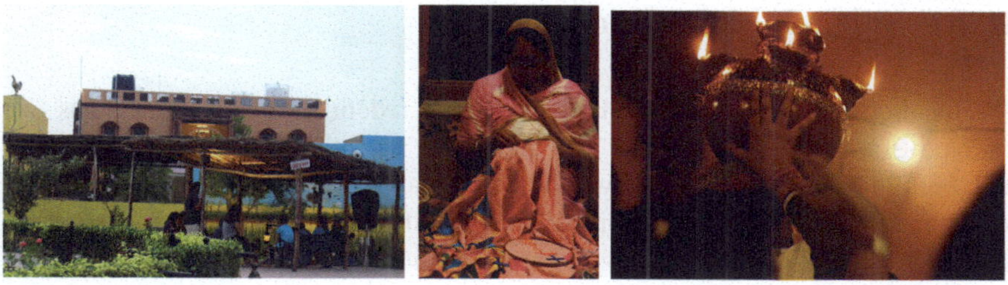

Fig. 9 *Sadda Pind* (from left): a magic show, artisan preparing an embroidered wrap, *jago*: a ceremony prior to Punjabi wedding

Fig. 10 Glimpse of Punjabi culture in *Haveli*

weaver, a blacksmith, a magician, a tailor, etc. The most beautiful part is that it showcases all traditional crafts, arts, cuisine, agricultural tools, music and dance; all within its boundaries beautifully showcased (Fig. 9). The young can reconnect with their roots, foundations and comprehend the customs and estimations of their fathers and progenitors. The elderly can recollect their more blissful times. Foreign tourists can test the genuine Indian provincial way of life. Tourists from other states can encounter true Punjab in one place. Shopping complex at *Sadda Pind* is an outlet for buying traditional Punjabi artefacts and souvenirs. Such destinations have revived many crafts which are now recreated on demand as souvenirs, hence also helping the artisans to produce and earn more, etc.

The popular *Havelis* are also well-liked stops and most visited destinations of almost all major cities of Punjab offering delightful mouth-watering food and a peep into the gorgeous culture and heritage of Punjab (Fig. 10). Such tourist places give a lot of boost to the sale of cultural and traditional artefacts of Punjab.

8 Revival and the Status of Artisans

The handicraft of Punjab today has immensely redefined the cultural status and to some extent the artisan lives. Till some time ago, the handicraft had touched rock bottom but a steady trend towards 'going back to roots' has led to the revival, reintroduction of crafts and material culture into the cultural lifestyle of the vibrant lives of the Punjabis. An artisan working for *paranda* making revealed:

> "These days Punjabis order *parandas* for all women who will be attending the *mehendi*/bangle ceremonies during a Punjabi wedding. Such practices are not only limited to the rural but urbanities too relive and revive the traditional culture still being modern."

This section details the status of various artisans involved in different crafts: *phulkari*, *paranda*, *jutti*, wood inlay and *durrie* making.

The survey revealed that artisan's age varied from 20 to 80 years. The crafts which involved lot of eyesight and precision especially *phulkari* where embroiderer has to carefully pass needle and thread through the motifs or in inlay work where the grooves are to be minutely worked, majority of the artisans belonged to the age group of 20–40 years, whereas in case of *durrie* making, majority of the artisans were in the age group of 40–60 years. There were master craftsmen or craftswomen in all the crafts who were above 70 years of age and pursuing the crafts. With regard to creating products, embroidery and hair accessory—*paranda*, purely women were involved in the making whereas contact with the market or selling of the same was handled by men of the family. In case of *phulkari*, women artisans reported to practice the craft as a free time activity not realizing the amount of hard work they put into the craft. Similar were the observations with *paranda* craftswomen. Both men and women were involved in the *durrie* making, but the work like spinning, warping or preparatory processes was primarily done by females and weaving by males. It was interesting to find that old fabrics/garment strips '*chindi*' were also used as a weft yarn to prepare *chindi durries*. The inlay work is dominated by male artisans, and no female artisans are involved. Artisans mentioned that this work is like a work of carpenter where one has to cut and recreate wood which involves a lot of energy and strength and hence cannot be handled by women. *Jutti* was prepared by male artisans. Usually, families are involved in this work. Further interaction with artisans disclosed that women are involved in the lighter part of craft like pasting and sticking beads and stitching other embellishments.

There were very less artisans who had pursued secondary or higher secondary education. Very few were graduates. This was one of the reasons that these artisans do not get to know about various schemes of the state regarding their welfare or they are dependent on others for filling many of their forms or they hesitate to participate in any marketing activity where they are supposed to go out and interact with the customers. As time and society witness changes, there is a change in the mindset of people. Even the artisans have changed over a period of time and have recognized the need for education and most of them were trying hard to educate their children. Some of the *durrie* artisans reported that they do not want their children to get involved in *durrie* making because this craft has lost its significance overtime and is now a difficult source to earn living. The artisans were interviewed in their living workshops which speak about their low standard of small-living house with all the necessities placed in one room. Another revelation by the artisans of various crafts was that despite their hard work and efforts they were not able to earn much. In case of *phulkari* and *paranda,* women consider this as a pass time work, but at the same time they are able to earn Rs. 1000/- per month according to the work and many a times amount is much less than this. These artisans were paid per piece. Moreover, it was informed that these artisans are not paid regularly; many a times their money is kept as a security so that the artisan's loyalty remains with the vendor. The *paranda* craftswomen also earn a meagre amount. In case of inlay work, the artisans earned more as compared to the other crafts: this craft contributed around rupees ten thousand per month but was reported as a small earning to sustain the family.

Lot of occupational discomforts, reported by the artisans from the field like back-ache, headache and strain on eyes, cervical pain, weak eyesight, needle pricking, regular cuts on hands with cutters, pain in knees, neck and legs, discourage artisans to further impart training to their family members as a profession. But many artisans impart training to their family members so that they could also add up as a source of family income. Artisans who were getting good orders, over a period of so many years, have started their own enterprises and feel that Punjabi heritage is in demand; hence, more and more people should be trained.

It was reported that in inlay work, elephant and floral motifs used to be popular in earlier times, but with design interventions and change in the market, new motifs are created to cater to the latest demands of the customer. There is a change in material too. Ivory has been replaced by plastic material. Cutting by hands and many tools have been replaced with laser cutting tools and techniques. Material changes were also seen in *jutti* from leather to synthetic and experimentation with a lot of accessories. Though leather footwear is durable, the market requirement has given a way to non-skin material. Earlier, needlework on leather footwear was very famous, but now along with embroidery other embellishments have also taken space on the surface of the traditional *jutti* such as beads, rhinestones and bells.

Raju Dati, 28 years old, is a migrant from Rajasthan and now lives and work in Patiala. He started making *jutti*s at the age of 14 years. Being in a joint family, he worked with his family members to make more and more number of footwear to earn income. Over a period of time, he started experimenting with decoration of the footwear with all the possible notions. His forte is *jutti* ornamentation. While experimenting with the decorations of *jutti,* he keeps himself updated on the prevailing trends. Hence, he has developed knowledge of the market, and his interest keeps him updated about the market trends. He has crafted a *jutti* of pure gold worth Rs. 9,00,000. Moreover, while sitting in a market place, he is able to sell well. There were other artisans who were in the interiors of the city/not located near the market place did complain about their sales. They voiced "It is easy to sell while sitting in the market whereas we further sell to the vendors who give us money at their own terms".

This statement was also reported by embroidery (*phulkari*) artisans. In case of *parandas*, single as well as multicoloured options are now available. In addition to cotton yarn and beads, artisans these days use silken yarn and woollen threads and accessorize it with mirrors, tassels, crystals, etc.

In case of *durrie* weaving, artisans reported that nowadays no one uses *durrie* on a bed. There is an entire shift in the usage and making of *durries*. Artisans were working with the old tools and traditional technology-*panja*, a metal beater made in the shape of the hand with its teeth made of metal, which is used to beat the weft yarns and such durries are called as *panja durries*. Earlier, *durries* were preferred in bold colours but now customers want durries in muted and subtle colours. *Durrie* patterns tend to be either geometrical or figurative. Sometimes motifs of flora and fauna are used.

Pushpinder, an 80-year-old artisan, is born and based in Nakodar. He did not have his formal school education and started weaving *durries* at a very young age as a help to his father who taught him this art. His living conditions are very bad which can be visualized in the photograph (Fig. 11): as he earns Rs. 19 for weaving 5 kg of yarn and there is no other source of income. According to him, *durries* are not very popular and the craft is slowly vanishing. Pushpinder says that machine-made *durries* are much cheaper and have variety vis-a-vis his handmade pieces. Moreover, he cannot work on product diversification because of his financial status and relies only on orders.

The mechanization of the crafts [29], decrease in the quality when there is an increased demand for more number of handcrafted products [19, 22] and lack of knowledge of new innovative ways to design [36] pose immense threat to the sustainable livelihood of the artisans. It is the middlemen or the vendors who decide the design and any personal variation in the same leads to the rejection of the craft and hence a loss for the artisan and making them vulnerably susceptible to exploitation by all those who are their only designates of access to distant markets.

There are several generic schemes introduced in the 11th five year plan for the promotion and development of handicraft sector providing financial assistance, support in acquiring raw material, health and insurance benefits and technological know-how. Many of these initiatives are already being administered by various agencies as part of promotion and protection of the craft in the last twenty years or more. Surprisingly when asked about the benefits of these schemes, hardly any artisan was aware of these. Though many government policies are available with regard to credit and loan facilities, they are not aware of it. Hence, artisans always remain small or invisible players. A report of the Steering Committee On Handlooms and Handicrafts Constituted for the Twelfth Five Year Plan (2012–2017) also emphasizes that "though the 11th Plan has helped in the revival of the handicrafts sector, certain concerns such as infrastructure gaps, inadequate social security for crafts workers, the largely unorganized nature of the sector and lack of consolidation of existing programs, continue to impede its growth"[1]. Jena [16] shows concern for the precarious condition of the artisans and recommends meticulous interventions: so the policies need to be implemented wholeheartedly by the government agencies and more categorically the state governments because these have not shown productive results. The researcher

Fig. 11 Artisan's living environment

stresses paramount reasons for why many government schemes have not been so fruitful is the lack of proper knowledge about the target group. A consummate cognizance on the target group could well distribute the developmental schemes and plans of the government. Those who have moved up in the craft sector and earning well, feel empowered. Though many crafts are revived and many are in the process of experimentation, many need more efforts to be revived and the welfare and status of artisan need more attention. There are many schemes, but the need of the hour is to make artisan aware of it and they are encouraged to avail the same. Finally, crafts will flourish if the artisans are in a good state.

References

1. A Report of the Steering Committee on Handlooms and Handicrafts Constituted for the Twelfth Five Year Plan (2012–2017). http://planningcommission.gov.in/aboutus/committee/strgrp12/str_handloom0305.pdf. Last accessed 10 Oct 2017
2. Abraham, T.M.: Handicrafts in India. Graphics Columbia, New Delhi (1964)

3. All India Handicrafts week. http://www.indiacelebrating.com/events/all-india-handicrafts-week/. Last accessed 12 Nov 2017
4. Annual Domestic Marketing Plan 2017–2018. http://www.handicrafts.nic.in/cmsUpload/ANNUAL_DOMESTIC_MARKETING_PLAN_2017_18.pdf. Last accessed 14 Nov 2017
5. Bagha, S.: 'Virasat-e-Khalsa' designed by Moshe Safdie becomes most-visited Museum in India within 5 years. World Architecture Community (28 Feb 2017). https://worldarchitecture.org/architecture-news/cvzmc/_virasatekhalsa_designed_by_moshe_safdie_becomes_mostvisited_museum_in_india_within_5_years.html (28 Feb 2017). Last accessed 11 Nov 2017
6. Black, S.: Eco-chic: The Fashion Paradox. Black Dog Publishing, London (2008)
7. Chattopadhyay, K.: Handicrafts of India. Indian Council for Cultural Relations, New Delhi (1975)
8. Crafts Council: Studying craft: trends in craft education and training. http://www.craftscouncil.org.uk/content/files/Studying_Craft_Report_single_pgs_.pdf (2013). Last accessed 11 Nov 2017
9. Crill, R.: Indian Embroidery. V&A Publications, London (1999)
10. Deepak, S.: Writings on the shirt: the secret behind the very Punjabi brand 1469 is hidden in its celebration of the basic. India Today. http://indiatoday.intoday.in/story/harinder-singh-owner-of-punjabi-brand-1469-situated-in-janpath-talks-about-his-ideas-and-secret-behind-his-brand/1/654955.html (2016). Last accessed 17 Sept 2017
11. Donkin, L.: Crafts and conservation: synthesis report for ICCROM. http://www.iccrom.org/ifrcdn/pdf/ICCROM_02_CraftsandConservation_en.pdf (2001). Last accessed 10 Nov 2017
12. Frater, J.: In the artisan's mind: concepts of design in traditional Rabari embroidery. In: Vatsyayan, K. (ed.) Embroidery in Asia: sui dhaga: Crossing Boundaries Through Needle and Thread, pp. 84–93. Wisdom Tree, New Delhi (2010)
13. Gale, C., Kaur, J.: The Textile Book. Berg, New York (2002)
14. Hargrove, C.M.: Cultural Heritage Tourism: Five Steps for Success and Sustainability. Rowman & Littlefield, London (2017)
15. Harvey, J.: Traditional Textiles of Central Asia. Thames and Hudson Ltd., London (2002)
16. Jena, P.K.: Indian handicrafts in globalization times: an analysis of global-local dynamics. Interdisc. Description Complex Syst. **8**(2), 119–137 (2010)
17. Kapur, H.: Handmade Tales: Sustainable Fashion Through Craft Connections. Master of design thesis, Massey University, Wellington, New Zealand (2016)
18. Kumar, S., Dutta, N.: Weaving a knowledge tapestry of traditional crafts for modern fashion designers: an Indian experience. Art Libraries Journal **36**(2), 17–24 (2011). https://doi.org/10.1017/S0307472200016874
19. Liebl, M., Roy, T.: Handmade in India: traditional craft skills in a changing world. In: Finger, J.M., Schuler, P. (eds.) Poor People's Knowledge: Promoting Intellectual Property in Developing Countries, pp. 53–73. The World Bank and Oxford University Press, Washington DC, USA (2004)
20. Lockwood Kipling: Arts & Crafts of Punjab and London (15 Jan 2017). The Heritage Lab. http://www.theheritagelab.in/lockwood-kipling/. Last accessed 15 Nov 2017
21. McIntyre, M.H.: Consuming Craft: The Contemporary Craft Market in a Changing Economy. Crafts Council, United Kingdom (2010)
22. Mohapatra, R.C.: Changing patterns in the applique crafts of Pipili. Orissa Review, 66–67 (2005). http://www.Paramparaproject.org/downloads/changing_patterns_in_the_applique_craft_of_pipili.pdf. Last accessed 11 Nov 2017
23. NSEAD: A manifesto for art, craft and design education. http://www.nsead.org/downloads/NSEADManifesto-2014.pdf (2014). Last accessed 11 Nov 2017
24. Paine, M.: Textile Classics. Mitchell Beazley Publishers, London (1990)
25. Parikh, R.U.: Nift students travel to study crafts. The Times of India, Ahmedabad (26 Jan 2014). https://timesofindia.indiatimes.com/city/ahmedabad/Nift-students-travel-to-study-crafts/articleshow/29379379.cms. Last accessed 14 Nov 2017

26. Philadelphia Museum of Art. http://www.philamuseum.org/exhibitions/857.html. Last accessed 15 Oct 2017
27. Reubens, R.: Bamboo canopy: creating new reference-points for the craft of the Kotwalia community in India through sustainability. Craft Research **1**(1), 11–38 (2010)
28. Sharma, M.G.: Rich in monuments, Punjab gets a tourism high: from 28th spot to 12th in eight years. Hindustan Times, Chandigarh (25 Feb 2017). http //www.hindustantimes.com/punjab/rich-in-monuments-punjab-gets-a-tourism-high-from-28th-spot-to-12th-in-eight-years/story-Jm2ybgRfm4kvydeGkFLUSI.html. Last accessed 15 Nov 2017
29. Singh, M.K.: Evolution in crafts: negotiating the handmade and the machine made. Craft Revival Trust (2005). http://www.craftrevival.org/voiceDetails.asp?Code=115. Last accessed 30 Oct 2008
30. Subrahmanian, K.K.: Introduction: an overview of the handicrafts industry. In: Subrahmanian, K.K. (ed.) The Handicrafts Industry in Kerala: Blending Heritage with Economics. Dannish Books, Delhi (2006)
31. The Director, Publication Division, Ministry of Information and Broadcasting: Indian handicrafts. The Director, Publication Division, Ministry of Information and Broadcasting, New Delhi (1972)
32. UNESCO World Report: Investing in cultural diversity and intercultural dialogue. www.un.org/en/events/cultural_diversity_day/pdf/Investing_in_cultural_diversity.pdf (2009). Last accessed 15 Nov 2017
33. Vatsyayan, K.: Introduction. In: Vatsyayan, K. (ed.) Embroidery in Asia sui dhaga: Crossing Boundaries Through Needle and Thread, pp. ix–xiii. Wisdom Tree, New Delhi (2010)
34. Wearing 'Punjabiat' through tradition and catchy slogans.: IANS (2016). http://www.sify.com/news/wearing-punjabiat-through-tradition-and-catchy-slogans-newsothers-qfdnagagchjea.html. Last accessed 15 Nov 2017
35. Winston, A.: Using traditional crafts in design is not "sentimental" says Sebastian Cox. Dezeen (30 Oct 2014). https://www.dezeen.com/2014/10/30/sebastian-cox-interview-british-craft-industry-coppicing-wood/. Last accessed 15 Nov 2017
36. Wood, S.: Sustaining crafts and livelihoods: handmade in India. Craft Plus Design Enquiry, 89–100 (2011). http://press-files.anu.edu.au/downloads/press/p211931/pdf/ch07-Sustaining-crafts-and-livelihoods.pdf. Last accessed 15 Nov 2017

Moving Lights as Moving Spaces: Reinterpreting Traditional Bamboo *Chik* Making

Shubhra Singh

Abstract The product is an innovative handwoven screen inspired from the craft of traditional *chik* making. Context played an important role; light and wind were crucial, owing to the orientation of the built mass. The product was designed to enhance the experience of the space through sensory associations. Experimentation with light culminated into the project idea: Moving Lights as Moving Spaces, where vertical screens were created to be used as a shading device with perforations to add playfulness to the space and make it usable for multipurpose activities. Technique and material turned out to be the most important factors. Hand weaving was explored employing the technique of hand leno. The use of six twisted warp threads worked the best for the scale. Combination of polythene bags, shrinkable films with bamboo sticks occurred to be the best combination owing to the feasibility, cost, context, sustainability, and social factor. And hence, the project ideology 'Weave your Sin' emerged, and the process furthered into creating explorations on small and actual scale prototypes. The product caters to small-scale industries and provides scope to empower our weavers through retained cultural practice.

Keywords Space · *Chik* making · Handloom · Plastic waste · Sustainability

1 Introduction

The grand semi-open space at SID, CEPT University, remained unusable during peak working hours of the day. The problem identified was the amount of light entering the space. The proposed design idea intended for managing the amount of light to provide shade, yet maintaining connect with surrounding nature. The design intends to cater to the aspect of light, in a way to extract maximum usability and possibility of the context with optimal use of the color, material and technique. The designed handwoven vertical screen is inspired by traditional bamboo *chik* making craft. It is purposed to be placed against light. It contains perforations in the form of geometrical

S. Singh (✉)
National Institute of Design, Ahmedabad 380007, India
e-mail: shubhrasingh28@gmail.com

A. Majumdar et al. (eds.), *Functional Textiles and Clothing*,
https://doi.org/10.1007/978-981-13-7721-1_32

patterns through which light penetrates and casts interesting shadow patterns based on the location of sun, still maintaining the structural stability of the created surface. The design is such that colors and pattern can be profoundly seen against light.

'*Chik* making' refers to the process of making bamboo *chiks* (screens). These screens are used to make beautiful vertical shades and blinds capable of providing respite from the scorching summer sun. *Chik* making is a traditional craft, simple yet elegant; *chiks* provide mesmerizing results.

1.1 History and Origin

Literature study of the *chik* making craft globally suggests about the craft drawing inspiration from various cultures and time zones for a variety of purposes. While *chik* making is the traditional craft of New Delhi, it is also practiced in Aligarh, Agra, and Gwalior. Highly ornamented screens may be inspired from the reed screens used in Kyrgyz yurts of Kyrgyzstan, silk-wrapped floral designs of bamboo screens used in the royal stable of Emperor Aurangzeb, with the change in technique.

- *Chik* making finds mention as the traditional craft of New Delhi as mentioned in the book, Handmade in India. Chiks are blinds or semi-rigid window panels created out of fine bamboo splits or rigid stems of *sarkanda* grass. It is held in place by a warp of cotton threads. The created *chik* can easily be rolled. It is edged with a woven tape called *nivar* and is sometimes lined with a fabric called *tirpal*, to make it opaque and waterproof. Bamboo *chiks* are usually given a waterproof backing as they are generally used in *veradahs*. The *chik* is an inexpensive, earthy window blind which succeeds in diffusing harsh light, while the geometrical patterns of the wrapped cotton threads contribute a certain elegance.[1]
- Traditionally, nomadic herders called Kyrgyz lived in circular felt tents known as yurts. Yurts were made using willow wood, reed screens, felt. They carried all their possessions with them, possessions that were beautiful, portable, less breakable, and provided for variety of purposes, such as the reed screens (plain/patterned). The reed used is called *chiy*. It is slender and stiff. Plain reed screens were created with reeds bound together using strong woolen cords, while for patterned reed screens, all reeds were first wrapped in dyed wool as per the pattern retained in the memory of the artisan. A patterned screen was woven on a weaving frame. The frame contained strong woolen cords wound onto heavyweights. It took around two to three months to prepare one 9′ long screen. Twinning process bounded the reeds together with the woolen cords. Each stem when placed in a sequence emerged as a pattern which is no less than a miracle. It can be compared with the tie-dyed and woven *ikat* textile traditions. Making a reed screen was a Kyrgyz women's art. It formed part of a Kyrgyz bride's dowry. Young girls learnt this traditional craft from their mothers and grandmothers. These reeds were used to furnish the yurt as a wall cover and as space dividers inside a yurt.

[1]Handmade in India.

The long *kanat chiy* measuring approximately 20 ft was used to cover the wall. Placed between yurt's lattice frame and the outer felt covering, it helped to insulate against cold. During warmer weather, the outer felt cover was removed, and therefore, the patterned reeds allowed for visual aesthetics, ventilation, and privacy. The shorter *ashkana chiy* functioned as a space divider. Plain reed screens *ak chiy* were also used as insulation from cold when placed beneath felt carpets, provided structural base for creating felt carpets acted as surface for food drying, windscreens, or coverings for newly threshed wheat.

The designs of reed screens share similarities with patterns on flat woven *kilims*, felted rugs, and silk *ikat* fabrics. Geometric motifs were popularly seen on reed screens, such as octagon, rhombus, square, and rectangle. This might be attributed to the ease in working with geometrical patterns, whereas curvilinear forms required higher artisanal skills. Animal forms or written inscriptions are rare to find on Kyrgyz reed screens.

- Another mention suggests about silk-wrapped bamboo screens probably dated from the late seventeenth century to early eighteenth century. These screens were used in Emperor Aurangzeb's royal stables. These screens prevented flies from tormenting horses. These screens were created using bamboo splits woven with twisted silk. Each fine bamboo reed was individually wrapped, in the manner of *ikat,* with multiple colors of silk yarn, dark blue-green, red, yellow, and white, to create a floral lattice enclosed within floral borders; the silk-wrapped reeds were held in place by a twisted 2-ply cotton cord warp.[2]

2 Traditional *Chik* Making

The craft of traditional *chik* making was explored as a part of process to enhance the experience of the space through sensory associations. The tools needed for creating a *chik* comprise of a temporary portable bamboo stand, a knife to split bamboo, warp yarns (cotton/nylon) to bind bamboo splits, *newar* to cover up the edges, and *tirpal* to cover up the entire surface to resist moisture (Figs. 1 and 2).

The technique employed is an up and down movement of the long continuous yarns (secured with a weight at the end) in alteration along the width of the *chik*. This helps in the creation of a long surface. The above description of *chik* is suitable for the non-decorative *chiks*. But at National Institute of Design (NID) highly decorative technical *chiks* are created on loom by the technically brilliant weavers (Fig. 3).

Woven on a primitive frame handloom, multiple designs are produced only with the use of two shafts. Usually, though basic plain weave is employed; another specialty is the use of hand leno technique. Hand leno is a technique where yarns are hand-picked, twisted, and retained through the use of a thin bamboo stick (*salli*). Hand leno technique provides tiny perforations for the play of light and wind. The

[2]Durbar: Royal Textile of Jodhpur by Rahul Jain.

Fig. 1 Traditional *chik* making in Ahmedabad (Courtesy: Shubhra Singh, 2015, Ahmedabad)

Fig. 2 Process of cleaning split bamboo (Courtesy: Shubhra Singh, 2015, Ahmedabad)

materials used are bamboo sticks (*salli*) and cotton yarns (as warp and binding weft); both materials are employed in natural shades. The above decorative *chiks* are functional and usable for various contexts.

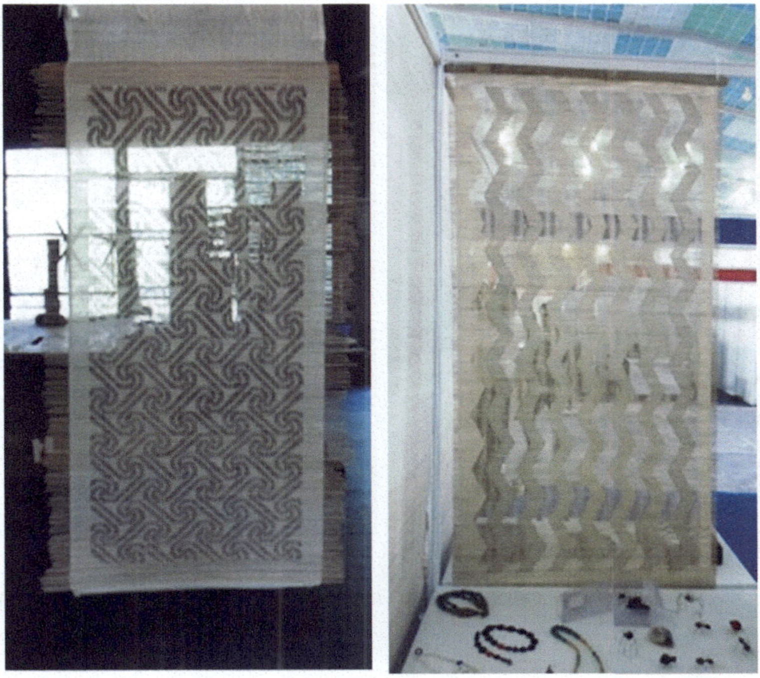

Fig. 3 Decorative woven *chiks* made on handloom at NID (Courtesy: NID Textile Department Archive, Ahmedabad)

3 Context

Designed for the grand semi-open space at SID, CEPT University, owing to the wide opening, the aspect of light and wind was considered to enhance the experience of space through sensory associations. Chiaroscuro formed the major opportunity area to design and innovate (Fig. 4).

4 Design Process

Brainstorming and explorations were carried out with material, color, form and technique, and significant insights were achieved for design intervention to get the desired results.

Fig. 4 Identified space at SID, CEPT University (Courtesy: Shubhra Singh, 2016, Ahmedabad)

4.1 Understanding and Learning the Technique

The technique of hand leno is practiced by handpicking four consecutive warp yarns and then twisting the two yarns on the extremes around the central yarns, and this twist in succession provides for the perforations to appear on the surface. In order to retain the twist, a long bamboo stick is inserted in continuation along the entire width with each twist (Figs. 5, 6 and 7).

4.2 Understanding the Space

Sun path and solar angle at various points of the day were analyzed through simulations carried out on software for the designated space (Figs. 8 and 9).

4.3 Opportunities for Design Intervention

Opportunities were derived on interaction with weavers at NID, through exploration and experimentation carried out with the variety of material on loom and learning hand leno technique from master weavers.

Fig. 5 Learning the technique of hand leno (Courtesy: Dhanashri Hase, 2016, NID, Ahmedabad)

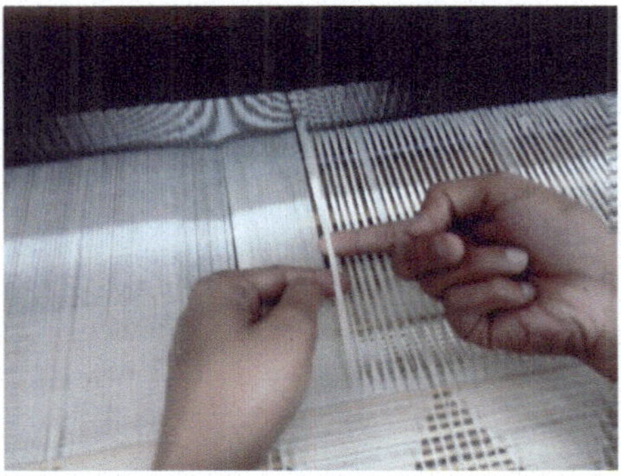

Fig. 6 Handpicking warp threads for twisting (Courtesy: Dhanashri Hase, 2016, NID, Ahmedabad)

Fig. 7 Hand leno technique (Courtesy: Shubhra Singh, 2016, NID, Ahmedabad)

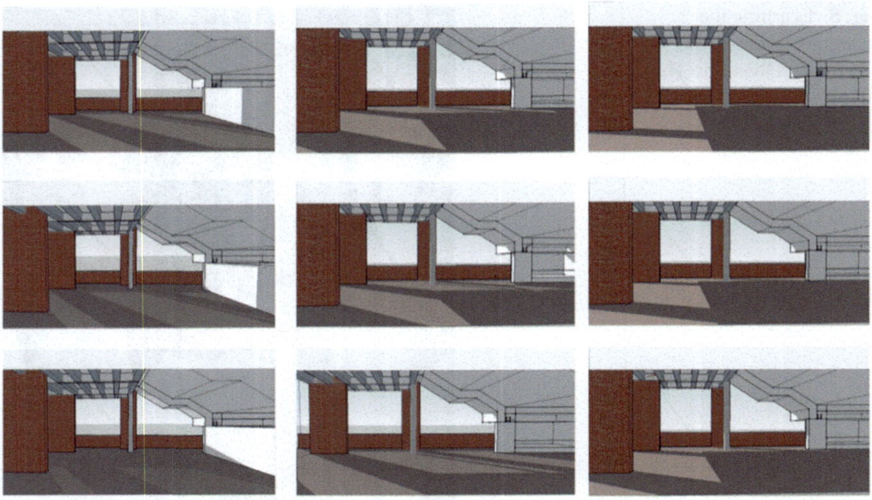

Fig. 8 Solar angle simulation in the space (Courtesy: 3D model made on SketchUp)

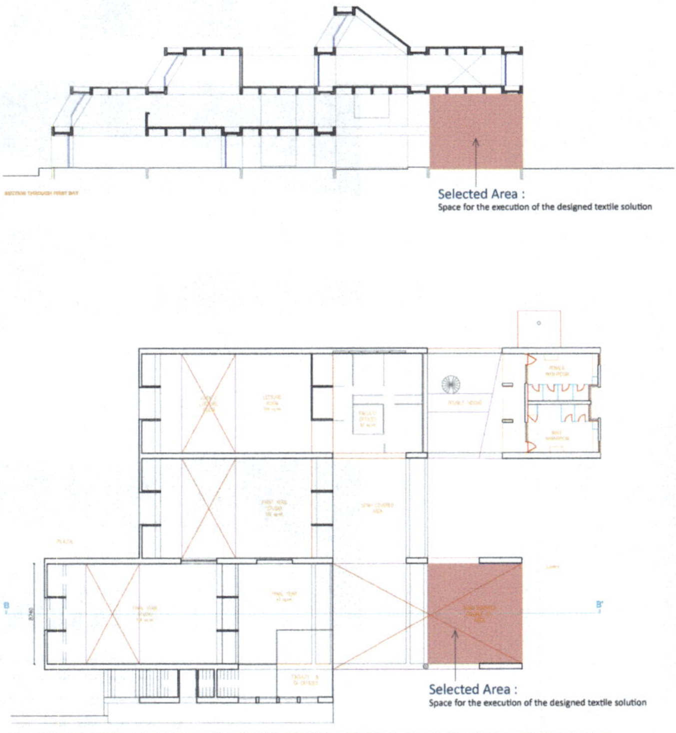

Fig. 9 Identified double-heighted volume for design intervention. Plan and section of SID, CEPT University (Courtesy: NID through SID, CEPT University)

Fig. 10 Exploration made on tabletop (sample) loom combining hand leno, extra weft and balanced weave together (Courtesy: Shubhra Singh, 2016, NID, Ahmedabad)

While traditional *chik* is a single-layered surface, this product is a layered entity. Perforations are integrated through the use of hand leno technique (−1 layer), extra weft technique facilitates layering (+1 layer), and basket weave acts as the base layer (0 layer) binding the layers together. Application of extra weft technique facilitates the introduction of patterns and colors, providing a new identity to the old (Fig. 10).

4.4 Innovation

In order to reduce cost, add colors, and address water resistance, environmental concerns, and sustainability through the product, unconventional materials such as polythene bags and shrinkable films were employed, along with conventional materials such as cotton yarns and bamboo sticks (*salli or tilli*) as the structural framework.

In the traditional bamboo *chik* making, cotton and bamboo sticks are used in natural shades, employing plain weave and hand leno (four threads) techniques. While this product utilizes, wastes such as polythene bags and shrinkable films, along with cotton and bamboo, employing techniques such as hand leno (six threads), extra weft, and basket weave technique. This process magnified the scale of perforation, added color to the schema, and provided visibility to the colors against light respectively.

Innovation took place at various levels such as:

Theme: Layering

Achieved through the combined use of hand leno technique, basket weave, and extra weft technique on hand loom (Fig. 11).

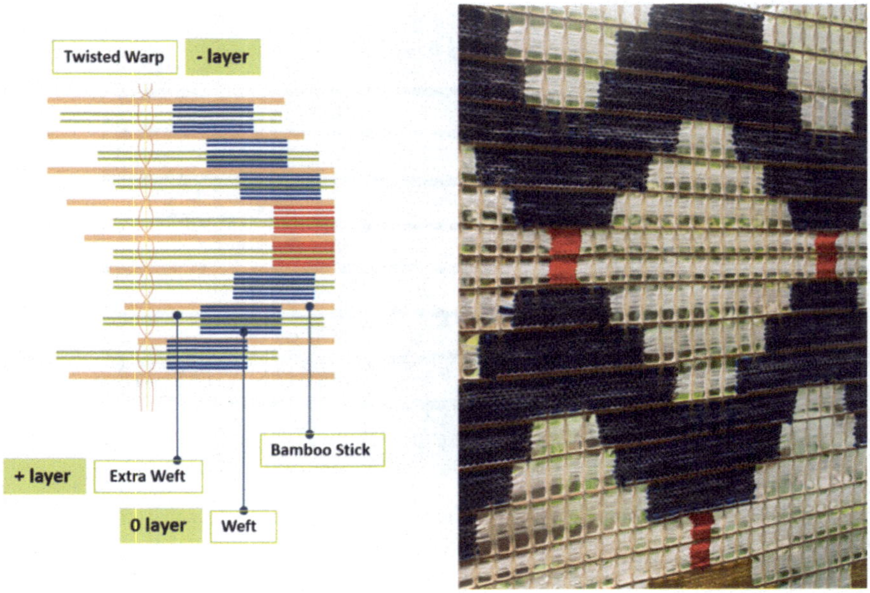

Fig. 11 Illustration showing the components of layering (left) and photograph (right) (Courtesy: Shubhra Singh and Ronak Parmar, 2016, NID, Ahmedabad, respectively)

Technique: Hand Weaving

Instead of twisting four warp threads, various other permutations and combinations were explored. The use of six twisted warp threads worked the best based on the scale for hand leno technique.

Extra weft facilitated the visibility of color and pattern against light, acting as the second skin to the existing layer. Pattern was generated by counting and handpicking warp threads and filling with color-coded extra weft strands (Fig. 12).

Material

A variety of materials were tried, namely aluminum, acrylic, cotton, polyester, jute, linen, MDF, bamboo, PVC pipes, plastic wires, polythene bags, shrinkable films, etc.

Combination of polythene bags (in assorted colors-cut into strips), shrinkable films along with bamboo sticks occurred to be the best combination owing to the feasibility, cost, context, sustainability, and social factor (Fig. 13).

Color, Form, and Composition

Constructed in red brick and concrete and surrounded by nature, the space comprised of earthy tones, gray and red. In order to break the monotony, colors were picked inspired from the color-coded bags (red, blue, yellow) used for waste segregation and disposal as derived through material exploration.

Stepped triangles were combined to be perceived as diamonds through the proximity of color and as waves individually. Patterns were worked out based on geometry

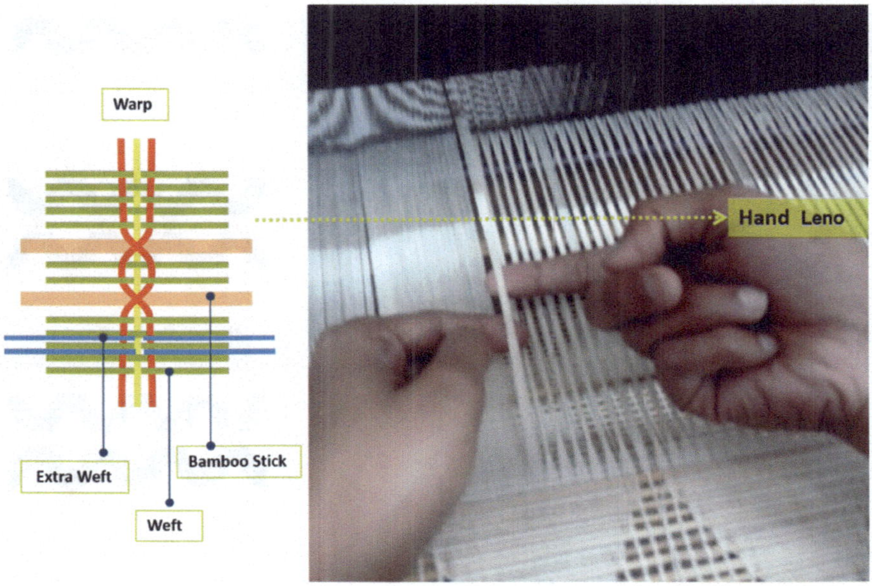

Fig. 12 Illustration (left) and photograph (right) showing the technique of hand leno (Courtesy: Shubhra Singh and Dhanashri Hase, 2016, NID, Ahmedabad, respectively)

Fig. 13 Materials used: cotton for warp; polythene bags, shrinkable films, and bamboo sticks for weft (Courtesy: Shubhra Singh, 2016, NID, Ahmedabad)

Fig. 14 Materials used: cotton warp, polythene bags, shrinkable films, and bamboo sticks (Courtesy: Shubhra Singh, 2016, NID, Ahmedabad)

to comply with the physical context, in a manner to reduce the number of bamboo sticks to optimize the cost (Fig. 14).

Processes

The process began with setting up the loom followed by preparing the weft, cutting waste polythene bags into strips. These strips were then wound to act as shuttle for extra weft. In a similar manner, shrinkable films were cut and wound around the shuttle to run continuously along the entire length of the *chik* in basket weave. Alongside, bamboo sticks were cleaned and smoothened to get rid of tiny fibers that may hurt the weaver or break the warp threads. With the preparation in place hand weaving began, the warp was managed with treadles controlled by feet and wefts were inserted and controlled by hands. At the end of the weaving process, the *chik* was removed from the loom and its edge was finished to convert it into a usable form (Fig. 15).

Fig. 15 Materials used: cotton warp, polythene bags, shrinkable films, and bamboo sticks (Courtesy: Shubhra Singh, 2016, NID, Ahmedabad)

5 Final Product

The product is a provocation with the design ideology 'Weave your Sin,' a reminder to use and throw lesser plastic into the environment and turn to natural substitutes (Fig. 16).

Suitable for use in the outdoors and indoors, the product can be used as space enhancers, space divider, canopy, etc. Against light, the appearance enhances and the shadows formed add playfulness to the space. This textile solution is quick to make, lightweight, easy to roll, pack, install, and transport. It is a sustainable and affordable textile design solution which utilizes handloom resource in an unconventional way.

Fig. 16 Display of product in space (Courtesy: Ronak Parmar, 2016, NID, Ahmedabad)

For safeguarding the *chik* from fire hazard, a fire retarding spray solution can be created by mixing equal amounts of ammonium phosphate and urea. The solution is sprayed over the surface thrice in succession as the previous coat dries up. This can provide with the required fireproofing.[3]

Alternatively, before creating the surface of *chik*, each individual ingredient could be treated for fireproofing separately. Fire retardant chemicals such as zinc borate, antimony trioxide, alumina trihydrate, and decabromo[4] could be added while manufacturing.

This process will help the product with fire-related hazards and increase its usability and market viability (Fig. 17).

5.1 Potential and Ability

- Innovative use of material and technique;
- Structural stability;
- Customizable sizes to suit various contexts;
- Innumerable design and compositions;
- Sustainable and green product;
- Water resistant and suitable for both outdoor and indoor;
- Empowering weavers and generating livelihood opportunities.

---

[3]As shared by Mr. Ashish Karnani from Acuro Organics Limited.

[4]http://www.acuro.in.

Fig. 17 Photographs of the final product (Courtesy: Ronak Parmar, 2016, NID, Ahmedabad)

6 Suggestive Sustainability Model for Livelihood Generation for Artisans (*Chik* Makers and Weavers) Presently Engaged in the Craft

The artisans who are working in the segment have many roadblocks to grow and work in this sector. Due to the lack of demand, often skilled artisans are bound to work as daily wagers and construction labors and are attributed as unskilled. Therefore, creating opportunities for self-employment is the best way to tackle the problem. However, engaging in a self-employed setup comes with its own struggles, such as

- Lack of access to capital due to inadequate banking facilities;
- Excessive paperwork required for getting loans from microfinance institution and banks;
- Lack of access to community organizations to work with;
- Lack of design abilities and capacity to make cutting edge products suitable for the urban markets;
- Mismanagement of funds available under government-sponsored self-employment schemes.

Traditionally, crafts were carried out inside a close-knit community of artisans. Segregation of labor had been based on gender. Works in the pre-production and postproduction stages were carried out by the women of the family. While the main works were carried out by the men, women have been deprived of their true credits as creators.

Scaling up women cooperative credit societies or an SHG will provide true worth to the artisans and also will provide a platform for them to sell their products at market prices, directly to buyers across the world. It would provide income generation for both skilled and unskilled artisans at the various stages in the entire process of the product.

It will help in working on regular monetary savings by a group of individuals who then lend money back to its members. Such a system ensures greater transparency as there is community involvement. It also enhances the access to capital required for self-employment. Further, the chances of loan defaults are reduced due to the role of peer pressure in such cooperatives. These activities can be funded under many government schemes for self-employment such as National Rural Livelihood Mission and many other state-sponsored schemes. The provision of funds is also available through KVIB, Mudra schemes, and PMEGP. These schemes may facilitate the artisans with finance, infrastructure, and equipment for producing *chiks* and come up with products suitable for the global demands. These cooperatives would help artisans to earn approximately 15–30% more than what they would normally get through their existing channels.[5]

By linking designers with the artisans through government-funded schemes like DDUGKY, BADP, Skill Development Mission, and many other state-sponsored

[5]Literature study on GoCoop.

schemes, a process-driven approach can be implemented focusing on the needs of urban clients, trends, and forecasts. The problem of design diversification, standardization, skill, and capacity building can be taken care of. During this process, designers will work on behalf of government to facilitate the artisans. It will help in marketing initiatives, design and technical innovation, and quality improvements. Also, techniques which are no more in practice can be revived under the supervision of the designer. An online application (app) can be created as a marketplace for artisans, where artisanal goods and services can be provided at ease. The designer will be instrumental with helping in photographing, cataloging, and compiling product-related information, etc.

After building the financial capacity and design ability, the forward linkages could be established to market the products through various online and offline channels (B2C and B2B models). This will help them to develop a sustainable and independent model to grow exponentially and provide livelihood opportunities. State government emporiums, crafts councils, revival trusts, museums, etc., might also help in marketing and spreading awareness about the craft. The cooperative with its elected heads can help market the products made by its members and distribute profits equally among them.

7 Conclusion

The product deals with the acute environmental concerns of non-biodegradable waste usage and decomposition. It addresses a way for the revival of traditional practices in the contemporary context. It also aims at providing livelihood opportunities to *chik* makers and handloom weavers.

With 44 lakh handloom weavers across the nation, immense magnitude and variety of patterns could be created, requiring lesser time and material to create the surface, adding the pop of color through the use of non-biodegradable waste solving environmental concerns by addressing the alarming condition of waste disposal for a better sustainable future.

Bibliography

1. Ranjan, A., Ranjan, M.P. (eds.): Handmade in India. Council of Handicraft Development Corporations (COHANDS), New Delhi (2007)
2. Rahul, J.: Durbar: Royal Textiles of Jodhpur. Mehrangarh Museum Trust, Jodhpur (2009)
3. https://kauffman.bethelks.edu/
4. http://www.acuro.in and conversation with Mr. Ashish Karnani from Acuro Organics Limited
5. Literature study on GoCoop, https://gocoop.com/

An Ergonomic Perspective of Uttarakhand Weavers

A. Goel and R. Garbyal

Abstract The study of the relationship shared between man, machine and environment is known as ergonomics. One of the most widely established cottage industry of India is handloom. In the early days, weaving was just a leisure time activity, but now it has changed into a full-time job for many weavers. Despite this positive change, the workstation designs remained the same. Some ergonomic hazards have been associated with weaving sector which may cause problem to the weavers, i.e., headache due to long and continuous work duration, shoulder and neck pain due to repetitious hand movements of throwing of shuttle and beating, etc. So the need was felt to study the current status and working conditions of handloom weavers. Data was collected through questionnaire cum interview schedule using random sampling method; total sample size of selected weavers was 40. It was observed that in Uttarakhand that weaving is performed by male and female both, using traditional handloom (table loom). So, under the present study weavers of both genders were selected from weaving units of Nainital and Udham Singh Nagar district. Ergonomic effects of weaving on the selected weavers while using their traditional handloom (table looms) was assessed, i.e., comfort level/pain, physiological cost of weaving, grip strength, etc. Then table loom was modified according to weavers need to reduce the physical pain and to make it ergonomically friendly. Weavers of both genders were asked to perform weaving on the modified looms, and their results were compared.

Keywords Ergonomics · Human factor · Weavers · Handloom · Uttarakhand

1 Introduction

The country's second largest employment provider is textile and apparel sector which has employed nearly 119 million people directly and indirectly in 2015–16 [1]. In

A. Goel (✉) · R. Garbyal (✉)
GBPUA&T, Pantnagar, U.S. Nagar 263145, Uttarakhand, India
e-mail: alkagoelp@gmail.com

R. Garbyal
e-mail: reenagarbyal@gmail.com

A. Majumdar et al. (eds.), *Functional Textiles and Clothing*,
https://doi.org/10.1007/978-981-13-7721-1_33

the developing countries, the traditional methods of weaving are still significantly practiced which turns handloom into an important cottage industry. The assessment of human workforce is vital for improving their working conditions. The ancient art of weaving is one of the crafts which have survived until the present times. Handloom sector comprises a majority of rural and semi-rural population which makes it one of the largest unorganized economic activities in the nation. There has been a major reduction in the number of people engaged in weaving and allied activities from 65.5 lakhs in 1995–96 to 43 lakhs people in 2009–10 according to the 2nd and 3rd handloom census. At present, approximately 15,468 numbers of weavers are engaged in the weaving industry in Uttarakhand [1]. This sector is gaining strength to withstand competition from the power loom and mill sectors due to effective government intervention through financial assistance and implementation of various developmental and welfare schemes. Nearly 19% of the total cloth produced in the country comes from the handloom sector which further adds substantially to the export earning of the nation. It is permitting experimentation and encouraging novelty due to its flexibility and versatility. Therefore, handloom being a part of nation's heritage glorifies the artistry, richness and diversity of our country [2].

Ergonomics is the science of fitting the work to the user instead of forcing the user to fit the work [3]. The study and application of ergonomics are gaining momentum in the industrial setup. Workers will get negatively affected by the ergonomic problems which will lead to decrease in productivity, non-achievement of the target value, etc. [4]. Repetitive action of weaving on handloom is closely related to musculoskeletal disorders. The performance of weavers gets affected by the high work demand during peak period. There are different ergonomic parameters which are to be considered to overcome these problems. According to OSHA, ergonomics is "the science of fitting the job to the worker." Musculoskeletal disorders (WMSDs) happen when there is a mismatch between the physical requirements of the job and physical capacity of the worker. In the handloom sector, these disorders have emerged as a major health problem among workers. Upper extremity pain and discomfort are closely associated with different workplace factors, such as repetitive work, awkward and static posture. According to studies in Iranian handloom carpet industry, constraints like poor design of loom, working postures, long working time, repetitive work and seating posture increase the possibility of musculoskeletal problems among weavers [5]. Weather has profound effect on human health, and some of such factors include temperature, humidity, light and noise, etc. The magnitude of activities is greatly influenced by such environmental variables.

Basically, both men and women participate in the cloth weaving process in India. There is existence of gender difference in the prevalence and severity of MSDs and perception of work as stressors despite similar occupational work. Their leisure time activity has now been changed into an eight hours job. But still the workstation design remains the same even with the increased weaving time spent on loom. So, present research focuses upon ergonomic effect of weaving on the selected weavers while using their traditional handloom (table looms), modification of handloom (table loom) with minimum expenditure to reduce the physical fatigue and study of weavers of both genders working on modified handloom.

2 Materials and Methods

1. **Locale of the Study**:
   The study was conducted in Kumaun woolens, Haldwani, Distt. Nainital and Mahua Dabra, Jaspur, Distt. Udham Singh Nagar, Uttarakhand. These are neighboring places to Pantnagar; thus, the selected areas were easily accessible to the researcher.
2. **Selection of Respondents**:
   Total sample sizes of 40 respondents involved in weaving activity were contacted. They were checked for their physical fitness; finally, 32 weavers (16 male and 16 female weavers) found physically fit were selected for the study.
3. **Data Collection**:
   Data was collected through questionnaire cum interview schedule and random sampling method was used. The data was collected in two stages: In the first stage, descriptive data through interview method was noted, whereas in the second stage the experimental data was collected through instrumental measurements.
   To study the comfort level or pain of the male and female weavers working on handloom, comparison was done, i.e., while working on unmodified handloom vs modified loom. Comparison was done on the basis of grip strength, energy level, percent increase in heart rate, energy expenditure, blood pressure and posture analysis (deviation of body). For the experimental work, measurements were taken at resting level (before starting work), during/after performance. Then through difference in rating/readings of both stages, results were analyzed and conclusions were drawn.

3 Results and Discussion

The weavers of Haldwani and Jaspur of Uttarakhand were selected for the present study. It was observed that in Uttarakhand, handloom weaving is done at regular basis in these places and also weavers of these two places were professional and progressive; therefore, due to their readiness to adopt changes in their traditional skills/methods, weavers of these places were selected. Another reason was that the selected areas were easily accessible to the researcher.

4 Assessment of Ergonomic Problems of Weavers

The weaver's problems related to the ergonomics were first assessed while they were using unmodified or control handloom of the selected locale. These observations were tabulated and analyzed. Results of the observations concluded that weaving is actually a highly repetitive task. Weaving and related activities of this sector may result in various musculoskeletal disorders. The risk factors noted were due to improper design of hand tool, lack of awareness about the appropriate postural conditions during work, lack of frequent rest breaks during weaving, adverse environmental conditions and poorly designed workstations, etc.

It was observed that most of the weavers of Mahua Dabra, Jaspur were using two types of handloom, i.e., table loom and pit loom; however in Kumaun woolens, Haldwani traditional table looms were used by the weavers. Therefore, handloom modifications were carried out on table loom only, to reduce the pain and postural discomfort of weavers. Again the above-mentioned properties with regard to ergonomic problems were assessed, where comparison was made to check the effect/benefit of loom modification.

1. **Assessment of Work-Related Musculoskeletal Disorders Using Unmodified Loom for Weaving**
   Work-related musculoskeletal problems and body pain as perceived by the respondents were determined by administering standardized body map questionnaire. Responses of all the selected 32 weavers were notified and analyzed.

   1.1 **Feeling of pain and discomfort**: It can be clearly envisaged from Table 1 that out of 16 male respondents, 37.5% respondents were having knowledge about the musculoskeletal pain/discomfort and 62.5% were ignorant about it. In case of female respondents, 25% respondents were acquainted with the muscular pain and 75% were not aware about it. It was observed that male respondents had better knowledge about musculoskeletal pain as compared to their female counterparts because of their higher literacy rate.
   It is also clear from Table 1 that nearly 31.25% male weavers reported that they used to avoid household activities, whereas 56.25% female weavers said that they were forced to stay away from normal work activities because of the pain/discomfort. 68.75% male and 43.75% female respondents revealed that they have no pain or negligible pain.
   It was expressed by weavers that the weaving is their traditional work which they used to perform throughout the year since their childhood. Due to this

Table 1 Pain and discomfort with the locomotion organs N-32

S. No.	Statements	Male (n-16)		Female (n-16)	
		Yes n (%)	No n (%)	Yes n (%)	No n (%)
1.	Knowledge about musculoskeletal pain/discomfort	6 (37.5)	10 (62.5)	4 (25)	12 (75)
2.	Unable to perform household work activities because of musculoskeletal pain/discomfort	5 (31.25)	11 (68.75)	9 (56.25)	7 (43.75)

Table 2 Subjective feeling of pain in different body parts of weavers (male) expressed through body map using unmodified loom N-16

Body part	Yes	No	Severity of pain			Frequency of pain		
			Acute	Less acute	Negligible	Always	Some time	Never
	n (%)	n (%)	n (%)	n (%)	n (%)	n (%)	n (%)	n (%)
Neck	7 (**43.75**)	9 (**56.25**)	–	5 (**31.25**)	2 (**12.5**)	–	7 (**43.75**)	–
Shoulder	10 (**62.5**)	6 (**37.5**)	6 (**37.5**)	4 (**25**)	–	8 (**50**)	2 (**12.5**)	–
Elbows	5 (**31.25**)	11 (**68.75**)	–	5 (**31.25**)	–	–	5 (**31.25**)	–
Upper back	12 (**75**)	4 (**25**)	8 (**50**)	4 (**25**)	–	5 (**31.25**)	7 (**43.75**)	–
Lower back	10 (**62.5**)	6 (**37.5**)	5 (**31.25**)	5 (**31.25**)	–	6 (**37.5**)	4 (**25**)	–
Hand	4 (**25**)	12 (**75**)	–	–	4 (**25**)	–	4 (**25**)	–
Palm	4 (**25**)	12 (**75**)	–	–	4 (**25**)	–	4 (**25**)	–
Thigh	8 (**50**)	8 (**50**)	6 (**37.5**)	2 (**12.5**)	–	6 (**37.5**)	2 (**12.5**)	–
Knees	6 (**37.5**)	10 (**62.5**)	–	4 (**25**)	2 (**12.5**)	4 (**25**)	2 (**12.5**)	–
Ankle	4 (**25**)	12 (**75**)	–	–	4 (**25**)	–	4 (**25**)	–

reason, their body has adapted to weaving activity, and so they do not feel pain. Now, weaving activity has become an integral part of their life.

1.2 **Feeling of pain in different body parts of weavers (male) expressed through body map**: It is evident from Table 2 that out of 16 male respondents, maximum pain was observed in shoulder and lower back, i.e., 62.5% weavers and minimum pain was found in hand, palm and ankle as reported by 25% respondents.

Severity of pain was divided into three parts, i.e., acute, less acute and negligible. It is clear from Table 2 that 37.5, 50, 31.25 and 37.5% male respondents had acute pain in shoulder, upper back, lower back and thigh, respectively. Less acute pain was felt by 31.25, 25, 31.25, 25, 31.25, 12.5 and 25% respondents in neck, shoulder, elbow, upper back, lower back, thigh and knees, respectively. Negligible or no pain was reported by 12.5, 25, 25, 12.5 and 25% respondents at their neck, hand, palm, knees and ankle, respectively.

Frequency/occurrence of pain is shown in Table 2 that 50, 31.25, 37.5, 37.5 and 25% male respondents always felt pain in shoulder, upper back, lower back, thigh and knees whereas 43.75, 12.5, 31.25, 43.75, 25, 25, 25, 12.5, 12.5 and 25% male respondents felt pain sometime in their neck,

Table 3 Percent distribution of data pertaining to subjective feeling of pain in different body parts of weavers (female) expressed through body map by using unmodified loom N-16

Body part	Yes	No	Severity of pain			Frequency of pain		
			Acute	Less acute	Negligible	Always	Some time	Never
	n (%)	n (%)	n (%)	n (%)	n (%)	n (%)	n (%)	n (%)
Neck	12 (75)	4 (25)	6 (37.5)	6 (37.5)	–	–	12 (75)	–
Shoulder	16 (100)	–	12 (75)	4 (25)	–	16 (100)	–	–
Elbows	6 (37.5)	10 (62.5)	–	6 (37.5)	–	–	6 (37.5)	–
Upper back	14 (87.5)	2 (12.5)	10 (62.5)	4 (25)	–	12 (75)	2 (12.5)	–
Lower back	16 (100)	–	10 (62.5)	6 (37.5)	–	10 (62.5)	6 (37.5)	–
Hand	8 (50)	8 (50)	–	8 (50)	–	3 (18.75)	5 (31.25)	–
Palm	6 (37.5)	10 (62.5)	–		6 (37.5)	1 (6.25)	5 (31.25)	–
Thigh	14 (87.5)	2 (12.5)	8 (50)	6 (37.5)	–	10 (62.5)	4 (25)	–
Knees	6 (37.5)	10 (62.5)	–	6 (37.5)	–	–	6 (37.5)	–
Ankle	6 (37.5)	10 (62.5)	–	4 (25)	2 (12.5)	–	6 (37.5)	–

shoulder, elbow, upper back, lower back, hand, palm, thigh, knees and ankle, respectively.

1.3 **Feeling of pain in different body parts of weavers (female) expressed through body map using unmodified loom**: It is apparent from Table 3 that in case of females, maximum pain was observed in shoulder and lower back, i.e., 100% and minimum pain was found in elbows, palm, knee and ankle, i.e., 37.5%.

It is evident from Table 3 that 37.5, 75, 62.5%, again 62.5 and 50% female respondents had acute pain in their neck, shoulder, upper back, lower back and thigh whereas 37.5% respondents had less acute pain in neck, elbow, lower back, thigh and knees.

The length of occupational exposure and age of the workers affect their working capacity. It was studied that musculoskeletal pains were increased with an increase in the length of occupational work exposure [6].

It was found that feeling of pain in different body parts of women was more as compared to male due to their feminine body structure. Men have been found to suffer less musculoskeletal disorders than women even when they perform identical work [7].

2. **Modifications in the Handloom**
Handloom (table loom) was modified according to the weaver's need and demand to make it ergonomically suitable. Ergonomics adequately takes care to minimize the factors governing physical and mental stress to the person. Benefits of ergonomics are higher productivity, higher quality, reduced operator injury and greater job satisfaction. The required modifications were applied to the handloom sector of Mahua Dabra, Jaspur (Distt. Udham Singh Nagar) and Kumaun woolens, Haldwani (Distt Nainital). The modifications of handloom were as follows

- **Noise control**
- **Shifting of cloth beam**
- **Addition of gear attachment with cloth beam**
- **Weaver's seat**
- **Warp beam belt**
- **Selvedge finishing clip**

- **Noise control**: Noise is an aspect of the working environment which has received much attention for many decades. Noise is conveniently and frequently defined as unwanted sound. In the weaving units, noise is caused by vibration of mechanical parts of the weaving machine. Noise level emitted by shuttle running strokes causes serious reduction in alertness, hearing loss and decline of some cognitive performance can be expected. Earplugs and other types of personal protective equipment (PPE) were suggested to be used by weavers to control exposure to noise of loom and equipment in the working area. In a handloom, slay is mostly consisted of plastic blocks which creates noise. To reduce this noise, rubber washer was placed between the two plastic blocks.
- **Shifting of cloth beam**: In weaving operation, the newly constructed fabric must be wound on a cloth beam; this process is called taking up. In case of more than 30 m cloth, the wrapped cloth touches the lower stomach of the weaver which creates problem/discomfort during the weaving process. To eliminate this problem, cloth beam was shifted to the lower part of the handloom. So that weaver can easily weave the cloths comfortably.

Shifted cloth beam

- **Addition of gear attachment with cloth beam**: Cloth rolling is to be repeated after every 20–30 min of weaving which is a very tiring physical activity in itself. At first, the weaver has to get out of the seat and go to other side of the loom to roll the woven cloth.

Gear system

Then, tightening and adjusting the fresh fabric onto the cloth roll are done after unlocking the cloth beam. After which the relocking of cloth beam is done followed by the weaver getting back to the seat. Whereas to avoid this whole process of getting up and adjusting the cloth beam, weavers lean forward and maintain the bending posture which leads to severe back pain. So, automatic cloth winding with gear attachment was introduced in the existing handloom (table loom). This additional gear system helps in wrapping up the woven fabric as weaving proceeds. This attachment reduces physical movements and fatigue of the weaver during winding up of fabric. Along with this, gear system also helps in adjusting the density of the woven cloth.

- **Weaver's seat**: During weaving process, good posture on the loom is essential. It is important to sit on hip bones, not on tail bone. Weaver's abdominal muscles should be tight, and back should be straight (no squatting). The next important thing is the height of weaver's bench. If weaver sits too low, weaver has to hunch his shoulders to raise arms while throwing the shuttle. This is very tiring and may lead to muscle cramps in the shoulders and neck of the weaver.

  In order to avoid physical stress and fatigue, weaver's seat (stool/chair) was modified ergonomically according to the weaver's need. Slightly tilted weaver's seat with cushion was designed. When weaver sits on ergonomically designed seat, with hips higher than knees, the workload is shifted to thighs. Thigh muscles are the biggest and strongest part of human body, so weaver was able to weave longer and more comfortably by using the ergonomically designed seat.

  The chair intervention plays an important role in reducing musculoskeletal symptoms among workers who are required to sit for prolonged period [8].

Unmodified seat

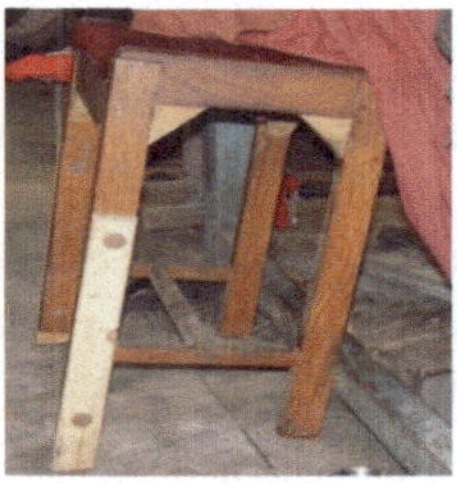

Modified seat

- **Warp beam belt**: Along with the primary weaving motions, i.e., shedding, picking and beating, secondary motions are also important in weaving process, these are: Let off and Take up Motion.
  These two processes are interconnected to each other and also affected by the wheel motion of warp beam which is attached with gear. Motion of warp beam is dependent on the chain which is responsible for the smooth motion and speed.
  Surface of the chain used in the traditional table loom was zigzag not flat therefore the prepared fabric was not even. To overcome this problem, smooth belt was used so that during beating process warp yarns were released evenly, and surface of produced fabric was smooth/better.

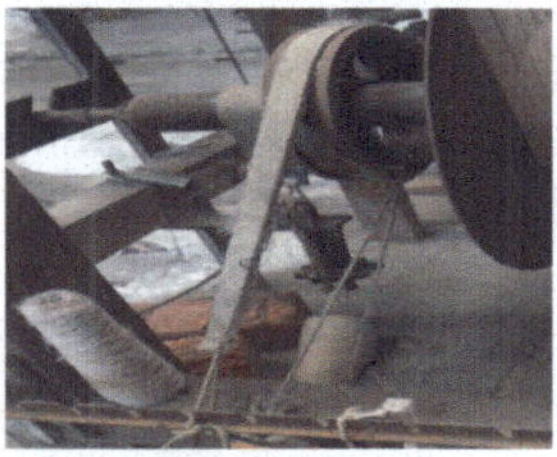

Warp beam belt

- **Selvedge finishing**: During weaving process, selvedge of the woven fabric may be unfinished; this was due to looseness of the warp threads at the edges/sides. Therefore, to overcome this problem clip was used. The clip holds selvedge yarns tightly so that selvedge of woven fabric can be finished in a straight manner.

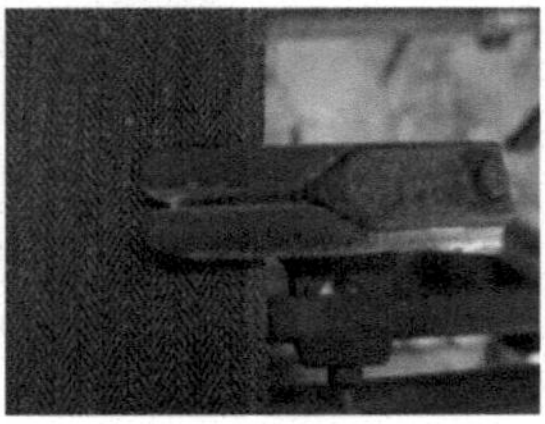

Selvedge Clip

- **Cloth meter**: The weaver needs to open the entire yardage of woven cloth and after measuring has to wrap it again. This activity increases fatigue and consumes time. It was observed during survey that a professional handloom weaver weaves around 12–30 m of cloth/day, depending upon various factors, i.e., yarn count, yarn strength, cloth yardage, density and design, etc. Here, the maximum length of cloth meter decided to be kept according to the length of warp.
  In order to check the yardage of the cloth woven, a cloth measuring device was also added in handloom. A required length of thread/tape was measured; when weaving was started, measured thread was inserted with the cloth. Thread was inserted in a manner so that thread was also wrapped with the woven cloth. This showed exact quantity/meters of prepared cloth (woven), while continuing weaving. This cloth meter will reduce fatigue, save time for measuring of cloth in between the weaving process.
- **Permanent setting of the treadle support**: During weaving process, the level of harness should be at equal level to each other. The treadles should be leveled evenly. Thus, proper separation of warp threads can be done (shedding), and then weft threads could be passed through it. Generally, to keep the treadles at an equal distance bricks are used as a supporting tool by most of the handloom weavers. During weaving process, these bricks are moved due to the treadle movements. It disturbs the equal level of the treadles and thus improper separation of the warp thread takes place, due to this reason the warp threads may break. As the weaver has to sit and stand up for each time to set the bricks, it causes extra fatigue and consumes time. So to overcome these problems, permanent wooden pieces were attached with the loom, as a supporting tool.

Modified treadle support

3. **Assessment of work-related musculoskeletal disorders of the weavers through body map method**
   Weavers of both the genders were asked to do weaving, and body map questionnaire was also administrated twice, i.e., using unmodified loom and then modified loom.

   3.1 **Feeling of pain in different body parts of male and female weavers expressed through body map using modified loom**:
   It is clear from Table 4 that after working on modified loom continuously for one month, 37.5% respondent felt maximum pain in lower back and 6.25% respondents reported pain in palm. It can be seen from Table 4 that 18.75% male respondents had acute pain in thigh followed by 12.5% respondent who reported acute pain in shoulder, upper back and lower back whereas 12.5% respondents had less acute pain in neck, elbows, upper back and thighs even after working on modified loom. It was found that level of pain in male weavers was decreased in case of modified loom as compared to unmodified loom. This is due to the ergonomically designed tools, i.e., sitting position, body posture and automatic fabric winding technique.
   Female weavers were also assessed for feeling of body pain while working on unmodified and then after that working on modified loom for at least one month continuously. It is apparent from Table 5 that in case of modified handloom, maximum pain was observed in female weavers, in their shoulder, i.e., 37.5% and minimum pain was found in elbows and ankle, i.e., 12.5% respondents.
   It is also evident from Table 5 that 12.5, 18.75, 18.75, 12.5 and 18.75% female respondents had acute pain in neck, shoulder, upper back, lower back and thigh, whereas 18.75% respondents had less acute pain in neck shoulder and knees, respectively.
   Frequency of pain as given in Table 5 showed that 37.5, 12.5, 6.25, 18.75, 6.25 and 25% female respondents had continuous pain in shoulder, upper back, lower back, hand, palm and thigh, respectively. But 31.25, 12.5, 12.5, 12.5, 6.25, 12.5, 6.25, 18.75 and 12.5% female respondents some time had pain in neck, elbow, upper back, lower back, hand, palm, thigh, knees and ankle, respectively. It was found that in modified loom, body pain was decreased as compared to unmodified loom.
   According to a study on "Gender differences, work stressors and musculoskeletal disorder in weaving industries," gender differences exist among the weavers in the prevalence of musculoskeletal disorders and the perception of work and psycho-social stress. Female weavers were more prone to musculoskeletal disorders in upper back in power loom and handloom [9].

4. **Physiological cost of work of subjects during experiment using unmodified table loom and modified table loom**
   Tables 6 and 7 show the effect of weaving activities performed on physiological cost of work. A total of 5 replications was conducted to observe physiological cost of work in terms of heart rate (HR) and energy expenditure rate (EER).

Table 4 Percent distribution of data pertaining to subjective feeling of pain in different body parts of male weavers expressed through body map by using modified loom N-16

Body part	Male (n-16)							
	Yes	No	Severity of pain			Frequency of pain		
			Acute	Less acute	Negligible	Always (continuous)	Some time	Never
	N (%)	N (%)	N (%)	N (%)	N (%)	N (%)	N (%)	N (%)
Neck	3 (**18.75**)	13 (81.25)	–	2 (**12.5**)	1 (6.25)	–	3 (**18.75**)	–
Shoulder	5 (**31.25**)	11 (**68.75**)	2 (**12.5**)	3 (**18.75**)	–	3 (**18.75**)	2 (**12.5**)	–
Elbows	2 (**12.5**)	14 (87.5)	–	2 (**12.5**)	–	–	2 (**12.5**)	–
Upper back	4 (**25**)	12 (**75**)	2 (**12.5**)	2 (**12.5**)	–	1 (6.25)	3 (**18.75**)	–
Lower back	6 (**37.5**)	10 (**62.5**)	3 (**18.75**)	3 (**18.75**)	–	2 (**12.5**)	4 (**25**)	–
Hand	2 (**12.5**)	14 (87.5)	–	–	2 (**12.5**)	–	2 (**12.5**)	–
Palm	1 (6.25)	15 (**93.75**)	–	–	1 (6.25)	–	1 (6.25)	–
Thigh	4 (**25**)	12 (**75**)	2 (**12.5**)	2 (**12.5**)	–	2 (**12.5**)	2 (**12.5**)	–
Knees	3 (**18.75**)	13 (81.25)	–	3 (**18.75**)	–	1 (6.25)	2 (**12.5**)	–
Ankle	2 (**12.5**)	14 (87.5)	–	–	2 (**12.5**)	–	2 (**12.5**)	–

4.1 **Physiological cost of weavers in terms of heart rate (beats/min)**: The results of experiments conducted on weavers were analyzed and presented in Table 6 for mean heart rate (HR) and percent increase in heart rate from resting level. It was seen that exerting human energy during performance of activities, resulted an increase in heart rate.

The mean resting heart rate of male and female weavers was 77.50 and 78.12 beats/min, respectively. While working on traditional (unmodified) loom, heart rate was increased up to 98 and 105.12 beats/min during the weaving performance; thus, the percent increase in heart rate was found to be 8.06 and 10.24 beats/min of male and female weavers, respectively. It was observed that heart rate (beats/min) of female weavers was higher as compared to male weavers.

The mean heart rate at resting level was 77.50 (male) and 78.12 (female) beats/min, while working on modified table loom, an increase in the heart rate of male weavers was observed as 95.25, and for female weavers, it was

Table 5 Percent distribution of data pertaining to subjective feeling of pain in different body parts of female weavers expressed through body map by using modified loom N-16

Female (n-16)							
Yes	No	Severity of pain			Frequency of pain		
		Acute	Less acute	Negligible	Always (continuous)	Some time	Never
N (%)	N (%)	N (%)	N (%)	N (%)	N (%)	N (%)	N (%)
5 (31.25)	11 (68.75)	2 (12.5)	3 (18.75)	–	–	5 (31.25)	–
6 (**37.5**)	10 (62.5)	3 (18.75)	3 (18.75)	–	6 (**37.5**)	–	–
2 (12.5)	14 (87.5)	–	2 (12.5)	–	–	2 (12.5)	–
4 (25)	12 (**75**)	3 (18.75)	1 (6.25)	–	2 (12.5)	2 (12.5)	–
3 (18.75)	13 (81.25)	2 (12.5)	1 (6.25)	–	1 (6.25)	2 (12.5)	–
4 (25)	12 (**75**)	–	4 (25)	–	3 (18.75)	1 (6.25)	–
3 (18.75)	13 (81.25)	–	–	3 (18.75)	1 (6.25)	2 (12.5)	–
5 (31.25)	11 (68.75)	3 (18.75)	2 (12.5)	–	4 (25)	1 (6.25)	–
3 (18.75)	13 (81.25)	–	3 (18.75)	–	–	3 (18.75)	–
2 (12.5)	14 (87.5)	–	–	2 (12.5)	–	2 (12.5)	–

99.60 beats/min during performance. The percent increase in heart rate was found to be 6.19 (male) and 7.65 (female) beats/min.

It was concluded from the above table that in case of working on modified loom, increase in heart rate of weavers was less as compared to weaving on unmodified loom. This may be due to the reason that weaving on unmodified loom requires more energy to pull and push the reed (Hatha) for beating the woven fabric as compared to modified loom.

5. **Physiological cost of weavers in terms of energy expenditure**

The mean energy expenditure rate kJ/min and increase in energy expenditure rate (EER) from its resting level for weaving activities performed by selected weavers are shown in Table 7.

In case of male weavers, energy expenditure rate at resting was 3.60 kJ/min. Energy expenditure during work was increased as 6.86 and 6.42 kJ/min while they worked on unmodified loom and modified loom, respectively. In case of female, the energy expenditure rate before work was 3.70 kJ/min which were

Table 6 Physiological cost of weavers in terms of heart rate (beats/min) N-32

S. No.	Mean heart rate (beats/min)								
	Unmodified table loom					Modified table loom			
		Resting	During work	Recovery	% increase in heart rate	Resting	During work	Recovery	% increase in heart rate
1	Male	77.50	98	83.75	8.06	77.50	95.25	82.30	6.19
2	Female	78.12	105.12	86.12	10.24	78.12	99.60	84.10	7.65

Table 7 Physiological cost of weavers in terms of energy expenditure (kJ/min) N-32

S. No.	Sex	Energy expenditure							
		Unmodified table loom				Modified table loom			
		Resting	During work	Recovery	% increase in energy expenditure	Resting	During work	Recovery	% increase in energy expenditure
1	Male	3.60	6.86	4.59	27.5	3.60	6.42	4.36	21.11
2	Female	3.70	7.99	4.97	34.40	3.70	7.11	4.65	25.72

Table 8 Physiological cost of work in terms of blood pressure N-32

S. No.	Sex	Blood pressure			
		Unmodified loom		Modified loom	
		Before work	After work	Before work	After work
		Systolic/diastolic	Systolic/diastolic	Systolic/diastolic	Systolic/diastolic
1	Male	123.12/79.21	**125.5/79.5**	123.12/79.21	125.0/79.5
2	Female	111.5/74.12	115/74	111.5/74.12	114.62/74.5

increased up to 7.99 kJ/min while they used unmodified loom for weaving, and an increase of 7.11 kJ/min was noted during weaving on modified loom.

It was concluded that energy expenditure was higher in case of female as compared to male weavers while using unmodified and modified loom. This may be due to the muscular structure of the male weavers. In general, physically male is stronger than female. It was also observed that energy expenditure was decreased while working on the modified loom. This may be due to the attachment of gear lead to easy beating process.

6. **Physiological cost of work in terms of blood pressure**

   The amount of force exerted upon the walls of the arteries is known as blood pressure. The change in blood pressure which happens throughout the day and from day to day is affected by many factors. The data revealed the information regarding the blood pressure of weavers during the rest period and after 15 min of work in both causes, i.e., unmodified loom and modified loom.

   It is clear from Table 8 that while using unmodified loom, mean blood pressure of male and female weavers were 123.12/79.21 and 111.5/74.12, which increased up to 125.5/79.5 and 115/74 during the performance. It was observed that an increase in the blood pressure of female weavers was slightly higher as compared to male weavers.

   In case of modified table loom, the mean blood pressure at resting level was 123.12/79.21 (male) and 111.15/74.12 (female), which increased up to 125/79.5 (in male) and 114.62/74.5 (in female) during performance.

7. **Percentage changes in grip strength (kg)**

   Tabulation of percentage reduction of grip strength of the respondents as envisaged in Table 9 shows that in case of unmodified loom, before starting the weaving work, average grip strength of left and right hand of weavers was 30.02 and 33.56 kg (in male), respectively. But after work on the same (unmodified) loom, average grip strength of left hand reduced to 28.5 and grip strength of right hand reduced to 31.0 kg (male weavers). It was found that in weaving, work percentage reduction in grip strength of the left hand was 5.06% whereas percentage reduction in grip strength of right hand was 7.62%.

   In case of female, grip strength of left and right hand before work was 22.0 and 24.01 which get decreased after activity to 20.6 (left hand) and 21.8 (right hand), respectively. Percentage reduction in grip strength of the left and right hand was 6.36 and 9.20%, respectively.

Table 9 Percentage changes in grip strength (kg) N-32

S. No.	Sex	Average muscular grip strength (kg)											
		Unmodified loom						Modified loom					
		Resting		After work		% reduction in grip strength		Resting		After work		% reduction in grip strength	
		Left	Right	Left	Right	Left	Right	Left	Right	Left	Right	Left	Right
1	Male	30.02	33.56	28.5	31.0	5.06	7.62	30.02	33.56	29.1	32.2	3.06	4.05
2	Female	22.0	24.01	20.6	21.8	6.36	9.20	22.0	24.01	20.9	22.3	5	7.12

Table 10 Angle of deviation on handloom N-32

S. No.	Sex	Angle of deviation (%)	
		Unmodified loom	Modified loom
1	Male	16.47	5.88
2	Female	18.07	10.84

In case of modified loom, male weaver's percentage reduction in grip strength of left and right hand were 3.06 and 4.05% and 5 and 7.12% in, respectively, case of female weavers.

It was found that the grip strength decreased more in right hand as compared to left hand. This is because the more rigid work, i.e., throwing the shuttle from one side to another side is mainly carried out with right hand of the weaver. In modified loom, percentage reduction in grip strength was less in both left and right hands of male and female respondents; this is due to the good sitting posture of the weavers and other modified features of loom. Grip strength was found to be a strong predictor of an individual's nutritional status.

8. **Posture analysis of weavers**

Table 10 shows the postural analysis of the male and female weavers in weaving activity. The angle of spinal cord was taken by using flexi curve. In case of unmodified loom, the angle of deviation in male weavers was found 16.47% in which decreased to 5.88% in modified loom, i.e., 5.88% while weaving on the modified loom, i.e., 5.88%.

In case of unmodified loom, angle of deviation in female weavers was recorded 18.07% which reduced to 10.84% in case of modified loom. In modified loom, angle of deviation found to be decreased as compared to unmodified loom. The higher the deviation of angle, the greater is the pain in the body parts. So it was concluded that the postural discomfort is more in unmodified loom; this may be due to the tools and sitting chair which were not according to the weavers need. It is clearly understood from the results that use of modified loom is beneficial for the weavers.

5 Conclusion

It can be concluded from the present study that ergonomics is significant for each industry. Handloom industry also requires ergonomically safe structure and environment. It is clear from the result that an awareness of the weavers regarding various ergonomic factors effected their comfort level and health. Therefore, such issues are required to be addressed. Some simple and easy to adopt modifications may tremendously reduce the health hazards associated with ergonomics. Loom was modified according to weavers need to reduce the physical pain. It was found that weavers working on modified loom had less pain and lesser muscular skeletal disorders. Due to the gear attachment on the loom and cushioned sitting tool, weavers worked on loom comfortably for longer period of time. Gear attachment improved the speed of weaving process so that the production of clothes was also increased as compared to unmodified table loom.

The modifications carried out on the table loom were simple, and little amount of money was used to make changes more adaptable to the weavers. Similar modifications can be used for other handloom as per need. The modified looms are used in the Jaspur and Haldwani regions successfully.

References

1. Anonymous: Handloom sector. http://handlooms.nic.in. Last accessed on 20 Nov 2017 (2015)
2. Sudalaimuthu, S., Devi, S.: Handloom industry in India. http://www.fibre2fashion.com/industry-article/1/10/handloom-industry-in-india1.asp. Last accessed on 15 Oct 2017
3. Adams, C.: Ergonomics. https://www.thoughtco.com/ergonomics-meaning-1206557. Last accessed on 29 Nov 2017
4. Balasundaram, K., Adugna, A., Kumar, A., Kumar, M.S.: Improvement of ergonomic factors in a textile industry: a case study. J. Recent Res. Eng. Technol. **4**(5), 01–06 (2017)
5. Choobineh, A., Lahmi, M.A., Hosaini, M., Jamali, M., Rezaee, G.Z.: Musculoskeletal problems and corrective measures in Iranian hand-woven carpet industry. In: Proceedings of the XVth Triennial Congress of the International Ergonomics Association and the 7th Joint Conference of Ergonomics Society of Korea/Japan Ergonomics Society, Seoul, Korea, Ergonomics Society of Korea, pp. 270–272
6. Metgud, D.C., Khatri, S., Mokashi, M.G., Saha, P.N.: An ergonomic study of women workers in a woolen textile factory for identification of health-related problems. Indian J. Occup. Environ. Med. **12**(1), 14–19 (2008)
7. Arvidsson, I., Arvinsson, M., Axmon, A., Hansson, G.A., Johansson, C.R., Skerfving, S.: Musculoskeletal disorders among female and male air traffic controllers performing identical and demanding computer work. Ergonomics **49**(11), 1052–1067 (2006)
8. Sjan-mari, S.V.N., Quinette, Q.L., Susan, S.H.: The effectiveness of a chair intervention in the workplace to reduce musculoskeletal symptoms. A systematic review. BMC Musculoskelet. Disord. **13** (1), 28–29 (2012)
9. Nag, A., Vyas, H., Nag, P.K.: Gender differences, work stressors and musculosketal disorders in weaving industries. Ind. Health **48**(3), 339–348 (2010)

The manufacturer's authorised representative in the EU is Springer Nature Customer Service Centre GmbH, Europaplatz 3, 69115 Heidelberg, Germany. If you have any concerns regarding our products, please contact ProductSafety@springernature.com

Printed and bound by CPI Group (UK) Ltd, Croydon, CR0 4YY

23/09/2026

02221532-0010